SOMETHING IS WRONG. IS IT THE PICTURE, OR IS IT ME? WHATEVER, IT IS JUST NOT RIGHT.

SHALL I IGNORE THAT WHAT I SEE IS WRONG AND SHALL I ACCEPT THAT WHAT I SEE IS RIGHT, AND THAT WHICH I THOUGHT IS WRONG?

YES. READ THAT TWICE, OR MORE.

WELCOME TO A WORLD WHEREIN THINKING CAN DETERMINE SURVIVAL AND/OR SANITY.

SOMETHING IS STRANGE ABOUT THIS PICTURE. **Or can't you see it?**

Bob's Legacy 4

It Isn't Al(l)right

OR

It ain't necessarily so!

NOT EVERYTHING YOU SEE, HEAR, READ, OR ARE TOLD, IS NECESSARIY SO, NO MATTER WHO OR WHAT IS THE SOURCE OF THE INFORMATION.

"Of course at this stage of human existence the individual human is actually totally unable to witness any such "universal expansion" of the cosmos, but merely observes expansion that appears to radiate away from him in all directions. He perceives "red shift" caused by retreating light sources. Thence by reason and logic he must conclude the same applies universally. Seeking an explanation he concludes that logically it must have once had a common source of origin, and thus must have began the observed expansion from one source. **He ignores the fact that the one inescapable source is the very "place" occupied by him.** He fabricates a "big bang" theory, and seeking evidence to validate that theory, will surely find what seems to "prove" it as correct." (Bob)

****

"Perhaps fairies, dwarfs, elves, leprechauns, dragons, monsters, vampires, werewolves, ghosts, poltergeists and flying saucers all exist. And perhaps the cynics who say that it is all in the mind are also right, ***because all of these things exist or are produced at the second or etheric level.***

The strange behavior of all apparitions suggests that they obey laws not quite like those of conventional physics, and that they probably belong to a reality with slightly different space-time references. ***The fact that those who come closest to these phenomena, usually receive information structured to support their own beliefs or fears,*** **suggests that these apparitions cannot be entirely independent of the minds of those involved**." *The Romeo Error, Lyall Watson.*

****

"To the naïve realist the universe is a collection of objects. To the quantum physicist it is an inseparable web of vibrating energy patterns in which **no one component has reality independently of the entirety; and included in the entirety is the observer."** ***Paul Davies "Superforce"***

****

SUNDAY (12th) 7.11pm ***I have learned that some times it is necessary to not only ignore things but to "not see them" at all.*** *It is beyond any powers of denial – it is for survival.*

Thus today I determined not to see nor dwell on floors, walls, clothes or sundry other things as have in the past been 'alive'. In consequence my sanity is preserved, *but the poor lady patient wheeled out last nite may have had experience none would enjoy – not the vilest horror movie sadist even. That lady did look calmer today, and is wearing an oxygen mask. (Bob)*

****

The game we play is let's pretend
and pretend we're not pretending.
We choose to forget who we are
and then forget that we have forgotten.
Who are we really?
(we are) the centre that watches
and runs the show
that chooses which way it will go
the "I AM" consciousness -
that powerful loving perfect reflection
of the cosmos.
But in our attempt to cope with early
situations we chose or were hypnotized
into a passive situation to avoid
punishment - or the loss of love.
We chose to deny our response/ability
pretending that things just happened
or that we were being controlled -
taken over.
We put ourselves down, and have become
used to this masochistic posture, this
weakness, this indecisiveness...
but we are in reality free,
a centre of cosmic energy.
Your will is your power - don't pretend
you don't have it - or you wont.
(Bernard Gunther)

“Question is this: Is there perhaps another (third) state of our existence of which we are generally and presently totally unaware, within (the) hyperspace? If we are indeed multidimensional entities as described earlier in this book, regardless of earthbound body of flesh and blood, then there is a distinct and definite probability that this mortal life in its entirety is no more than the equivalent of that higher dimensional self's rambling and perhaps ultimately meaningless dream. Just as meaningless as are the events and experiences of our mortal body's dream world and life meaningless to the mortal self.” (Bob)

****

“That has become the way Darwinists handle any and all challenges to their pet theories: **if they can no longer defend one, they don't talk about it, or they talk about it as little as possible. If forced to talk about it, they invariably try to "kill the messenger" by challenging any critic's "credentials". If the critic lacks academic credentials equal to their own, he or she is dismissed as little more than a crackpot. If the critic has equal credentials, he or she is labelled as a "closet Creationist" and dismissed.**”

****

Bob Maddison Upper Hutt, New Zealand

Name: Bob Maddison Location: Wellington, NZ I was born in Queensland in 1945 and have lived in New Zealand since 1973. My life has been enriched by many children and now grandchildren. I have always sought answers to life's questions, and finding answers generate more questions. I would like to share some of these experiences and answers, trusting they may educate, entertain, or at least amuse or cause one to think, and in thinking, ask questions. Join the journey of Life, living, and the richess freely offered. Contact is available at maddison.nz@gmail.com

ALSO BY THE SAME AUTHOR

Bob's Legacy
Bob's Legacy 2, Reclaiming Life
I Can See Clearly Now

I believe that you may really enjoy this book. It may even cause you to question things you never thought could be questionable.

DON'T PANIC

This is quite normal and you are not asked for any agreement, but rather leave for you the potential delicious pleasure of discovering an entire new paradigm that may re-awaken your self awareness, and the possession of your own life and being.

THIS BOOK IS ENTIRELY FOR AND DEDICATED TO YOU.

Reformatted and edited November 2013

TABLE OF CONTENTS:

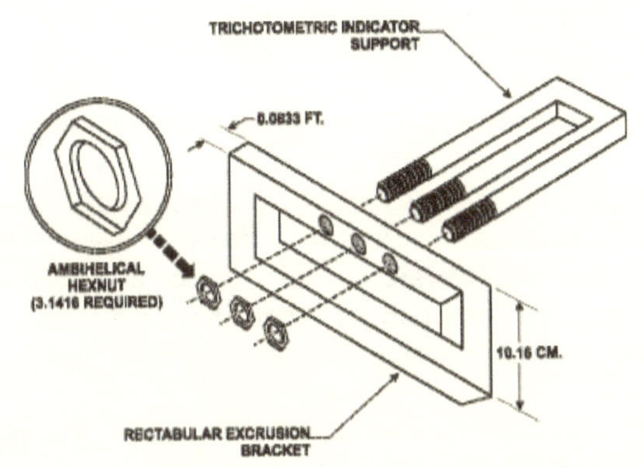

PART THREE

THE "GOD" FAMILY, OR,

WAY BACK THEN, WHEN....

(I don't really think you can make up this stuff.)

(And the more you look at it, the more confusing it can become.)

300 – THE "STAGE" – WHO, WHAT AND WHY.

"All truth goes through three stages. First it is ridiculed. Then it is violently opposed. Finally, it is accepted as self-evident." (Schopenhauer)

Although I had intended to place most of this material in an appendix at the rear of this book, I am inserting it now, as this is indeed the logical place to locate this initial material. By "initial material" I mean exactly that, as now the resource materials have become so vast that one could spend almost a lifetime researching this subject of Anunnaki alone. I would not expect the reader to be satisfied with the small amount of materials presented here, but to do the research him or herself.

"Now" is the appropriate time by reason of my having just dismissed the Jehovah (and possibly all other) based religions as unreasonable and incapable of sustaining the claims made for them, and then immediately thereafter dismissing both evolution and standard creationist paradigms.

Easy to tear down, but now let's see if we can establish something in their place. First we will look at material directly off the Internet.

PART ONE - ANNALS OF EARTH

© 1995, 2003 Dan Sewell Ward

Episode V -- The Anunnaki

In our last Episode, there was the hint that we were about to welcome Man and his history in these Annals of Earth. Moreover, Man's arrival on the scene, supposedly in a starring role, promised to be a dramatic event, complete with controversy and wonderment. It was a tense moment, mitigated only by the realization that there were a lot of unanswered questions still on the books. The good news is that it's now time -- in this very episode -- to welcome Man onto the scene, begin his history (and include a fair amount of herstory as well), and at the same time, answer all the piddling little questions.

In order to accomplish in one Episode all of these tasks, it is necessary to backtrack once again in our dating (but probably for the last time), in order to bring the relevant histories of the rest of the major characters into our drama. In doing this we will initially resort to an assortment of Sumerian texts, coupled with just enough Biblical quotes to add spice to the arguments. We will also recall to the stage, a major character who has received little mention in the latest Episodes: Nibiru.

For while life has been merrily evolving along on the planet Earth circa 1 million years B.C.E., back on the ranch on the planet Nibiru -- that currently distant member of our solar system -- evolution has marched and/or accelerated to the beat of a different (and possibly hyper) drummer. Nibiru has, according to the texts of Sumeria, not only developed an intelligent civilization, but has even managed to develop blood feuds and space flight, and inevitably to acquire imperialistic ambitions.

c. 485,000 B.C.E. In a Hittite version of a Sumerian text entitled by modern scholars Kingship in Heaven, the story is told of the descendents of LAMA. The text, after elaborating on their ancestry, "the fathers and mothers of the gods", tells the tale of Anu and Alalu, "the mighty olden gods, the gods of the olden days". When the time of succession had arrived on Nibiru, it was not Anshargal, Anu's father and the heir apparent, who had ascended the throne. Instead, a relative named Alalu (or Alalush in the Hittite version) became the ruler. Then in a gesture of reconciliation (or possibly by custom), Alalu appointed Anu to be his royal cup-bearer, an honored and trusted position. But after nine Nibiruan years (known as Sars, where 1 Sar = 3600 Earth years), Anu "gave battle to Alalu" and deposed him (so much for the honored and trusted bit).

> *"Once in the olden days, ALALU was king in Heaven, ALALU was seated on the throne; The mighty ANU, first among the gods, was standing before him: He would bow to his feet, set the drinking cup in his hand.*
>
> *For nine counted periods [Sars], ALALU was king in Heaven.*
>
> *In the ninth counted period, ANU gave battle to ALALU. ALALU was defeated, he fled before ANU -- Down he descended to the dark-hued Earth.*
>
> *ANU took his seat upon the throne."*

Noted the phrase: "Descended to the dark-hued Earth." Ah yes, the critical connection.

Mind you, the texts did not say that Alalu came from Nibiru, but rather from Heaven -- what we now recognize -- at least in these Annals -- as being periodically located in the environs of the asteroid belt. Was Alalu a fallen angel? By no means. Man, at the time, was only a gleam in the eye of Homo erectus. He was not ready for gods yet. And Alalu was not destined to be one of the major characters in our drama anyway. But his arrival on Earth began the process, the initiation of all the events that would be detailed within the Sumerian texts.

The House of Anu had put an end to the rival house by deposing its ruler Alalu, who in the best tradition of deposed rulers escaped the planet of his birth and chose exile. Nibiru, apparently, was not big enough for the both of them. Alalu was thus forced to find greener pastures (and what greener pastures can one find in the solar system, than Earth?). By descending to dark-hued (Nibiruan for greener pastures) Earth, one assumes that Alalu made the trip in a spaceship, probably with a retinue of aides and supporters, and thereafter found refuge on an unexplored planet.

c. 450,000 B.C.E. After many, many moons, or roughly ten orbits of Nibiru, during which Alalu, et al made the best of it in their exile domicile, a momentous event occurred. According to the Sumerian texts, Alalu ***discovered gold****, a commodity apparently in great demand on the planet Nibiru. Why the gold was so valuable to Nibiru's inhabitants is not made clear in the texts, but as the narrative continues we will soon realize that this was a discovery of immense importance. Furthermore, Alalu, even though deposed, was sufficiently concerned for Nibiru's need for gold, he advised Anu that he had found the precious metal in Earth's oceans. This momentous discovery apparently had a major impact, in that a reconciliation of sorts then ensued with Anu appointing KUMARBI, a grandson of Alalu, to be his royal cup-bearer. Alalu then, it is assumed, was allowed to return to Nibiru for a more genteel exile.*

All the evidence suggests that at this juncture the Nibiruans launched into a space program of some magnitude [pardon the pun]. It was probably a crash program (or rather a hurried up program where it was hoped there would be no crash). But the process nevertheless took some 7,000 years. This may sound a trifle long, but keep in mind that it was two orbits of Nibiru, or only two of their years.

As will become apparent in reading this Episode, the inhabitants of Nibiru were extremely long-lived. If one is to take the Sumerian texts literally, one would have to suspect they had a Nibiruan-normal life-time in excess of some 500,000 Earth years, about 140-150 orbits of Nibiru. But they were not immortal! As time will tell, they were quite capable of dying young, having their life taken by dismemberment and other bloody deeds, and other forms of accidental or homicidal death. Of course, from the viewpoint of any human with a life expectancy of less than 100 years, such long-lived creatures would appear immortal.

A related and highly speculative corollary to their long lives is the question as to whether or not such creatures would progress at an equally slow rate. As we will see, different factions of the Nibiruans created and maintained a family feud worthy of the Hatfields and the McCoys, and one which lasted for more than 400,000 years. This would hardly be considered progress from any species' point of view, where the species had less tenure than a tenth the length of the feud.

Furthermore, all progress, improvements, or advances in the Nibiruan technology seem to also proceed at an incredibly slow rate. The fact that it took them 7,000 Earth years to take advantage of Alalu's discovery would attest to this rather laid-back way of life. One could also speculate on whether or not their Earth-daily lives were quite leisurely by human standards, and ***whether or not a good night's sleep for them was on the order of Earth weeks or months.*** *The Sumerian texts do not address this question (nor for that matter the personal traits of the Nibiruans which do not affect the human race), but one wonders if extreme longevity might be coupled with extremely slow transformational rates, both personal and cultural. And vice versa for the very short-lived species. E. g. not cleaning a toilet for a month might allow the bacteria therein to reach the state of a civilization advanced enough to be capable of achieving spaceflight. A thought to ponder.*

But eventually, the Nibiruans began their imperialistic conquest of the Earth.

c. 443,000 B.C.E. The Sumerians considered Nibiru to be a twelfth member of the solar system -- after the Sun, Moon, and the nine other planets that modern science recognizes. The texts also revealed that Nibiru's orbit took the planet to a "station" in the distant heavens, then brought it back to Earth's vicinity, crossing between Mars and Jupiter.

In the latter position, the planet obtained its name Nibiru ("crossing") and its symbol, the cross. The inhabitants of the planet, the "Righteous Ones", were known as the Anunnaki ("Those Who From Heaven to Earth

Came"). The inhabitants of Nibiru who stayed home (supposedly the overwhelming majority of the population) were still Nibiruans, but the astronaut-wing were known by a very specific name -- the one we will hereafter use. Keep in mind also that Heaven has been identified as the asteroid belt located between Mars and Jupiter, the exact location of the "crossing" point, the perihelion (closest approach to the Sun) of Nibiru in its highly elliptical orbit. The Anunnaki of the planet Nibiru, then, quite literally came from Heaven.

For reasons we will discuss in future Episodes, the Biblical Deluge, the torrential covering of the Earth with water in ancient times, has been dated to circa 11,000 B.C.E. This date is not without controversy, but is well supported by evidence from several diverse sources. In any case, the date of the arrival of the Anunnaki on Earth is calculated by back dating from the Deluge. ***According to the Sumerian texts, the all-important arrival of the Anunnaki occurred 120 Sars, or -- inasmuch as one Sar equals 3,600 Earth years -- 432,000 years before the Deluge. 432,000 years! Hmmmmmm.***

For those who were wondering where the Hindu traditions (Episode III) came up with their figure for the 432,000 year Yuga... Well, now you know! *And even if you're not convinced about which came first, the Anunnaki or the Hindus, there nevertheless is a clear connection. And if you consider the fact that 432,000 years is a very long time for Hindu humans to keep track of, then perhaps the Anunnaki have the best claim to originating the dating. In any case...*

Exactly one ***Yuga (432,000 years)*** *prior to the Deluge, the DIN.GIR ("Righteous Ones of the Rocketships") came down to Earth from their own planet. ["Rocketships" might be pushing the translation a bit, but as Zecharia Sitchin continues his interpretation of the Sumerian texts, the term makes more and more sense.]* ***The first major expedition of the Anunnaki was let by E.A ("Whose House is Water"). After landing and establishing a base at Eridu, located at the headwaters of the Persian Gulf, this eldest son of Anu, assumed the title: EN.KI ("Lord of Earth"). According to his first-person report:***

*"When I approached Earth, there was much flooding.
When I approached its green meadows, heaps and mounds were piled up at my command.
I built my house in a pure place... My house -- its shade stretches over the Snake Marsh."*

Ea / Enki was well chosen for the mission. He was a brilliant scientist and engineer whose nickname was NU.DIM.MUD ("He Who Fashions Things"). His plan was to extract gold from the waters of the Persian Gulf and the adjoining shallow marshlands. In this regard Sumerian depictions showed Enki as the lord of flowing waters, sitting in a laboratory and surrounded by interconnected flasks. Enki was also destined to be one of the two major players in the history of our race. Keep him in mind. His place in the early parts of Genesis are of the highest significance.

The extraction by Enki and his followers of gold is of critical importance. The whole purpose of the imperialistic intentions of the Anunnaki was to accumulate gold. But apparently, not for any vain purposes. It is noteworthy, for example, that at no time during the millennia that followed were the Anunnaki ever shown trading in gold or wearing golden jewelry. There was no hint of a monetary role either. The Anunnaki apparently had much higher purposes for the precious metal. (There is, however, absolutely no evidence for the Anunnaki having higher porpoises.)

Gold may have been required for their space craft (as does our modern day space program), and as noted in Hindu texts, references to the celestial chariots of the gods refer to them being covered with gold*. But there is little need for spacecraft except as a means to acquire gold from Earth, and thus this idea is probably not correct. Instead, there appears to be a much more important need of the metal, with its unique properties, possibly some vital need back home on Nibiru, a need conceivably affecting the very survival of life on Nibiru. One theory Sitchin seemed fond of was that the Nibiruans needed the precious metal in order to suspend gold particles in Nibiru's waning atmosphere and thus shield it from a critical dissipation. Sitchin, however, has not been willing to defend this idea to any great extent. The question must remain mute for a bit longer. But we shall return to this all important detail a bit later in the narrative.*

Another factor worth mentioning is the genealogy of the leaders of the Anunnaki and their relationship to the planets. First of all there is the matriarch LAMA, who is responsible for both lineages of the warring houses of ANSHARGAL and ALALU. Anshargal is identified with Anshar (Saturn). His mate is KISHAR (Jupiter). Their son is ANU (Uranus). By NAMMU or ID, a concubine of Anu's, Anu "begot his twin and equal" NUDIMMUD, EA or ENKI (Neptune).

In the Sumerian traditions, Anu is generally considered the Sky God, the head of the Sumerian pantheon of gods and goddesses. Enki is considered the God of Wisdom and the God of Waters (Nammu, his mother, was the Goddess of the Watery Deep --"Deep Space"?). At the same time, Enki is Anu's "twin and equal", and Enki's planet Neptune is at the gateway to the solar system, such that the solar system is Ea's (Enki's) "abode".

As already mentioned, Enki will become one of the most important gods in the saga of Man. He is also the father of Marduk, the latter being Enki's first born son. This makes sense, inasmuch as it was Neptune's gravitational pull that first brought Nibiru (Marduk) into the solar system. Marduk also plays a very major part later on, particularly during the time of Babylon. Meanwhile, at this stage, Enki, the equal of Anu is on Earth, prospecting in the Persian Gulf waters for gold.

The Sumerian texts also note that Nibiru (or Marduk) made for himself two abodes: One in the "Firmament"; the other, "in the Deep" -- the latter being called the "Great Distant Abode", as well as E.SHARRA ("Abode/Home of the Ruler/Prince"). The abode in the Firmament is not entirely clear as to its exact name, location, address, and/or zip code, but there is a possibility this abode was on the planet Mars. This rather outrageous suggestion will be revisited more than once as we progress in these Annals.

The sea we today call the Arabian Sea, the body of water between the Persian Gulf and the Indian Ocean was called in antiquity the Sea of Erythrea, from whence we derive the word, Earth. The first settlement of the Anunnaki on Earth was at a place called E.RI.DU ("Home In a Faraway place"). The Sumerian term for Earth's globe and its firm surface was KI. Note, for example, that the word, Anunnaki breaks down into ANU, N, NA, and KI. ANU is thought of as heaven and KI, Earth. In the same fashion, EA, after having established the first five of the seven original settlements on Earth, was given the title of EN.KI ("Lord of Earth"). KI also conveys the meaning "to cut off, to sever, to hollow out." Its derivatives illustrate this. For example: KI.LA meant "excavation", KI.MAH "tomb", and KI.IN.DAR "crevice, fissure." In Sumerian astronomical texts, the term KI was prefixed with the determinative MUL ("celestial body"). Thus MUL.KI meant "the celestial body that had been cleaved apart", a reference to the creation of Earth from Tiamat. Over the years, the pronunciation of KI change to GI, and ultimately to "geo" (as in geo-graphy, geo-metry, geo-logy, etcetera).

The cosmology of the Sumerian gods and their related planets has its counterpart in the Greek version. In the eighth century B.C.E., Hesiod began the divine tale of events that ultimately led to the supremacy of Zeus with:

"Verily, at first Chaos came to be, and next the wide-bosomed Gaia,
She who created all the immortal ones; Who hold the peaks of snowy Olympus:
Dim Tartarus, wide-pathed in the depths, and Eros, fairest among the divine immortals...
From Chaos came forth Erebus and black Nyx; and of Nyx were born Aether and Hemera."

It seems apparent that Gaia was the Greek equivalent of Tiamat, and the divine pairs of Tartarus and Eros, Erebus and Nyx, Aether and Hemera were the Romans' Mars and Venus, Jupiter and Saturn, Uranus and Neptune, just as in the Sumerian version. At the same time, Ouranos ("Heaven") came about, according to Hesiod's Theogony, in a similar way as well:

"And Gaia then bare starry Ouranos -- equal to herself --
To envelop her on every side, to be an everlasting abode place for the gods."

Later on after the battle, Hesiod spoke of Gaia as being the half equivalent to Heaven: on one side she bore Urea, who "brought forth long hills, graceful haunts of the goddess-Nymphs"; and on the other side "she bore Pontus, the fruitless deep with its raging swell." The former would be on the side the firm lands had formed from the crust of Tiamat; while on the other side there was a hollow, an immense cleft into which the waters of the erstwhile Tiamat must have poured. In effect, the fruitless Pontus was the Pacific Ocean (where no fruit grows in the extensive salt water). Apparently, the authors or compilers of the Book of Genesis were not the only ones accepting the Sumerian cosmogony (and also editing it to outline the derivation of their own gods and Olympic dynasties).

c. 429,000 B.C.E. Four Sars later, more Anunnaki arrive on Earth, among them a Chief Medical Officer, Enki's half-sister Ninharsag, (also referred to as NINTI). The Anunnaki came in groups of fifty, as attested to by one of these groups led by Enki's firstborn son,

MAR.DUK. *Marduk describes to his father an "attack" on his spacecraft by one of the solar system's larger planets (possibly Jupiter):*

"It has been created like a weapon;
It has charged forward like death...
The Anunnaki who are fifty it has smitten...
The flying, birdlike Supreme Orbiter it has smitten on the breast."

It's relevant that these "gods" were not exactly omnipotent... *They could certainly get smitten, and space travel was not to the level of being routine. In fact, as will become more and more apparent in these Annals, just getting from place to place on the Earth had a degree of adventure.*

There is also the suggestion from the Sumerian artifacts that Marduk may have landed on Mars on the way to Earth, that Mars may have been a way station between Nibiru's crossing in the asteroid belt and Earth. This clue comes from a depiction on a Sumerian cylinder seal. It also makes sense in that Nibiru making its perihelion approach to the Solar System may have been a lot closer to Mars at the time in its orbit than to Earth (which might easily have been on the other side of the Sun at the time).

Other more indirect evidence includes photographs by the NASA's Mars probe, Mariner 9, which shows what has come to be called "The Face on Mars." The photo includes the "Face" and what appears to be an extensive area of pyramids and other allegedly artificial structures, all faintly reminiscent of ancient cities and major religious areas on Earth. Any one of the latter might easily be attributed to the Anunnaki, at least in terms of their design and layout.

c. 414,000 B.C.E. Four Sars later -- about 15,000 Earth years -- it finally becomes apparent to the Anunnaki that the gold production from seawater is failing to live up to expectations. More significantly, these primeval prospectors were now faced with a decision to either abandon the project -- essentially out of the question -- or try to obtain the gold by other means, for example, by mining. The Anunnaki had by this time become aware that gold was available in abundance in the AB.ZU, "The Primeval Source". Scholars, such as Sitchin, interpret this to mean South Africa. But such deep mining in "the place of the shining lodes", would necessitate expanding the processing facilities and fabricating ore vessels (MA.GUR UR.NU AB.ZU -- "Ships for Ores of the Abzu"). It was major decision time!

And as in all major decision times, the ones on the front lines, Enki and the Gang, were not going to be allowed to make the decision. This was one for the Big Guy. And so, eight Sars and 28,800 Earth years after Enki's landing, the Supreme Anu arrived on Earth for a closer inspection. His gold train had been threatened! It was time to step in and act authoritative.

Anu arrived on the scene, however, with considerable baggage. His retinue included his heir apparent, ENLIL ("Lord of the Command"), and the young Kumarbi (the grandson of the same Alalu whom Anu had deposed). One can rationalize Anu's decision to include Kumarbi on the basis that it might have been unwise to have left the young contender back home, close to the throne. As for Enlil, Anu may already decided there needed to be a shake-up in the Earth bureaucracy. As heir apparent, Enlil was the son of Anu and his half-sister ANTUM (or KI or URASH, later considered an Earth goddess). Enlil was someone whom Anu could trust, while Enki was the disenfranchised son who just might have a grudge or genteel upset toward Anu and the powers that be. Note in this regard that Enki was Anu's first-born son, but that Enlil was Anu's first-born son by his half-sister, and thus the legal heir apparent! This is the same situation on which is based the saga of Abraham, Sarah, Ishmael, and Isaac. In some respects, the Ishmael/Isaac rivalry was just a rerun of the Enki/Enlil one. But in the latter case, as we shall see, the stakes were much greater. It was the Earth that was up for grabs!

We might note in passing a fundamental factor in the succession code of the Anunnaki. The heir apparent is first and foremost the first-born son by a half-sister. Only if no such son exists do we drop down to the subsequent levels on which Enki found himself. Later on, we will see that Enki finds himself strongly attracted to Ninti, his half-sister, whose union could produce an heir that would have even more clout than Enlil's son by any other woman. The epic of this romantic interlude, we will return to in a later episode. For the moment, suffice it to say that the half-sister-wife rule of the Anunnaki surfaces in all of its glory with Abraham, who went to some lengths to emphasize Sarah's status as his half-sister and wife. From Genesis 20: 12:

> *"And yet indeed she is my sister; she is the daughter of my father, but not the daughter of my mother; and she became my wife."*

Curiously, scientists in 1980 found that given a choice, female monkeys preferred to mate with half-brothers. These preferred half-brothers shared the same father, but had different mothers. Other reports [Discover, December 1988] showed that "male wasps ordinarily mate with their sisters, but preferentially mate with half-sisters, those with the same father but different mother." One might wonder if the succession code of the Anunnaki is more than just whim. Or if these older and wiser gods have determined in their near omniscience that: Incest really is best!

The importance of lineage and genealogy with the Anunnaki cannot be stressed enough, as it will be intimately involved in the struggles for succession and supremacy in the coming millennia. Much of the ferocity of these later wars stemmed from a code of sexual behavior based not on morality but on considerations of genetic purity; that-is-to-say, sexual acts were judged not by their tenderness or violence, but by their purpose and outcome. Furthermore, while the code prohibited marriage (but not lovemaking) between full brother and sister, marriage with a half-sister was not only allowed, but the male progeny by a half-sister even had precedence in the hierarchical order. And while rape was condemned, sex -- even irregular and violent -- was condoned if done for the sake of succession to the throne. Finally, the same code which condemned rape did not prohibit extramarital affairs per se.

Meanwhile back at the gold-producing crisis, Anu took stock of the situation, showed why he was on the throne, and made the command decisions: Enlil would take command of the Earth operations and organize the gold deliveries back to Nibiru. At the same time, Enki was to be demoted as Lord of Earth, and given a lesser command, the actual mining in the Abzu; and would be allowed to keep his abode in Eridu (housing prices, at the time being at an all time low). But the latter gesture wasn't sufficient to prevent heated arguments from Enki, who threatened to return to Nibiru. More than the loss of the command of Earth, there was the fact Enki had lost it to Enlil!

The choice of Enlil for command of the Earth mission might have been a necessary one, but it greatly sharpened the rivalry and jealously between the two half-brothers. Enki had already had to deal with being disenfranchised as heir apparent when Anu's half-sister wife Antum had bore Enlil. Anu, thus, had to deal with Enki's outrage at having to step down in favor of Enlil, as well as being chastised for the lack of gold-production. As the scientist-engineer in charge, Enki could easily have taken it as a professional insult to have his plans be universally acknowledged as failing to meet expectations. Ultimately, it was decided to draw lots; allow chance to determine how it would be.

"The gods clasped hands together, then cast lots and divided: Anu to heaven went up; To Enlil the Earth was made subject; That which the sea as a loop encloses [South Africa?], They gave to the prince Enki.

To the Abzu Enki went down, assumed the rulership of the Abzu."

Unfortunately, for Anu, the drama was not yet over. Kumarbi had been left by Anu on the space platform orbiting the Earth. When Anu returned "up to heaven" (or at least, enroute), the two "gave battle" to one another. As Kumarbi momentarily bested Anu in the wrestling (the Anunnaki's preferred method for settling differences), "Anu struggled free from the hands of Kumarbi". But then Kumarbi managed to grab Anu by his feet, and "bit between his knees", hurting Anu in his "manhood". Ouch, that must have hurt. (This, if you can believe it, was a typical "hold" in Anunnaki wrestling.) Anu then took off for Nibiru, disgraced and in pain, leaving Kumarbi behind with the IGIGI manning the space platform. Thus was delivered in the classic fashion of the Anunnaki the first blow that would ultimately pave the way for the "War of the Olden Gods." In the interim...

c. 400,000 B.C.E. The Anunnaki begin arriving on Earth in larger numbers. Ultimately as many as six or seven hundred resided at any one time somewhere on Earth. Some were assigned to the "Lower World" to help Enki mine the gold, some manned the ore ships between the Abzu and Mesopotamia, some remained on the space platform orbiting the Earth, and some stayed with Enlil in Mesopotamia -- the latter controlling the all important spaceport. According to the Sumerian texts, Enlil established settlements according to a master plan:

"He perfected the procedures, the divine ordinances; established five cities in perfect places, called them by name, laid them out as centers,

The first of these cities, Eridu, he granted to Nudimmud, the pioneer."

(Sounds like Enlil is doing a bit of revisionist history himself, claiming his generosity in "granting to Nudimmud" the city of Eridu, when Anu had already done it! Ah well, such is imperial life.)

Meanwhile, each city was given a specific function. E.RI.DU ("House in Faraway Built"), continued as the gold-extracting facility by the water's edge (and which for all time remained Enki's home away from home). The others included BAD.TIBIRA ("Bright Place Where the Ores Are Made Final"), the metallurgical center for smelting and refining; LA.RA.AK ("Seeing the Bright Glow"), a beacon-city to guide the landing shuttlecraft; SIPPAR ("Bird City"), the Landing Place; and SHU.RUP.PAK ("The Place of Utmost Well Being"). The latter was considered the medical center and was placed under the control of SUD / NINHARSAG / NINTI ("She Who Resuscitates"), a half-sister of both Enki and Enlil.

Later, other cities were established including LA.AR.SA ("Seeing the Red Light") and NIBIRU.KI -- Nippur in Akkadian -- ("The Earth-Place of Nibiru"), essentially the mission control center. As such, Nippur was complete with a DIR.GA ("Dark, Glowing Chamber") where space charts ("the emblems of the stars") were displayed and where the DUR.AN.KI ("Bond Heaven-Earth") was maintained. Finally, the IGI.GI ("Those Who See and Observe") remained on the space platform in constant Earth orbit. Refined metal from the smelters at Badtibira was sent aloft to the Igigi, until such time as the gold was transferred via spaceships periodically to Nibiru. Were these Anunnaki organized, or what!? Of course it took them 10,000 to 20,000 Earth years to get things running smoothly, but hey, it worked! But then, just when it looked as if things could become routine, things became very complicated again.

c. 380,000 B.C.E. In the quintessential example of how the Anunnaki and Nibiruans take forever to progress in any meaningful way (at least in terms of Earth years), the lingering and bitter struggle between the House of Anu and the House of Alalu broke out in the "War of the Olden Gods". This war, upon which the Greek "War of the Titans" is undoubtedly based, pitted "the gods who are in heaven" against the "gods who are upon dark-hued Earth". (What was going on back on Nibiru is not addressed by the Sumerian texts -- the texts almost always restricting themselves to only those portions of the conflicts and events which directly affect the Earth.)

Gaining the support of the Igigi (who had been going around in circles in their space platforms and were probably bored, as well as possibly atrophying from the weightlessness), Alalu's grandson Kumarbi tossed aside his effectively tenured job as cup-bearer, and attempted to seize mastery over Earth. Kumarbi first attempted to enlist Enki in his cause -- thinking, obviously, that Enki just might still be a little bitter about the "recent" demotion. The Anunnaki do, after all, know how to carry a grudge... For millennia!

Subsequently, Kumarbi attempted to seek help from Lama, "mother of the two gods". But as we all know, the plans of mice, men and Anunnaki oft times go astray, and Anu got wind of the activity. Deciding once and for all that enough was enough, Anu ordered Enlil's son to find Kumarbi and kill him. Ferocious battles then ensued between the terrestrial gods led by Enlil's son and the sky-borne gods led by Kumarbi. In one battle, no less than 70 gods participated, all riding in "celestial chariots". It was Star Wars in the very neighborhood of Earth! In the end, the son of Enlil (and grandson of Anu) prevailed against Kumarbi.

There is some question as to which of Enlil's son was the avenging god. In the Hittite text entitled by modern scholars, The Kumarbi Cycle, the avenging god is identified as Enlil's youngest son, ADAD (also known as the Storm God Teshub and the principal Hittite deity). However, in the Sumerian version, in the tale known as The Myth of Zu, the hero is NINURTA, Enlil's first born son by his half-sister Ninharsag (and therefore the Heir Apparent to Enlil). Even in the Hittite version, though, Adad is assisted by his older brother.

For our purposes, it appears that the Sumerian version is more pertinent, in that it provides additional detail to the war, in particular to the attempts by Zu / Kumarbi to take control of Enlil's Mission Control Center in Nippur. For it was there in the DIR.GA room, the most restricted, holy-of-holies room, where the vital celestial charts and orbital data panels -- the "Tablets of Destinies" -- were installed and maintained. Control of this sacred chamber could conceivably be used to control the fate of the Anunnaki on Earth as well as on Nibiru. In getting there, Zu apparently had help from Enki (who was never loath to put a bee in Enlil's bonnet). Enki, aware of Zu's ancestry (the grandson of Alalu), suggested to Enlil that Zu be allowed into his service.

"Your service let him enter, in the sanctuary, to the innermost seat, Let him be the one to block the way.

To the words that Enki spoke to him, the god [Enlil] consented. At the sanctuary Zu took up his position at the entrance to the chamber."

As Zu stayed by his post, he "constantly viewed Enlil, the father of the gods, the god of the Bond-Heaven-Earth [communications post?]... his celestial Tablet of Destinies Zu constantly viewed." Soon a scheme took shape. "The removal of Enlilship he conceives."

"I will take the celestial Tablet of Destinies; The decrees of the gods I will govern;

I will establish my throne, be master of the Heavenly Decrees; The IGIGI in their space I will command!"

"His heart having thus plotted aggression," Zu saw his chance one day as Enlil went to take a cooling swim. "He seized the Tablet of Destinies in his hands" and in his Bird "took off and flew to safety in the HUR.SAG.MU ("Mountain of the Sky-Chambers"). No sooner had this happened than everything came to a standstill:

"Suspended were the divine formulas; The lighted brightness petered out; silence prevailed.

In space, the IGIGI were confounded; The sanctuary's brilliance was taken off."

At first, "father Enlil was speechless", the latter a possible reference to communications being cut. But when Anu on Nibiru was informed of the coup, the order to capture Zu was clear. (It was a Zu coup!) There were, however, few volunteers to go chasing after Zu. The apparent reason was that Zu had also taken "Enlil's brilliance", a powerful weapon. It's not altogether clear what the weapon was, but one must suspect it was a bit more than a polished sling shot, something that the average Anunnaki might be loath to challenge.

But then, in the true spirit of a futuristic Hollywood script, the young hero, Ninurta stepped forward, ready to do battle for god Anu, country Earth, and Mom. Of course, he was given the slightest of motivations by his mother who pointed out Enlil's loss of the throne was ultimately Ninurta's loss! Mom armed Ninurta with weapons equal to Zu's stolen weapons, including a few of her own design. (There's nothing like having a mother who keeps a lethal arsenal in her broom closet!)

The ensuing battle had all the earmarks of nuclear weapons, guided missiles, and fighter aircraft. Had Homo erectus occasioned to look up, he would undoubtedly have been impressed -- even if he hadn't had the slightest idea of what was going on. Ultimately, Ninurta defeated Zu, and in a subsequent trial was given the right to cut Zu's throat -- which he did. Forthwith.

However... As a condition for his volunteering to go after Zu, wily old Ninurta had extracted from the other Anunnaki, a promise to ensure that the vanquisher of Zu was appropriately rewarded:

"Thy name shall be the greatest in the Assembly of the Great Gods;
Among the gods, thy brothers, thou shall have no equal;
Glorified before the gods and potent shall be thy name."

While this is all well and good for the war hero, it had an unintended, undesirable effect. It planted the seed for future conflict by establishing Ninurta higher in the hierarchical order than even Enki. And while Ninurta was indeed Enlil's Legal Heir on Nibiru, having been born there, he was not necessarily the next in line on Earth! Thus the son of Enlil and a son of Enki were ultimately destined in the future to battle for control of the Earth. But such is still in the future. Try to be patient.

Meanwhile, back at the zoo, Zu (Kumarbi?), prior to his demise, had managed to impregnate a goddess of the mountain. It's the old soldier-off-to-the-war trick -- works almost every time. And in typical fashion, this dalliance led to the birth of a possible avenger, the "Stone God ULLIKUMMI.

Ullikummi grew up in secret, but eventually, the Sumerian Sun God UTU, a grandson of both Enki and Enlil, saw Ullikummi roaming the skies one day and informed Ninurta. Enlil's son promptly attacked Ullikummi, but at first to no avail. Even when his brother Adad joining the battle, Ninurta continued to be unsuccessful. The two then went to Enki in the Abzu, to seek an oracle according to "the old tablets with the words of fate". (Essentially they wanted some advice from the uncle known as the God of Wisdom.)

Enki realized that Zu / Kumarbi had created a monster, and Enki, himself, went to Enlil to warn him of the danger. Enki also brought a solution: "let them bring out the Olden Metal Cutter, and cut under the feet of Ullikummi the Stone God". Crippled, the Stone God was still defiant, even to the end when Ninurta caught up with him at sea and engaged him in a final battle. Ultimately, the Enlilites won this final phase of the War of the Olden Gods, what the Greeks called the Battle of the Titans. The meaning of the Greek word "Titan", incidentally, may appear obvious, but it is worth noting that the word, TI.TA.AN, in Sumerian means "Those Who in Heaven Live". This is rather precisely the designation of the Igigi led by Zu / Kumarbi.
****

Up to this point, other than references to Anunnaki colonies in the Tigris-Euphrates valley and South Africa, much of these Sumerian tales could apply to any number of locales. There is, for example, only minimal mention of any activity that might definitely connect the Anunnaki antics to the Annals of Earth. However, that is about to change. Suddenly the lives of Anunnaki and mankind are about to become irrevocably entwined. Suddenly, our very concept of what it means to be human and from whence we came, is about to be severely challenged. It all began with a mutiny. Not on the Bounty, mind you, but at the source of all the activity, the gold mines.

c. 270,000 B.C.E. Something on the order of 30 Sars (110,000 years) later, or about 40 Sars (144,000 years or 40 "counted periods") after the mining of the Abzu began, the Anunnaki toiling the minds did something quite extraordinary: They mutinied! From The Alta-Hasis Epic, we find:

"Inside the mountains, in the deeply cut shafts, the Anunnaki suffered the toil; excessive was their toil for forty counted periods.

[The Anunnaki] suffered the toil day and night, they were complaining, backbiting, grumbling in the excavations."

The Sumerian texts go on to relate in vivid detail the fact that the miners mutinied, and marched on Enlil's quarters in the middle of the night (apparently receiving night visitors being one of the joys of being Top Gun on Earth). Taking Enlil prisoner, the mutineers then demanded that Anu be sent a message and asked to come to Earth to negotiate. Enlil quickly obliged them, but upon Anu's arrival, Enlil called for a court-martial before the Great Anunnaki to be convened. Enlil, rather clearly, was not one of those diplomatic facilitators with consummate skills of negotiation and compromise.

At the court-martial, according to the texts, "Enki, Ruler of the Abzu, was also present." Enlil took this opportunity to accuse Enki of being the instigator of the mutiny. This didn't fly with the Great Anunnaki, even though it was certainly not beyond Enki to have planted a few seeds of discontent in order to get Enlil's goat (Enlil's goat being an award-winning goat, one that all the Anunnaki coveted). But then, not getting the support of Anu, Enlil offered his resignation: "Noble one, take away the office, take away the power; to Heaven will I ascend with you."

Anu, however, did not bite. Instead, he calmed Enlil, while at the same time expressed his understanding of the miner's hardships. Anu was, of course, playing the role of ruler/diplomat, attempting to offend no one and managing to take both sides at the same time. Not unexpectedly, he thus created a situation with little or no hope of a solution: Someone had to mine the gold, it was really tough on the Anunnaki miners, the mining process was indeed arduous, and yet, still, someone had to mine the gold. Ye olde vicious circle. The problem would simply not go away. Then, into this quandary, rushed Enki.

Enki "opened his mouth and addressed the gods." He and his Chief Medical Officer, his half- sister Ninti, had a plan: a brilliant, and simultaneously, brazen and outrageous plan:

> *"Let her create a Primitive Worker; and let him bear the yoke...*
> *Let the Worker carry the toil of the gods, let him bear the yoke!*

In the following one hundred lines of the Alta-Hasis text, and in several other "Creation of Man" texts that have been discovered in various states of preservation, the tale of the genetic engineering of Homo Sapiens has been told in amazing detail. Man, Homo sapiens sapiens, was to be created through the wonders of genetic engineering for the sole purpose of mining gold for the Anunnaki. The yoke was on Man! *(Somehow, we always suspected something like this.)*

To achieve this feat, Enki suggested that a "Being that already exists" -- Apewoman -- be used to create the Lulu Amelu ("The Mixed Worker"). This would be done by "binding" upon the less evolved beings, "the mold of the gods." It was time to mold the mitochondrial and nucleic DNA of Homo erectus (or possibly Archaic Homo sapiens) to evolve the Ape-man into a gold miner, complete, no doubt, with a jaw full of chewing tobacco and a pickax.

Where have we heard of this scenario before?

> *"And God said, Let us make man in our image, after our likeness...*
>
> *"So God created man in his own image, in the image of God created he him; male and female created he them." [Genesis 1:26, 27]*

If you've ever wondered why God said, Let "us" make man in "our" image... Now you know. Suddenly, the verse which makes absolutely no sense, becomes crystal clear. The "us" was Enki and Ninti, while the "our" was that of the Anunnaki. The Biblical version, therefore agrees with the Sumerian. And on the scientific front, the agreement is similarly precise. The Sumerian timing, for example, was perfect, the evolution of mankind beginning some 270,000 years ago. This, of course, is an agreeable compromise date between the different geneticists and paleoanthropologists immersed in the mtDNA debate.

Sumerian, Biblical, and Scientific sources agree on the beginning of mankind!

But before you identify the God of Genesis with Enki (whose "abode" is our solar system), note that it was the goddess Ninti who purified the "essence" of a young male Anunnaki and mixed it into the egg of an Apewoman." It was the goddess that implanted the fertilized egg into the womb of a female Anunnaki, and then when the "mixed creature" was born, it was Ninti who lifted him up and shouted: "I have created it! My hands have made it!" Not surprisingly, Enki and Ninti called their creation, Adapa, similar to the biblical, Adama. Eve, apparently, was the female Anunnaki! Maybe.

Irregardless of the exact identity of Adam and Eve, the "primitive worker" -- Homo sapiens -- had come onto the scene through a feat of genetic engineering and embryo-implant techniques. In effect the Anunnaki had taken a hand in the long process of evolution, jump-starting it and "creating" Man sooner than he might otherwise have evolved on his own. In effect the "missing link" in man's evolution turns out to be a feat of genetic engineering performed in an ancient laboratory of the Anunnaki ("Those who from heaven to earth came"). Suddenly we can explain why the progenitor of both Neanderthal and Cro-Magnon Man (Homo sapiens sapiens) appeared so soon on the scene, a mere million years after Homo erectus' initial foray into the wilds of Africa, Asia, and Australia.

It is, of course, worth mentioning that the creation of the "primitive worker" was easily 100,000 years before Neanderthal Man. As for Cro-Magnon Man's appearance, circa 90,000 B.C.E., the time frame is even later. The mitochondrial DNA dating is right on target, but the fossil remains of both Neanderthal and Cro-Magnon Man simply do not go back that far.

This variance, however, does not imply a serious problem, in that the "primitive worker", potentially the first in the Neanderthal-line, would not have left a lot of fossil remains for the benefit of paleoanthropologists in a later era, in that they would have been working (and probably dying) in the deep mines of South Africa. Furthermore their remains might have been disposed of by the Anunnaki in a fashion not conducive to becoming fossils later on. For these reasons, it is reasonable to assume that the "quantum leap" from Homo erectus to Neanderthal may have been due to the intervention of Enki, Ninti, and the Anunnaki.

One should also keep in mind that it is unlikely that God-Enki-Ninti would get it right on the first try. The first experiments in producing the "primitive worker" may very well have failed, but yielded enough information to continue the process, improving on each new attempt. According to the Sumerian texts, several kinks did in fact occur in the process. But then, after considerable trial and error, a mass-production process was launched, with fourteen at-a-time "birth goddesses" (female Anunnaki) being implanted with the genetically manipulated eggs of the Apewoman. Eventually, the process provided sufficient workers in the mines for the Anunnaki to retire (except for supervisory roles), and then later to work the land in Sumeria, the land between the Tigris and Euphrates rivers.

At the time, the "primitive workers" were still pretty crude: "When Mankind was first created, they knew not the eating of bread, knew not the dressing of garments, ate plants with their mouth like sheep, drank water from the ditch..." (Reminds me of some people I know in New York City.) The workers were good enough for the purposes of the mines, but they were not a thing of beauty and joy forever. They still lacked the all-essential style and grace of later years.

But Enki was not yet finished. It was clear to him and Ninti that the use of young female Anunnaki to perform the roles of "birth goddesses" was not a long term solution. The Anunnaki were not eager to work in the mines, but giving birth in an assembly line fashion did not have a great deal more glory or desirability. Accordingly, unbeknownst to Enlil, and with the connivance of Ninti, Enki contrived to improve upon the "primitive worker", and give the new creature one more genetic twist: granting to the hybrid beings -- incapable of procreating, as all hybrids are (mules, e.g.) -- the ability to have offspring, the sexual "Knowing" for having children. And while the original Sumerian text of this tale has not yet been found, a number of Sumerian depictions of the event were discovered.

"And God blessed them, and God said unto them, Be fruitful and multiply, and replenish the earth, and subdue it..." [Genesis 1:28]

There is the suspicion that Enki continued the work of improving (i.e. evolving) the primitive worker over a long period of time. It is entirely conceivable that his and Ninti's first efforts produced mule-like

Neanderthals (hybrids who were not procreating without the aid of the Anunnaki laboratory workers and female birth goddesses), then Neanderthals with the capability of procreating, and then in a second quantum leap, Cro-Magnon Man. The superior model might have then caused the first industrial lay off in human history with the Cro-Magnon Man replacing the Neanderthals in the mines, and with the early version of Homo sapiens sapiens discovering the distinct lack of unemployment benefits in the archaic world.

The Neanderthals (and later the Cro-Magnons) might also have occasionally left the mines of their own accord. In any case, the Neanderthals were soon roaming free in Africa, migrating, and then popping up all over in the fossil records, as they proceeded to "conquer" the rest of the world. And with their infusion of Anunnaki talents, skills, and genes (and quite possibly, tools), they probably had a much longer life expectancy than the other Homo erectus. The tools could easily have brought them into the Upper Paleolithic (blades, etc.), or just the Middle Paleolithic (when the newly freed workers would be aware of blades, and attempt to fashion a substitute). The paleontological evidence of Homo sapiens sapiens "outliving" and outperforming Homo erectus thus becomes clear.

It is also clear that the issue of "no intermixing" between the conquering Neanderthal (and later the Cro-Magnon invaders) -- with their superior genes, tools, etcetera -- and the conquered Homo erectus is quickly resolved as well. The Anunnaki had caused in the evolution of mankind a quantum leap, a clear distinction between what had gone before and a new breed. And this distinct difference was sufficient to keep any sexual intermixing to a minimum -- conceivably, even zero. Let's face it: How many humans develop a sexual yearning for chimpanzees?

The differences could, at least in the eyes of the Neanderthals and Cro-Magnons, have been as great. Furthermore, the lack of evidence of warfare between the species also makes sense, in that the vast superiority of the invaders made it clear to them that there was no competition, no fear of the Homo erectus as competitors, and thus no need for warfare. Moving onto an island prolific with monkeys is not likely to motivate a modern human inhabitant to declare war on the monkey population, i.e. no monkey holocaust. But the monkey population will just as surely go into serious decline as the human development of the island proceeds.

As this theory of an Enki/Ninti/Anunnaki genetic manipulation of the human species begins to register in your mind, you might take note of how many questions it answers. For example, in the twelve questions proposed at the end of the last Episode, note that virtually all are answered! (1) The Sumerians could have known about the cosmology of our solar system, because at some later date, the Anunnaki told their miners their history. Simple. (2) The concept that Nibiru somehow planted the "seed of life" on Earth is not as clear, but one at least recognizes that Nibiru did have life on it at the time. (There is also the realization that the Anunnaki may have had no motivation -- and there's no record that they ever did -- to confide in the Sumerians as to their thinking and private lives). Had the Nibiruans deliberately planted the seed of life on Earth, they may have chosen not to say why.

(3) The unusually slow mtDNA mutation rate of humans (as compared to other related species) could come from an infusion of Anunnaki genes. Furthermore, we have already commented on the longevity-implying-slow-progress theory as directly affecting the Anunnaki, and this would suggest a slow mtDNA mutation rate for the Anunnaki. It all falls together rather well. One can even quickly visualize the meaning of the phrase in Genesis 6:2: "the sons of God". In effect, this is a literal statement! And, as we will soon discover, most of Genesis becomes literal in this way!

(4) The question of the Neanderthal and Cro-Magnon "outliving" the Homo erectus they replaced is answered (see above). (5) Homo sapiens sapiens migrating Out of Africa (sounds like a movie title) and "denying their ancestry" becomes clear. One can easily visualize a "primitive worker" visiting the local career counselor and learning that mining is just not his bag -- no real chance for advancement or meeting his innermost needs. And at the same time, dashing off to become a big-game hunter, a prolific gatherer, or real estate tycoon, such a creature might have been embarrassed at his ancestry and former employment. Small wonder he or she kept it a secret (even from the fossil hunters of a later era).

(6) The no-intermixing issue has also been resolved (see above), and (7) one can even give credence to man evolving the ability to speak. Certainly, the genes for the physical equipment to be able to speak could have come from the Anunnaki, and there is also the possibility that speech would be encouraged in the mines (at least, discussions on the conduct of work, and the dire need for the development of the world's first employee suggestion box).

(8) The Anunnaki creating Man for work in the South African mines clearly locates mtDNA Eve in southern Africa (Botswana, for example), and allows for any subsequent migration from this central point. Shortly, we will point out that a secondary, and on a time scale, later, focal point will be generated in the Middle East -- just as the modern scientists of such topics would find evidence for. (9) There is even the possibility that Becky Cann's statement that the migrating Homo sapiens sapiens in leaving on a jet plane for Australia "knew exactly where they were going", can be explained. The Anunnaki could easily have told them.

The question (10) on races has not quite been answered, but there is the possibility that the different models of "primitive worker" created by Man may have resulted in significant variations, which later became type cast. There is also the possibility that some races did not make it (not to mention the fact that many offshoots might have been eliminated by the Deluge -- the latter subject to be discussed in great detail in a subsequent Episode). There is, finally, the "Tower of Babel" biblical episode which may account for a great deal of linguistic variations.

(11) Also, the potential for Cro-Magnon having superior evolutionary staying power over the Neanderthal may easily have derived from their acquiring such superiority from Enki and Ninti's tweaking of the "primitive worker". As to exactly what happened some 40,000 to 35,000 years ago to account for the rather abrupt disappearance of Neanderthal Man... Well, we can't give all the answers away. Besides, it's still in the future. There much more of a story to tell.

(12) But before we proceed, let us not forget that the half-sister rule of Abraham and Sarah's era clearly derives directly from the Anunnaki's rules of succession. As to the why of such rules, the Anunnaki were not into explaining themselves to their slaves.

Meanwhile, back at the mines, the Anunnaki who had been laid off from their work, were a bit at loose ends. Enlil, recognizing that an idle Anunnaki mind is an authority-questioning Anunnaki mind, took immediate action. Inviting the unemployed Anunnaki back to Mesopotamia, a virtual vacation paradise after the Abzu, Enlil thoughtfully put them to work digging ditches, raising dikes, and deepening canals. Theoretically, this was a fast improvement over their lot in Africa. However, it didn't take the Anunnaki long to yearn for another version of "undocumented workers" for whom they could assign all the dirty jobs (and at the same time, avoid paying social security taxes). Enlil (no longer a fan of house arrest) was quick to agree with the

Mesopotamian Anunnaki that the creatures with the thick black hair (man) were sorely needed in Mesopotamia as well as in the Abzu.

Naturally, Enlil was not allowed to just take the "primitive workers" from Enki's mines without a fight from the ruler of the Abzu. But eventually Enlil did his own version of a pre-emptive strike, and took many of the workers back to his place in Mesopotamia. You'll never guess what Enlil (and the Sumerian texts) called their little paradise-away-from-home! You guessed E.DIN? Okay, so you're beginning to catch onto the strange logic of all of this. But!!! If you're so smart...

Did you ever wonder why there are two creation of man epics in the Bible? If not, why not? Obviously, in addition to the latter verses of the first chapter of Genesis, there follows another version in the second chapter of Genesis, after God had taken to resting on the seventh day. It would appear that the first creation was for the miners of the Abzu, those created by Enki and Ninti in the image of the Anunnaki. The second creation of man was based on another motivation:

"And every plant of the field before it was in the earth, and every herb of the field before it grew: for the Lord God had not caused it to rain upon the earth [apparently spending all his time irrigating where it didn't rain], and there was not a man to till the ground." [Genesis 2: 5]

Rather precisely, the Bible makes it clear in this verse that Enlil's E.DIN did not have the slave labor for which the Anunnaki craved. The solution? A little raiding party, a man-rustling foray into the Abzu, to acquire someone other than the Anunnaki to "till the ground" of E.DIN.

"And the Lord God planted a garden eastward in Eden; and there he put the man whom he had formed [i.e. stolen from Enki's camp]. And out of the ground made the Lord God to grow every tree that is pleasant to the sight [with man tilling the soil, obviously it takes man and God to make a tree], and good for food; the tree of life also in the midst of the garden, and the tree of knowledge of good and evil." [Genesis 2:8-9]

This raises one of the most important aspects of the Sumerian and Genesis versions of the earliest times of man. In the Sumerian version, it was the Lord God, Enki (with help from his half-sister Ninti), who created man and woman in their image, and it was the Lord God, Enlil, who planted a garden eastward in Eden! There are, as a minimum, two gods, whose adventures are described in Genesis. The Hebrews, in their quest to attribute all creation to a single god, condensed the activities of the GODS into the character of a single deity.

As will become even more apparent in the continuing of these Annals, there are really TWO gods of Genesis. Only by understanding this can one hope to understand the early chapters of Genesis. The simple truth is that if one demands that there be only one god of Genesis, then one is faced with a picture of that same god, who is continually reversing himself.

For example, according to the Bible, God made man and placed him in Eden. Then he threw him and his spouse out for no good reason (more on this later). Then God repented "that he had made man" [Genesis 6:6] and determined to destroy him and his progeny. Then God changed his mind and decided to let Noah and his brood survive. Then God became upset with Noah's descendants and felt compelled to confuse his language. And so forth. As one progresses through the Bible, one is forced to question if the Biblical god has a multiple personality. Or is just psychotic. For the described Biblical god flips back and forth, is notorious for taking sides, and is radically unpredictable.

BUT! If there is more than one god in the proceedings, then things begin to make sense. Different peoples might choose one god as their favorite deity, and then defend his alleged superiority to the death. Which is precisely what the Hebrews did. They picked one God (as we shall see, Enlil and his heirs), attributed all of the good deeds of other gods to their god (in effect, ignoring the contributions of Enlil's rival, Enki et al), and then even went to the extreme to deny that any other gods ever existed.

The single-god hypothesis proposed by Genesis just doesn't make it! Once we let go of this artifact of a life-long brain-washing by the single-god fanatics, we find ourselves able to derive from the Biblical accounts a wealth of information. There's nothing like a basic truth to lighten our load.

Keep in mind one thing, however: A belief in a single deity, a supreme, divine being who created the universe, is NOT being questioned here. What is being questioned is that the "God of Genesis" is a single god. In fact, what is being strongly suggested (as a means of understanding the Book of Genesis) is that the God of Genesis is really an edited condensation of the "gods": Enki, Enlil, and possibly Ninti. Other Anunnaki also occasionally play a role, but primarily, it is the rival camps of Enki and Enlil that get all of the press in Genesis and the later books of the Bible.

Before we press on, we might note that nestled between Genesis 2:9 (when God made Eden) and Genesis 2:15 (when God "took the man, and put him into the garden of Eden to dress it and keep it.") there is the following very important and puzzling group of verses:

"And a river went out of Eden to water the garden: and from thence it was parted, and became into four heads. The name of the first is Pison: that is it which compasseth the whole land of Havilah, where there is gold; And the gold of that land is good; there is bdellium and the onyx stone. And the name of the second river is Gihon; the same is it that compasseth the whole land of Ethiopia. And the name of the third river is Hiddekel; that is it which goeth toward the east of Assyria. And the fourth river is Euphrates." [Genesis 2:10-14]

Let's see if we've got this right:

God created the Heaven and the Earth, created the flowers and the herbs, established night and day, created the creatures, formed man "of the dust of the ground", planted a garden in Eden, and then announced with great fanfare: "There's gold in them thar hills!"

Does this strike you as God having a strange sense of priorities? Does the God of creation seem overly concerned with the yellow metal? Is mentioning the rivers a means by which God suggests that Man pan for gold? Why is it that gold seems to play such an important part in the Annals of Man?

Of course, it all makes sense if one buys into the Sumerian texts and recalls that the "Prime Cause" for the Anunnaki to be on the Earth in the first place was for the location, mining, processing and shipping to Nibiru of gold in large quantities. What exactly "bdellium" constitutes is not clear from the text, but I would suggest it is the Elixir of Life, while the onyx stone is the Philosopher's Stone (the most sought-after elements of the Alchemists' trade). Bdellium" might also include the "rare earth elements" closely associated with gold in the Periodic Table of Chemistry, i.e. Ruthenium, Rhodium, Palladium, Silver, Osmium, Iridium, and Platinum. It is these seven additional elements, found in the same ores as gold, which may constitute the all important quest for gold. This is the same "gold fever" initiated by the Anunnaki, and carried on down through the ages by Man.

Stop and think about it: Why is gold so incredibly important to Man and his history? No other precious metal has the appeal, even though elements such as silver may have improved industrial uses. The fact remains that gold is the premier substance on which most of history revolves. The Sumerian texts do not explain why the Anunnaki were so gold hungry, but the evidence is prolific that they were after it in spades (the preferred means of mining gold). Furthermore, the Abzu was the place to get it, and the current residents of South Africa were not the first to work the mines.

In fact, realizing that sites of abandoned ancient mines might indicate where gold could be found, the Anglo-American Corporation, South Africa's leading mining corporation, in the 1970s (of the current era, C.E.) engaged archaeologists to look for such mines. Discoveries in Swaziland and other sites in South Africa indicated extensive mining areas with shafts to depths of fifty feet. Stone objects and charcoal remains established dates of 35,000, 46,000, and 60,000 B.C.E. for these sites! The archaeologists and anthropologists who joined in dating the finds came to the conclusion that mining technology was used in southern Africa "during much of the period subsequent to 100,000 B.C.E.!" In September 1988, a team of international physicists came to South Africa to verify the age of human habitats in Swaziland and Zululand. The most modern dating techniques indicated an age of 80,000 to 115,000 years for those areas with evidence of ancient mining.

If this doesn't fortify your resolve in believing the Anunnaki and their gold quest were for real, consider the fact that Zulu legends of the most ancient gold mines of Monotapa in southern Zimbabwe claim these mines were worked by "artificially produced flesh and blood slaves created by the First People." These slaves, the Zulu legends recount, "went into battle with the Ape-Man" when "the great war star appeared in the sky." [Indaba My Children, by the Zulu medicine man Credo Vusamazulu Mutwa.]

It doesn't take a creative or imaginative rocket scientist to conjecture Nibiru to be "the great war star", that the "artificially produced flesh and blood slaves" were the jump-started Homo erectus, or that "the First People" were the Anunnaki. Everything just fits!

There are, however, despite our honest endeavor to answer all of the questions prompted by the first four Episodes of these Annals, additional questions arising from our newly found knowledge:

1. What happened some 40,000 to 35,000 years ago to account for the demise of the Neanderthals and the unchallenged ascendency of the Cro-Magnons? Were the Neanderthals part of an Anunnaki factory recall? Is there any possibility of another recall in our own immediate future?

2. Will the black haired humans enjoy their sojourn in E.DIN? Will they pine away for the mines back home in Swaziland? Will they find farming and the glories of dressing and keeping E.DIN to their taste? Will they freely eat of every tree of the garden, save one? Will they get their forest merit badge?

3. Will Enlil find out that his newly acquired slaves, the advanced models of Homo sapiens sapiens, have the ability to create little sapiens of their own? Can Enlil possibly miss the fact that there are all these little sapiens running about? Will Enlil insist on being called their Godfather? Will the art of puns be introduced in time for the reader to understand the last joke? (*)

4. Will Enki and Ninti get into trouble for their creativity creating creative creatures such as man and woman? Will Anu ground them? Will Enlil have a cat? Will the female get blamed for everything?

5. Why is it Man and Woman can eat of every tree of the garden save two, "the tree of life and the tree of knowledge of good and evil"? Why are the trees with the most intriguing names, off-limits to the gardeners who are maintaining them? And why is it, immediately after mentioning for the first time these wondrous trees, that the Book of Genesis then launches off into a discussion of rivers and the fact that "thar's gold in them thar hills!"? Is there a connection here? Or are we barking up the wrong tree?

6. Why was gold so important to the Anunnaki? Could mankind similarly benefit by wise use of the gold? Are the Anunnaki likely to share it with us?

7. And in looking ahead to the next exciting Episode, who in the world are the Nefilim of Genesis 6: 4? Giants!? What about the Yankees? Don't they deserve some mention as well? What in the world am I talking about!?

8. Will there be no end to these questions?

For the answers to these and other exciting possibilities, stay tuned for the next thrill-packed Episode. Rest assured that if you thought the above was intriguing, the next Episodes will be even more astounding! To paraphrase John Paul Jones, "We have not yet begun to rattle the cages of outrageous fortune!"

(*) "Hanging is too good for anyone who makes puns. They should be drawn and quoted." [Fred Allen]

The Library of ialexandriah

The following articles are off the Internet, and some are copyright as per notices with the articles. They are a sample of what is available on these matters.

PART TWO - GENESIS

The first book of the Bible is a fascinating document. On the one hand, it is the basis for many religions -- either as a matter of faith or one of historical precedence. At the same time, it is a complete history spanning eons, but still capable of telling unique stories of individuals. In the original Hebrew, it is a masterpiece of Sacred Geometry -- wherein it incorporates the Geometry of Alphabets while recreating the story of mankind. But it is also the subject of a massive number of interpretations (and probably misinterpretations as well). In this latter respect, it is often more that of "a puzzle wrapped in an enigma."

There is, for example, in the King James version a distinction between "God" and the "Lord God". Chapter One is exclusively the province of "God", the creator of the heaven and the earth, while beginning at Genesis 2:4, the Lord God is supposedly in charge. It is almost as if "God" was the universal creator, while the "Lord God" was an earth-based deity. In fact, it was when (or after) God was resting on the "seventh day" that the Lord God began the "generations" on a whole new regimen. God created man in his image, and the Lord God used a dusty clay model. It's possible they're the same being, but the evidence is not strong in that regard. On the contrary...

There are the paradoxes, the apparent contradictions, the strange twists and turns of first one thing, and then, curiously, the seemingly opposite. Consider one the most obvious:

"And God said, Let us make man in our image, after our likeness..." "So God created man in his own image, in the image of God created he him; male and female created he them." -- Genesis 1: 26-27 [emphasis added]

Why the plural tense? Are we talking about the "royal we"? If so, then why was the proposal to make man in the plural tense, but the actual act, in the singular? Basically, we must ask, "What do you mean, "we", white man?" All of which is also applicable to:

"And the Lord God said, Behold the man is become as one of us, to know good and evil; and now, lest he put forth his hand, and take also of the

tree of life, and eat, and live for ever;" -- Genesis 3: 22 [emphasis added]

Become one of us, become a God? Man? Eat of the Tree of Life and live forever? It might be justifiable as punishment to send Man packing, but why bring up the tree of life thing? Obviously the tree's fruit was an eye opener for the naked couple, but...?

"And it came to pass, when men began to multiply on the face of the earth, [and presumably to do long division as well] and daughters were born unto them, That the sons of God saw the daughters of men that they were fair; and they took them wives of all which they chose." -- Genesis 6: 1-2

I can appreciate the daughters being fair, but "the sons of God" is not at all clear -- at least in the context of there being only one god. But if there were two or more... Then the "us" is clear, even if only one "he" was the final version of created man. And if more than two, why not have sons of God? (And presumably daughters?) A pantheon!

Biblical Scholars, including the Jesuits of the Catholic Church, have reluctantly had to admit that there must have been at least two gods in the story of Genesis. In fact, if the role model for Genesis was the Sumerian Epic of Creation, then it's pretty much of a done deal that there was a whole flock of Gods and Goddesses in the time before man. Not necessarily before the Chapter One Creator God, but by the time of the Garden, Adam and Eve, and so forth. This group of "lesser" gods and goddesses -- as distinct from the Creator God -- may indeed have been troubled by the possibility of man inadvertently joining their ranks by eating of the Tree of Life. They might have also been concerned about man's attempts to build new and wondrous things, such as:

"And the Lord said, Behold, the people is one, and they have all one language; and this they begin to do; and now nothing will be restrained from them, which they have imagined to do. Go to, let us go down, and there confound their language, that they may not understand one another's speech." -- Genesis 11: 6-7

If the local gods and goddesses are not omnipotent, but simply superior beings, then the concern for man's getting "uppity" makes sense. Otherwise, why would a tower built to reach heaven give any concern to a true Supreme Being. Did God actually think the Tower of Babel might reach the heavens? If your kid tells you of his or her decision to build a

Giza size pyramid in the backyard, is this going to cause you a great deal of worry?

Alternatively, was it simply a "tower" or something a bit more? Like a means of lifting off the planet? Ah, then, now there's a case of concern! Because the key phrase is: "nothing will be restrained from them, which they have imagined to do."

This latter phrase hasn't received a lot of press, even though it has been echoed in the words of Jesus Christ, i.e.

"Therefore I say unto you, What things soever ye desire, when ye pray, believe that ye receive them, and ye shall have them." -- Mark 11:24

But the idea is probably noteworthy. [The latter statement is the Grand Prize Winner for the Biggest Understatement within this website! Congratulations to the winner!]

Obviously, much of the early chapters in Genesis are perplexing in their interpretation. If one ignores the two creation of man stories -- Genesis 1:27 (the image thing) and Genesis 2:7 (the dust one) -- you've still got a lot of reversals and inconsistencies. For example:

[God placed Man in the Garden of Eden -- then threw him out. In the process, He blessed Man (Genesis 1:28), and then cursed him (Genesis 3:17-19), as in:

"...cursed is the ground for thy sake; in sorrow shalt thou eat of it all the days of thy life; Thorns also and thistles shall it bring forth to thee; and thou shalt eat the herb of the field; In the sweat of thy face shalt thou eat bread, till thou return unto the ground; for out of it wast thou taken; for dust thou art, and unto dust shalt thou return."

Hey man, you're dirt! On the other hand...

"Unto Adam also and to his wife did the Lord God make coats of skins, and clothed them." Genesis 3:21

Just as God show Adam and Eve the door, He gives them a door prize!?

[He cursed Cain...

"Now art thou cursed from the earth... When thou tillest the ground; it shall not henceforth yield unto thee her strength; a fugitive and vagabond shalt thou be in the earth." -- Genesis 4:11-12

And then He protected Cain!!!

"...Therefore whosoever slayeth Cain, vengeance shall be taken on him sevenfold. And the Lord set a mark upon Cain, lest any finding him should kill him." -- Genesis 4:15

[Man fell out of favor...

"And it repented the Lord that he had made man on the earth, and it grieved him at his heart." -- Genesis 6:6

But then...

"But Noah found grace in the eyes of the Lord." -- Genesis 6:8

We're either talking about a very inconsistent God -- even one possibly schizophrenic. Either than or something else.

The simplest answer is often the best. Therefore, how about the idea that there was more than one local god? This does not discount the possibility of a truly Supreme, Singular Being from creating the heaven and the earth, but the down-to-earth activities are inevitably the work -- and possibly the conflict -- of two or more gods.

Over the last fifty years or so, there have been found and interpreted, libraries of clay tablets from the Sumerian Civilization, circa 4,000 to 2,000 B.C.E. The Sumerian texts, specifically the Epic of Creation is essentially the long version of Genesis (or Genesis is the edited, condensed Reader's Digest version, or the executive summary of the Sumerian account. The Sumerian texts are on six tablets, with a 7th glorifying God -- akin to the 7 days of Genesis. The Epic details the creation of the planets (aka the firmament), and the separation of the "waters". (Keep in mind also that the Sumerian texts described the creation and some characteristics of Uranus, Neptune and Pluto -- with modern astronomy finding these planets only in the last 150 years or so.)

The latter brings up the critical question of: "How was the Sumerian knowledge obtained (who told them!)?

According to the texts: the Anunnaki (i.e. "Those who from Heaven to Earth Came").

Are there any more questions?

Okay. But pay attention here. This is a crash course. According to the Sumerian texts, a deposed leader of a planet called Nibiru discovered gold on Earth. This might not sound like a big deal for a species obviously capable of long range interplanetary travel, but gold figures heavily into the situation. Recall that:

In Genesis, after God made the heaven (the firmament) and the Earth, divided the waters, created grass, herbs and trees, placed lights in the firmament, created the fishes, the fowl, the great whales and animals, made man in his image, rested on the seventh day; after which he watered the earth (Genesis 2:6), formed man (again?) of the dust of the ground and breathed life into him (such that he became a living soul -- Genesis 2:7), created the Garden of Eden and placed man in it (Genesis 2:8), then grew the tree of life in the midst of the garden, and the tree of knowledge of good and evil (Genesis 2:9), arranged for four rivers, one to irrigate the garden (Genesis 2:10), and immediately thereafter (Genesis 2:11) announced, "Thar's Gold in Them Thar Hills"!

Strange. There is no mention of copper, aluminum, or carbon... Just... gold. Gold, apparently, is very important! Biblically, as well as to the Anunnaki -- the latter apparently spacefarers from the planet, Nibiru.

Again, according to the texts, thousands of years after the earth gold discovery, the gold rush began in earnest. The Anunnaki in charge was named EA (Nudimmud), or Enki ("Lord of Earth" -- EN, lord; KI, earth). Enki was an engineer/scientist. He established a base at the headwaters of the Persian Gulf, at Eridu (now considerably inland due to the sediment build up), in order to acquire gold from sea water.

[The fact the Anunnaki put an engineer in charge should give us considerable pause.]

Roughly 29,000 years later (greater than the time for the precession of the equinoxes -- and a strong implication of the Anunnaki extremely long lifetimes), it became apparent that acquiring gold from sea water was not living up to expectations. Anu (the head honcho of Nibiru) and his heir-apparent, Enlil ("Lord of the Command") arrived on the scene to set things straight.

Therein begin the earth portion of the continuing sibling rivalry saga of Enki and Enlil. One was the first-born son of Anu, and the other the first-born son of Anu and his half-sister. Remember the tale of Abraham, Sarah, Ismael, and Isaac? Same drill. Enlil had it over Enki (just as Isaac had it over Ismael), and their rivalry would last for some 450,000 years -- give or take an eon. Talk about carrying a grudge!

The Compromise Plan was to mine the gold. Enki was placed in charge of mining South Africa, while Enlil took over the administrative duties in Sumer. Thousands of years later, more Anunnaki began arriving, opening mines, and creating boom towns. (See the movie: Paint Your Spaceship.)

This was not a sterling plan, however. [pardon the pun] There were wars and mutinies. Turns out the Anunnaki are not into tens of thousands of years of working the mines -- even for gold. Enlil was then taken hostage, whereupon he blamed Enki (naturally!). But Anu knew better (it's the "Father knows best" syndrome). The problem was that mining was hard work -- and no fringe benefits for the last 130,000 years or so.

But Enki had a solution. He proposed to cross breed the Anunnaki with some local beasts known as Homo erectus, and make them do the work. Everybody agreed. (Well, all of the Anunnaki.) Enki's proposal was to combine DNAs -- i.e. create man in the image of the Anunnaki. Enki and his half-sister, Ninhursag (Ninti) began a program of genetic engineering and created ADAPA ("the mixed worker") or ADAMA. Ninti, given the job of carrying the creation to term, was able to announce, "I have created it!"

After Ninti's first born, the team resorted to 14 Birth Goddesses to begin an assembly line birthing operation for the new mixed workers. Unfortunately, the ADAPA was a hybrid -- i.e., he could not procreate. So this plan worked for a while.

But the assembly line goddesses found the program somewhat arduous. Assembly line birthing has never been what it's cracked up to be. For the goddesses it was less than appealing -- not a whole lot better than mining. Thus in a second act of creation of man [Ah, so! Two creation stories!], Enki and Ninti created a man and woman who could procreate. Enki just didn't tell Enlil about the new models. [The plot thickens.]

Meanwhile, Enlil had decided he wanted his own undocumented workers to do the ditch digging and crop raising in Mesopotamia. So he placed some of Enki and Ninti's creations in a place called in the Sumerian texts, E.DIN. To tend the garden, and the trees. Enlil, however, still thought they were hybrids and incapable of procreating, whereas Enki had sold him the new models. Boy was Enlil surprised! For, of course, the new state of (love) affairs eventually became obvious! These two had been eating of the Tree of Life! And for this crime, Enlil threw them out of the Garden -- the ultimate act in party pooping.

This then is the key! The solution to the Biblical paradoxes is that there were (at least) two gods -- and for the most part, working from contradictory agendas. In fact, most of the following history is based on the rivalry between Enki and Enlil, with the rest of the Anunnaki being split in their agendas as to who to support next.

According to the Sumerian texts, the Number One God was Anu (MARDUK in the Babylonian version), while in the creation of Man, Enki was God. The Lord God referred to either Enlil ("Lord of the Command") or Enki ("Lord of Earth"). Enki made man, Enlil created the Garden of Eden. Enki was the "serpent" who genetically engineered man so he could procreate (eat of the tree of knowledge of good and evil). Enlil is the God who threw man out of the garden, and Enki who clothed him. Enlil took Abel's offering, but ignored Cain's, cast Cain out and cursed him; Enki gave him his passport to freedom. The other players are identified in the Sumerian Family Tree.

The "sons of God" (not the Lord God), the "giants in the earth", were the Anunnaki who found the female half-breeds to their liking. Sumerian texts talk about the Anunnaki sons breeding with human women, and creating "mighty men of renown." Enlil hated it and vowed to kill off man. As luck would have it, he got help. The Deluge and Flood.

Roughly 11,600 B.C.E., Nibiru had a close encounter of the most important kind with Earth, triggering the ice cap of Antarctica to slide off and swamp the place. The Anunnaki had seen it coming and hauled ass while Enlil demanded that the humans not be forewarned! Enki by subterfuge saved Noah, the latter also known as Ziusudra (Sumerian) or Utnapishtim (Babylonian). Noah also saved the animals, the fowl, his family, and a fair number of laborers on the Ark. Enki told Noah to tell the city folk that he's building a boat to journey away from Enlil, who is mad at him and is causing untold misery -- and the town folk are only too eager to help build the Ark. ("Noah," by the way, means Respite. Things had been altogether too dry on earth prior to the Deluge and Flood.)

And of course, it was Enlil who did the Tower of Babel gig (about 3400 B.C.E.). Enlil is definitely not a fan of man! Even the stories of Abraham begin to make sense. Isaac and Ishmael are simply a reprise of the Enlil and Enki drama -- the drama which about 2000 B.C.E. flourished into all out war.

Abraham (as noted in the Chronicles of Earth) was the commander of an elite military, calvary force. In rescuing Lot, with his 318 well trained and armed men, Abraham was in the employ of Enlil. And with such credentials, Abraham, upon arriving in Egypt was able to immediately go before the Pharaoh -- which was not the privilege of most shepherds!

But the dramatic climax came at Sodom and Gomorrah, where Enlil's son went over the edge and used nuclear weapons to obliterate two cities in conflict with Enlil (a "grievous sin" from one point of view). The result was not only their fiery end, but anyone exposed to the blast was either incinerated or turned into a pillar of salt. Unfortunately, the fallout was even more grievous, in that the Sumerian Civilization met its end as a result of its being downwind from ground zero.

Now... Is any of the above, legitimate?

Basically, yes. The entire scenario is based on the Sumerian texts, and the thoughtful and insightful consideration of numerous scholars, including Zecharia Sitchin, Laurence Gardner, and many others. It is also based on hard science. In the latter category there is the evidence concerning:

The advent of Neanderthals and Cro-Magnon man (particularly in the timing of their appearances and evolutionary timescales);

The identification of the ancestry of Adam and Eve, from mitochondria DNA and the male equivalent;

An understanding of the varying lifetimes of The Adam's Family and their descendants, and the Sumerian King List ("Before the Flood" and thereafter); Comparative Religions

The identification of Adam as "Adama"; E.DIN as Eden, Enoch as Enmendaranna; Lamech as Ubar-tutu; Noah as Ziusudra, etceteras;

The fact of the Sumerian Civilization suddenly having all of the firsts: (aka The Me, which included: writing, law, proverbs, priests, animal husbandry, and genetic engineering of crops), and agriculture returning after the Flood (around 11,500 B.C.E.) in the highlands instead of the -- flooded -- valleys).

The gods of the ancient Egyptian Civilization: Ptah, Ra, Shu and Tefnut, Geb, Seth and Nephtys, Osiris and Isis, Horus, Thoth -- all being identical with their Sumerian counterparts, and the dating of Ra after the flood, Ptah's rebuilding, the creation of the Sphinx circa 10,000 B.C.E. (the Great Pyramid having been constructed earlier); and even

The early MesoAmerican Civilization, where the precursors to the Incas (the latter being the race immediately preceding the Dinkas and the Dos) created in what must have been a hell-of-a-place to start one, a civilization. The Andes was not an agricultural region, such as the Tigris, Euphrates, Nile, Indus valleys, but on the other hand, there was in abundance: gold and tin (the latter a critical ingredient in bronze). The Andes location was also at a high elevation in case of high water!

There is just too much evidence not to believe the plausibility of the above. But if you want more detail, simply refer to the Annals of Earth. The evidence is astounding!

But there is also an unanswered question in all of the above. What is so all-fired important about gold? (Besides the fact that it's "god" with an "l" inserted.)

In a word (or two), gold is the source for the ORME, the Star Fire, the "What is it?" of the Egyptian Book of the Dead, the "white powder of gold" of the Ha Qabala, the key ingredient in the long lives and powers of the Anunnaki -- and by implication of their step-children, the members of the human race. Gold is one route to the Tree of Life, as well as The Tree of Knowledge of Good and Evil. Gold is the premier example of the Precious Metals (gold, silver, rhodium, iridium, platinum, palladium, osmium, and ruthenium).

Genesis, therefore, is the story of such magnitude and majesty as to stun the imagination.

Genesis is, truly, "the origin, or mode of formation or generation," of humans for whom "nothing will be restrained from them, which they have imagined to do."

306 PART THREE – ANUNNAKI

Updated 22 August 2003

Genesis 6:1-4 reads:

"And it came to pass, when men began to multiply on the face of the earth, and daughters were born unto them, That the sons of God saw the daughters of men that they were fair; and they took them wives of all which they chose... There were nephilim in the earth in those days; and also after that, when the sons of God came in unto the daughters of men, and they bare children to them, the same became mighty men which were of old, men of renown." [emphasis added]

Nephilim is often translated as "giants", a legitimate and appropriate interpretation, but one which may be only partially accurate. A better definition might be "those who came down", "those who descended", or "those who were cast down." The Anunnaki of ancient Sumerian texts is similarly defined as "those who from heaven to earth came". Sitchin [1], Gardner [2], and Bramley [3] have all identified the Nephilim as the Anunnaki, more specifically, essentially the rank and file.

Virtually all open-minded historical and theological scholars agree the Old Testament's book of Genesis was extracted from the older Sumerian records, if only because of the similarity in their Comparative Religions. The Enuma Elish, the Sumerian Epic of Creation, and Genesis have a variety of common elements. Stories of a Great Flood and Deluge, among other stories, are also common to both Sumerian and Biblical accounts. An inevitable conclusion is that the Anunnaki were as real as Noah, Moses or Abraham.

Laurence Gardner [2] has written: "Every item of written and pictorial attestation confirms that the ancient Sumerians were absolutely sincere about the existence of the Anunnaki, and those such as Enki, Enlil, Nin-khursag and Inanna fulfilled earthly functions with designated community duties. They were patrons and founders; they were teachers and justices; they were technologists and kingmakers. They were jointly and severally venerated as archons and masters, but there were certainly not idols of religious worship as the ritualistic gods of subsequent cultures became. In fact, the word which was eventually translated to become 'worship' was avod, which meant quite simply, 'work'. The Anunnaki presence may baffle historians, their language may confuse linguists and their advanced techniques may bewilder scientists, but to dismiss them is foolish. The Sumerians have themselves told us precisely who the Anunnaki were, and neither history nor science can prove otherwise."

The Sumerian records recorded in great detail the stories of the Anunnaki, and among these, that of Enki, Enlil, Ninki, Inanna, Utu, Ningishzida, Marduk, and many others. Chief among these stories was the continuing conflict between Enki and Enlil, the sons of the supreme god of the time, Anu. Much of ancient human history, and the Biblical Genesis, can be explained as the militant differences between these two half-brothers, and how they affected the life of all sentient beings on Earth.

But the Anunnaki were more than just a pair of squabbling half-brothers. They were the council of Gods and Goddesses, who periodically met to consider their future actions with respect to each other, and probably as a smaller, nondescript item on their agenda, the fate of mankind. The Anunnaki, depending upon the context, were the Nephilim, the gods that Abraham's father, Terah, (according to the book of Joshua) was reputed to have served, the fallen angels, the lesser individuals of the race from which Anu, Enki, Enlil, Inanna and the other notables had sprung, and the "judges" over the question of life and death. They were in fact the bene ha-elohim, which translates as "the sons of the gods", or equally likely, "the sons of the goddesses." For example, from Psalm 82:

> *"Jehovah takes his stand at the Council of El to deliver judgment among the elohim." "You too are gods, sons of El Elyon, all of you."*

The Anunnaki have also been equated with the "Watchers" (who are also mentioned in the books of Daniel and Jubilees), i.e. "Behold a watcher and an holy one came down from heaven." -- Daniel 4:13

According to Zecharia Sitchin [1] and his interpretation of ancient Sumerian texts, the Anunnaki were extraterrestrials (aka "angels"?), who were an extremely long-lived race, potentially living as long as 500,000 years. Laurence Gardner [2] reduces this to more on the order of 50,000 years, and notes specifically that the Anunnaki were not immortal. He point out that no records are currently extant which relates to their natural deaths, but the violent deaths of Apsu, Tiamat, Mummu, and Dumu-zi are provided in some detail. (Sitchin and Gardner also disagree on the date of the Great Deluge/Flood; Sitchin assuming a time frame of 11,000 B.C.E., while Gardner assumes one of 4,000 B.C.E.)

Sitchin's book, The 12th Planet, published in 1976 was the first modern volume to begin to describe the Anunnaki, their arrival on Earth supposedly some 485,000 years ago, and from where they had come -- a planet called Nibiru. Sitchin believes Nibiru to be in an orbit about our sun, but in a strongly elliptical orbit which requires 3,600 Earth years to make a complete orbit. Nibiru's perihelion (closest point of approach to the Sun) is thought to be within the main asteroid belt between Mars and Jupiter, at a distance from the Sun of approximately 2.75 A.U. (an A.U. being the distance from the Sun to the Earth). (the Annals of Earth include a detailed description of how Nibiru created the asteroid belt by destroying a planet, Tiamat, in roughly the same orbit, and which created the Earth in the aftermath, the Earth being a remnant of the greater, destroyed planet.)

Nibiru is not known to modern astronomy primarily due to the extreme elliptical nature of its orbit and the fact its aphelion (furthest point in the planet's orbit from the Sun) is more than eight times the distance from the Sun to the planet Pluto (the latter being some 40 A.U. away, and thus the former, some 320 A.U. distant). Furthermore, Nibiru may be now far out in deep space and unlikely to be detected. (Or close by, e.g. Planet X.)

While Sitchin and Gardner may disagree with the extent of the long lives of the Anunnaki, it is clear that these gods and goddesses, baring accidents or "Anunnaki-cide", lived a very long time. It has also been theorized that because of their long lives, they do not quite move in "the fast lane" -- at least to the extent humans do.

This could be fundamentally important in that, quite possibly, the human life span, while enormously brief as compared to the Anunnaki gods and goddesses, might nevertheless be compensated by the humans possessing the ability to achieve a great deal in a relatively short time. The creativity of a shortened, and thus highly motivated lifespan is likely to be enormously greater than that of a god or semi-god resting on their laurels. This may also relate to the idea of why the gods and goddesses of the Anunnaki even bother with mankind. Humans may, on the one hand, act as workers to accomplish the Anunnaki's agenda, but an accelerated creativity may be well worth the trouble for the Anunnaki to manage a crew as motley as the human race.

But the connection between humans and the Anunnaki is much more profound than that of masters and slaves. All the evidence strongly advocates the concept that Adam and Eve and their ancestors, cousins, and what-have-you were created by genetic engineering and mixing the

DNA of Anunnaki with that of Homo erectus, the reigning progenitor of man at the time. Fundamentally, this was because the Anunnaki needed someone to work the mines in search of gold and other Precious Metals, and in all likelihood the ORME.

http://www.vibrani.com/Anunnaki.htm provides what just may be an insider view of the Anunnaki -- but from the perspective of Enki. The advantage of this link is that it provides extensive details on pre-Anunnaki history. While such channeled information is always speculative, it is nevertheless worthy of serious consideration.

The most fundamental question with respect to the Anunnaki is whether or not they're still on Earth! Sitchin [1] has pointed out that he never said they left (and there is no evidence that they did). There was, however, an apparently fundamental Anunnaki policy shift circa 600 B.C.E. wherein the overt, day-to-day interference in human affairs by the Anunnaki disappeared. There is also the scenario encapsulated in Richard Wagner's classic opera The Ring of the Nibelung, which included Night Falls on the Gods and the Entrance of the Gods Into Vahalla -- titles which are suggestive of possible changes in status of the Anunnaki. Finally, there is evidence to suggest that this state of affairs may be temporary, and may be scheduled to end with the end of the Mayan Calendar on or about 2012. A.D.

From mankind's point of view, the dysfunctional nature of the Anunnaki family, and the continuing rivalry of Enki and Enlil, may still be ongoing and having enormous effects on the quality of our physical, emotional, mental and spiritual lives. It's a very important question, and one that needs to be answered by each of us.

References:

[1] Zecharia Sitchin, The 12th Planet, 1976, The Wars of Gods and Men, 1985, Genesis Revisited, 1990, Divine Encounters, 1995, Avon Books, New York.

[2] Laurence Gardner, Genesis of the Grail Kings, Bantam Press, New York, 1999.

[3] Bramley, William, The Gods of Eden, Avon Books, New York, 1989, 1990.

The Library of ialexandriah

PART FOUR – ADAM AND EVE

The story of Adam and Eve is often treated as an allegory, but in reality is quite likely a great deal closer to factual history, a story of genetic manipulations at the Dawn of Man, by a group of extraterrestrials who commit the ultimate in "Prime Directive" Violations.

According to ancient Sumerian texts, as interpreted by Laurence Gardner [1], Zecharia Sitchin [2], and others, the Anunnaki ("those who from heaven to earth came") are extraterrestrials who arrived on the planet Earth after the discovery of gold by a deposed ruler of their race, named Alalu. The discovery eventually led to an Anunnaki mission to Earth to recover as much of this noteworthy example of the Precious Metals as possible.

The initial effort was led by Ea; whose title, Enki, meant "Lord of Earth", and who was the son of the new ruler, Anu. Enki (or Ea) set up shop at Eridu, near the northwest end of the Persian Gulf at the point where the Tigris and Euphrates Rivers meet the Gulf -- and at a time long before the silting of the two great rivers had extended the shoreline many miles to the southeast. At Eridu, Enki began to recover gold from sea water.

After this initial effort failed to produce the expected quantities of gold, an enlarged effort was commenced, this time under the command of Enlil ("Lord of the Command"), another son of Anu, and a half-brother to Enki. The new plan was to shift the operation from Mesopotamia to southern Africa (referred to in the Sumerian texts as Ab-zu), ship the gold back to Mesopotamia, and then lift it off the planet for trans-shipment to the home planet of the Anunnaki, Nibiru. (Nibiru is also a member of Earth's solar system, but has an extremely elliptical orbit and only reaches perihelion every 3600 years or so.)

At this juncture, there were supposedly 600 Anunnaki in the Netherworld (i.e. working the mines) and 300 in the heavens (doing the trans-shipments and minding the store). After a long period of time (Sitchin reckons the date as 300,000 B.C.E.), the Anunnaki who were laboriously mining the gold from the South African mines, mutinied! The mutiny and the difficulties the Anunnaki had encountered in working the mines were resolved, however, when Enki proposed to create a "primitive worker" to work the mines in lieu of the Anunnaki. Enki's proposal was, "Let us make man in our image, after our likeness."

Given the go-ahead, Enki and his half-sister, Ninki (Nin-khursag) created man, Homo sapiens, using genetic manipulation. They did so, allegedly, for the sole purpose of having workers to mine the gold for the Anunnaki, and thereby to quell the mutiny! The actual creation of Homo sapiens, as depicted in the ancient Sumerian texts in detail, was done by cross-breeding Homo erectus with that of the extraterrestrial Anunnaki! In other words, we're all half-breeds! (Except for a few notable personalities such as described in the Epic of Gilgamesh, who might better be described as one-third breeds. No kidding.)

Ninki, the Lady of Life (now we know where she received her title!), carried the first "mixed worker" to term and gave birth to a being she called the "Lulu." Later, fourteen "birth goddesses" (female Anunnaki) were used to produce additional workers. This solution was only a moderate success in that the Lulu was a hybrid and incapable of procreation (just as the mule, a cross between horse and donkey, cannot reproduce itself).

In addition, the "birth goddesses" had become weary of continually being pregnant!

Accordingly, it was back to the drawing board for Enki and Ninki. More work had to be done in genetically engineering a self-replicating, humanoid creature to be used as a slave.In the process, they may have encountered more than one dead end. It is possible, for example, that the legends of strange creatures and mythological monsters (from a Cyclops to a Hydra) may have arisen from the early experiments in genetic engineering, those which did not quite work quite so well. Cyclops, for example, was listed as the son of Neptune, the God of the Ocean -- another likely name or title for Enki (who was given the oceans as his bailiwick, and who was identified in the Sumerian texts with the planet Neptune. There is also the possibility that both the Neanderthal and Cro-Magnon species were simply different "models" of the genetic engineering project, and created for different forms of work -- ostensibly the Neanderthal being the mine workers, and Cro-Magnon being the new, improved version of Homo sapiens for the purposes of housework and domestic servants. The latter may also have been the precursor of Homo sapiens sapiens.

Eventually, Enki and Ninki were able to modify the genetic structure of the Lulu in order for it to be able to produce itself. Yea! They thus created the "Adama".

Significantly, they did so without Enlil's knowledge or approval! (Enki and Enlil did not really get along, a conflict stretching over eons.)

Once the interbreeding began to show some significant results, several Lulus were taken to Mesopotamia to work in agriculture (in Enlil's backyard, so to speak). There, they aided the Anunnaki efforts to raise food for themselves and their new workers. According to the Sumerian texts, Enlil created an E.DIN, a special place where new strains of edible crops could be developed and later implemented. Lulus were placed inside Edin, but without Enlil having learned of the Lulu's recently acquired talent for procreation.

When the obvious result becoming obvious even to Enlil -- when the Lulu's had in effect eaten from the Tree of Knowledge of Good and Evil (e.g., "sexual knowing"), the result was the Adamas' expulsion from Edin. At the same time, Enki, their paternal "creator", became the god who clothed the now homeless Adam and Eve on their way out the door.

The apparent duality of God in the book of Genesis is thus explained by the often opposing actions of Enki and Enlil. Enlil expels Adam and Eve; Enki clothes them. Enlil gives the old heave ho to Cain; Enki protects him. Enlil brings about a flood; Enki assists a Noah in building an ark. And so forth and so on. Even the Jesuits of the Catholic Church have begun to acknowledge the reality of at least two gods in the story of Genesis. Considering what we know of the Anunnaki and the elohim, there are a whole slew of Gods and Goddesses in the story of Genesis.

The scientific confirmation for the creation of mankind from Homo erectus comes in part from two sources. The first is the result of research by Cann, et al [3], where it was shown that mitochondria DNA (a form of DNA transmitted only maternally] could be shown to postulate a single woman living in Africa approximately 250,000 years ago who became the mother of every human being now living on the planet. Later, Dorit, et al [4] found no intraspecific polymorphism whatsoever in a gene paternally inherited, and concluded a date of the last common male ancestor to be roughly 270,000 B.C.E. These dates tie in well with Sitchin's argument, and tend to conflict with a purely evolutionary theory of humans evolving naturally from a survival of the fitness type scenario.

The conclusion is that Adam and Eve were real, possibly Lunatics, and were created by the genetic manipulation of cross-breeding Homo erectus with extraterrestrials from the planet, Nibiru. As such, they began a dynasty (i.e. The Adam's Family) of beings who lived to a much riper age than Homo sapiens sapiens are generally expected -- just part of their genetic heritage from the Anunnaki, who apparently live for hundreds of thousands of years. (There are, apparently, advantages to being a half-breed.)

References:

[1] Gardner, Laurence, Genesis of the Grail Kings, Bantam Press, NY, 1999.

[2] Sitchin, Zecharia, The 12th Planet, 1976, The Wars of Gods and Men, 1985, The Lost Realms, 1990, Avon Books, New York.

[3] Cann, R. L., Stoneking, M., and Wilson, A. C., "Mitochondrial DNA and human evolution", Nature, Vol 325, January 1, 1987.

[4] Dorit, R. L., Akashi, H., Gilbert, W., "Absence of Polymorphism at the ZFY Locus on the Human Y Chromosome," Science, Vol 268, May 26, 1995.

The Library of ialexandriah

PART FIVE – GODS AND GODDESSES

"We let them think of us as angels, or gods. It suits our purpose." "We must have them fear us and give reverence and adoration. It protects and conceals us. Yet we are no more divine than they."
(Anunnaki thinking, Bob, Jan.2010)

Gods and Goddesses

It should be readily obvious that the "God" who has been identified as the "Creator of Heaven and Earth" is by definition an extraterrestrial -- inasmuch as creating Earth implies both predating the Earth and not being "of Earth" at the time it was created. This Creator God, however, is not necessarily the God of Genesis, nor the gods and goddesses of the other religions of the ancient world. In fact, mankind may have very little traditional literature which describes this Creator God.

As for the "local gods and goddesses", numerous authors have discussed in great detail the stories of their lives (Mythologies, Ancient Myths, Descent into the Underworld), personality characteristics (Archetypes), the dysfunctional relationships between them, and to a lesser degree, their origin. Zecharia Sitchin [1], on the other hand, has argued in seven exhaustively researched books that the so-called gods and goddesses of the ancient world (including the god(s) of Genesis) were almost certainly extraterrestrials -- more specifically: mortal beings from another planet. Sitchin's work explains the particulars of the arrival of the Anunnaki ("those who from heaven to earth came"), and why these extraterrestrials are believed to come from a 12th planet called Nibiru.

Sitchin also describes the genetic experiments to create mankind, the story of Adam and Eve, the Garden of Eden, the events surrounding the biblical Flood and Deluge, and the early history of the world up until the time of Abraham (circa 2000 B.C.E.), including the reality of what happened at Sodom and Gomorrah and the Tower of Babel. Laurence Gardner [2] has independently described many of these same events, and also discussed the importance of Monoatomic Elements, the ORME, Star Fire, and related subjects. In Lost Secrets of the Sacred Ark, Gardner, in fact, goes into considerable detail about how the Anunnaki are connected to the all important subject of Gold. [This book is really a must read!]

The work of Gardner, Sitchin, and others makes it clear that the extraterrestrial "gods and goddesses" were intervening and lording over the human species with a vengeance! The Anunnaki, specifically Enki and his half-sister Ninki, were responsible for the genetic experiments which combined Anunnaki DNA with that of Homo erectus in order to create Homo sapiens sapiens, thus placing the human evolution far ahead of schedule and with the added ingredient of extraterrestrial DNA.

There is also the possibility that these experiments may have been responsible for the creation of Neanderthals (perhaps the early version), and even certain mythical creatures (in terms of genetic mishaps). While some of the later generation of Anunnaki were born on Earth (thus making them "terrestrials"), the "Prime Directive" Violations by these Anunnaki were extensive and far-reaching.

One of the best examples of interventions is the long lifetimes of The Adam's Family, the Biblical Patriarchs from Adam to Abraham. Prior to the Flood and Deluge, all of the patriarchs lived for roughly nine hundred years. (One exception is Enoch who did not die, but was taken up to heaven by "God". The other is Lamech, the father of Noah who lived a "mere 777 years".)

After the Flood and Deluge, however, there was a step wise reduction in the patriarch's lifetimes, from Shem (600 years) to roughly 450 years for the next three patriarchs, to an average of 222 years for the next six -- including Abraham. A curious result is that the five generations between Eber and Abraham were already dead and presumably buried, at a time when the four generations beginning with the son of Noah, Shem, were still alive (and presumably kicking). [Waiting for an inheritance must have been a lost art!]

Today, of course, average lifetimes are far less than a hundred years. This suggests a very directed and specific extraterrestrial intervention. This intervention is likely due to denying humans the ORME, Star Fire, or Monoatomic Elements, which may have been the key ingredients in keeping the patriarchs and their kin living to very ripe old ages.

One might also mention that there is nothing in Gardner's, Sitchin's or any other author's works to suggest that these extraterrestrials ever left the Earth! Just so you'll know.

On the other side of the planet, American Indian traditions have asserted that there are hundreds of extraterrestrial races which have routinely involved themselves in the affairs of Earth (and who are apparently continuing to do so). Specifics include: Jesus Christ allegedly being a Star Man [3], White Buffalo Calf Woman being an extraterrestrial -- who incidentally is returning in the immediate future [4], extraterrestrials being heavily involved in Iroquois traditions for tens of thousands of years [5], and 12 extraterrestrial cultures currently interested in the Earth, including the Pleiadians, Sirians, and Orions [6].

Other authors, such as Graham Hancock [7], William Bramley [8] and Erich Von Daniken in Chariots of the Gods have provided extensive documentation of advanced technologies in the Chronicles of Earth being held or wielded by peoples, who are now considered members of pre-historical and ancient civilizations -- and who supposedly had no such technologies. All of these authors have concluded, either mankind's previous civilizations were remarkably well advanced (and then later dissolved for some unknown reason), or that Earth was being visited on a regular basis -- and quite probably interfered with -- by extraterrestrials of one or more different cultures.

The evidence from ancient histories of extraterrestrials representing themselves as gods and goddesses is massive and virtually without alternative explanations. A review, for example, of such works as The Egyptian Book of the Dead [9] makes it clear that strange-looking, non-human beings were actively involved in terrestrial affairs. The traditional view that such depictions were based on mythological archetypes or fantasies of the allegedly backward ancients is nothing more than mainstream science burying its head in the traditional sand. Quite deeply, in fact.

Tricia McCannon [6] has stated explicitly, "All the ancient pictures are of real beings!" Beings from the constellation Pegasus, for example, are supposedly flyers. The Egyptian goddess, Isis, was, according to McCannon, a Sirian, while those from Sirius B had their mouths on top and looked to be a cross between a human and a fish. Drunvalo Melchizedek [10] places the Hathors (Hippo-like beings) as being from Venus. McCannon [6] also believes that Jehovah was definitely an extraterrestrial, but suspects that Yawyeh was more likely the "true God".

In effect, we already have picture galleries of extraterrestrials, and ample evidence of multiple "Close Encounters" of the most far-reaching kind. As Barbara Marcinak [11] once said, "The best secrets are carved in stone or on the ceilings."

References:

[1] Sitchin, Zecharia, The Wars of Gods and Men, Avon Books, New York, 1985; Genesis Revisited, Avon Books, New York, 1990; and others.

[2] Gardner, Laurence, Genesis of the Grail Kings, Bantam Press, New York, 1990; Lost Secrets of the Sacred Ark, HarperCollins, London, 2003.

[3] Standing Elk, Lecture at the Star Knowledge Conference, Lakota Indian Reservation, South Dakota, June, 1996.

[4] Hand, Floyd, Lecture at the Star Knowledge Conference, Lakota Indian Reservation, South Dakota, June, 1996.

[5] Underwood, Paula, Lecture at the Star Knowledge Conference, Lakota Indian Reservation, South Dakota, June, 1996.

[6] McCannon, Tricia, Lecture at the Star Knowledge Conference, Lakota Indian Reservation, South Dakota, June, 1996.

[7] Hancock, Graham, Fingerprints of the Gods, Crown Publishers, New York, 1995.

[8] Bramley, William, The Gods of Eden, Avon Books, New York, 1990.

[9] Faulkner, Raymond, The Egyptian Book of the Dead, The Book of Going Forth by Day, Chronicle Books, San Francisco, 1994.

[10] Melchizedek, Drunvalo, Lecture at the Star Knowledge Conference, Lakota Indian Reservation, South Dakota, June, 1996.

[11] Marchinak, Barbara, Lecture at the Star Knowledge Conference, Lakota Indian Reservation, South Dakota, June, 1996.

The Library of ialexandriah

In recent history, the Sphinx was The Sphinx was covered at least twice to its neck in sand, which means there was no time for the rain water to do this kind of damage, unless we go much farther back in time. When was the last time there was enough rain in the Sahara desert to cause such an extensive amount of erosion? It wasn't just a couple of years to create the depth and detail we see here. Computer models and geologists say that 10,000 - 12,000 years ago was the last time the Sahara had that much rain.

■■

In an earlier section I wondered where the gods are now, and put forth the idea that the Jehovah of Israel did not and could not surely be more than just a "local god", and one that has now either long departed or abandoned his godly duties. It all comes together now and starts to make big sense. Of interest is the fact that the human race were obliged to give tribute to the gods, and this will come up again later as a major issue that is still ongoing, or so it would seem. The world's gold reserves seem to be vanishing with no accounting why, and replaced with bogus bullion. Simple internet research will establish this.

Now is certainly the time to forget and abandon the whole dogma of established religion, open the eyes and the mind, lose that awe and fearful reverence that has been engendered for millennia by those proclaimed as priests and leaders.

EGYPTIAN PREHISTORY

Prior to the creation of dynasties in the prehistory of ancient Egypt, god-Kings reigned. (Or in some cases, just misted.) These beings were undoubtedly the same Anunnaki of the ancient Sumerian texts, but with different names. For example, Ptah of Egyptian fame is the Sumerian Enki, Isis the same as Inanna, and Ra, the same as Marduk.

But Egypt was distinct from Sumer -- even if the players were essentially the same. In Egypt, Ptah/Enki held sway, while in Sumer, it was his half-brother Enlil. This is a major difference -- and accounts for massive differences in their histories, cultures, and those traditions brought down to us today. (6/1/05) The fact that the symbols of Egyptian jewellery and that of other cultures might have included common elements does not dispute the fact that the manner in which the Egyptian culture focused its energies -- i.e. created their own realities -- may have been enormously different from those other races and cultures which may have been in existence at the time.

The Sinai Peninsula was essentially the neutral zone between the opposing forces of Enki and Enlil. Unfortunately for the humans in the zone, especially those living in Sodom and Gomorrah, for example, neutrality turned out to be a serious illusion, when nuclear weapons began flying about. But that's another story -- or another link.

For the moment (i.e. here) we will content ourselves with a brief description of the reigns (and the snow jobs) of the pre-historical Kings of Egypt.

Beginning circa 18,420 B.C.E. (possibly November 15th, a Tuesday), Ptah became the first king of Egypt and was known as the "Creator God, "A very great god who came forth in the earliest times." He undertook great works of land reclamation and dyking -- thus explaining Egypt's nickname, "The Raised Land".

Ptah/Enki was a "God of Heaven and Earth", and considered to be a great engineer and master artificer. His base of operations, according to legend, was on the island of Abu (now called Elephantine on account of its shape), located just above the first cataract of the Nile, at Aswan. **His symbol was the serpent** (i.e. the other God in Genesis was Enki). He and the other gods came from Ur (Sumer). The name, Ptah, has no meaning in Egyptian, but in Semitic, it means "he who fashioned things by carving and opening up."

After 9,000 years (give or take a fortnight), Ra, a son of Ptah became the ruler.

In the interim of Ptah's reign, there was a need for a whole lot more dyking. For roughly a thousand years after the Great Flood/Deluge (which may have occurred circa 10,500 B.C.E.), Ptah was back into land reclamation. [His reputation for dyking and what not, might, in fact, have been based purely on Ptah's post-flood work. Before that, he might, for all we know for sure, have been operating a casino on Elephantine Island.]

Circa 9,420 B.C.E., Ra ("the Complete, the Pure One") began a reign of roughly a 1000 years -- except that no one called it a "reign", what with the Deluge having somewhat recently wiped out most of the population. Feelings were still a bit tender.

Ra is reputed to have come to Earth from the "Planet of Millions of Years" in a Celestial Barge, which was later kept at Anu (biblical On, Greek Heliopolis).

[Ra's home was likely Nibiru -- which with a "year" equivalent to 3600 Earth years, would indeed seem to be a planet of "Millions of Years". For example, many of the Patriarch's lifetimes would, when multiplied by 3600, easily translate into millions of Earth years. Also, the "Celestial Barge" is very likely a space ship of some kind, but probably very dissimilar to what our narrow, current technological paradigm might suggest.]

Ra gave birth to Shu (male, "Dryness") and Tefnut (female, "moisture"). The two set the example for mortal Pharaohs in later times, i.e. **the brother married his half-sister**. Shu and Tefnut promptly set up housekeeping as King and Queen circa 8,420 B.C.E., and did their thing for some 700 years -- until something called the "700 year itch" came along...

Circa 7720 B.C.E., Geb ("Who Piles Up the Earth"), along with his sister Nut ("The Stretched-out Firmament") began a 500 year stint as King and Queen of Egypt. [This was one of the few occasions in all of history, where a husband could call his wife a "Nut" and get away with it.] Geb and Nut were so named based on activities related to the periodic appearance of the Bennu bird, from which the Greeks obtained the legend of the Phoenix.

The Bennu was an eagle with feathers of red and gold, and which died and reappeared at intervals lasting several millennia. It was for that bird -- whose name was the same as that of the contraption in which Ra landed on Earth -- that Geb engaged in great earthworks and Nut stretched out the firmament of the sky. These feats were carried out by the gods in the "Land of the Lions". [The Bennu bird might have been a "celestial barge", and its periodic appearance might have been related to the 3600 year Nibiru Cycle. I.e., every time Nibiru approaches perihelion, the Bennu bird/Phoenix again rises.]

Geb and Nut turned over the direct rule of Egypt to their four children: Asar ("The All-Seeing", whom the Greeks called **Osiris), his sister-wife, Ast (Isis),** Seth, and his wife, Nephtys (Nebt-Hat, "Lady of the House"). This is when things really got interesting!

With two brothers married to their own two sisters, the gods confronted a serious problem of succession. The only plausible solution was to divide the kingdom: Osiris was given the northern lowlands ("Lower Egypt") and Seth was give the southern, mountainous region ("Upper Egypt"). Seth, however, was not satisfied with the division of sovereignty.

But it's a bit more complicated than that. Plutarch (1st century A.D.), for one, based his writing on Egyptian sources believed at the time to have been the writings of the god Thoth himself (who as Scribe of the Gods, had recorded for all times the histories and deeds upon the Earth). In this version, Nut mothered three sons: Osiris (first born), a son with an unknown name, and Seth (the youngest son). Nut also gave birth to Isis and Nephtys. But! Only Seth and Nephtys were fathered by Geb. Osiris and his second brother were fathered by Ra (who came to his granddaughter Nut in stealth). Meanwhile, Isis was fathered by Thoth (the Greek Hermes).

Therefore, the firstborn was Osiris, who having been fathered by the great Ra himself, had a significant claim to succession. The legitimate heir, however, was Seth, because of his having been born to the ruling Geb by his half-sister Nut. This is why Egypt was split into two kingdoms, and each brother given a limited sovereignty. But this also led in turn to a highly charged competition between the two brothers to assure that their son would be the next legitimate successor of the whole of Egypt.

To achieve this goal, Seth would have to father a son by his half-sister Isis, whereas Osiris could achieve his aim by fathering a son by either Isis or Nephtys (both being half-sisters to him). The key was that Seth (or Osiris) would have the greater claim for their son to be the next king of a united Egypt, only if he fathered a son by his half-sister. However, Seth and Nephtys had identical parents, and thus this union would afford no advantage to Seth. Osiris, on the other hand, could utilize either Isis and/or Nephtys to stake his claim.

Osiris then proceeded to deliberately block Seth's chances to have his descendants rule over Egypt by Osiris taking Isis as his spouse. Seth then married Nepthys; but as she was his full sister, none of their offspring could qualify. In this manner the stage was set for Seth's increasingly violent rage against Osiris, who had deprived him both of the throne (of the combined Upper and Lower Egypt) and of the succession of his son.

Seth then used the old party trick of "fit the coffin" in order to get back at Osiris. Making grandiose drunken wagers and other means, Osiris was motivated to lie down in the coffin to prove that it fit him -- or that he was fit to be the coffin's occupant. Seth, however, had the coffin quickly sealed and then dumped it into the sea at a point where the Nile flows into the Mediterranean (at Tanis). This seemed to complete Seth's revenge against Osiris.

However. Seth had not accounted for Isis, who promptly went searching for her boxed lover. She found the chest near Lebanon, but before she could figure out how to resurrect Osiris, Seth found out that she had the chest, seized it, and cut up the body of Osiris into fourteen pieces, which he scattered all over Egypt.

Apparently being exceptionally talented in scavenger hunts, Isis found all of Osiris' parts... except for his phallus. Bummer! But not to be denied, she managed to extract from the body of Osiris its "essence", and thereafter self-inseminated herself with his seed. This led to her conceiving and eventually giving birth to Horus. For what were apparently obvious reasons, she hid him in the Nile delta far from the eye of Seth.

Meanwhile, what with Osiris having apparently died without an heir, Seth promptly initiated the next step in his master plan and kidnapped Isis. This portion of the plan was to allow Seth to father a legitimate heir via his mating with Isis). Seth held her until she consented. It was starting to look good for Seth, when Thoth turned up to help Isis escape. Which she did. The only problem, however, was that when she returned to the swamps where Horus was hidden, she found him dying from a scorpion's sting. [Aw, the trials and tribulations of motherhood!] But then she got some real help:

"Then Isis sent forth a cry to heaven and addressed her appeal to the Boat of Millions of Years. And the Celestial Disk stood still, and moved not from the place where it was. And Thoth [Isis' daddy!] came down, and he was provided with magical powers, and possessed the great power which made the word become deed. And he said: 'O Isis, thou goddess, thou glorious one, who had knowledge of the mouth; behold, no evil shall come upon the child Horus, for his protection cometh from the Boat of Ra. I have come this day in the Boat of the Celestial Disk from the place where it was yesterday. When the night cometh, this light shall drive away [the poison] for the healing of Horus... I have come from the skies to save the child for his mother."

Horus was thus revived and some say immunized forever. Educated and trained in martial arts by goddesses and gods who sided with Osiris, Horus was groomed as a Divine Prince worthy of celestial association, and eventually appeared before the Council of the Gods to claim the throne of Osiris. This was not good news for Seth, but he still had an ace (or a seed) up his sleeve. Using trickery, Seth attempted to plant his own seed in Horus (and thereby claim that Horus could only succeed Seth, not precede him).

Horus, however, had managed to catch the seed in his hand, and thus nullified Seth's claim. Meanwhile, Isis had taken Horus' seed into a cup and then sprinkled it on Seth's salad, such that after a healthy meal, Horus' seed was now in Seth. Thoth then checked that the semen Horus had caught in his hand was that of Seth, and at the same time that Seth indeed carried Horus' seed. Thus Seth could only succeed Horus, not precede him. Horus had turned the tables on Seth, and won the day! Yea.

Seth, of course, was not about to admit defeat. A typical male, he went to war to settle his differences. However, in a subsequent battle, Horus ended up with Seth at the stake, ready to be impaled. But Horus' mother, Isis, relented and released Seth. [What's a son to do with a mother like that?] Horus thereupon cut off Isis' head. But Thoth put it back on. Apparently, Isis' head still worked, but she did thereafter resort to the old guilt trip.

In a later, major battle Horus defeated Seth, and in the process cut off his testicles (a clear violation of the Geneva Convention, but then there was no Geneva at the time -- much less a Convention -- so it didn't really matter). The Lord of Earth, Geb, ultimately gave to the heritage of Horus the whole of Egypt. Seth was awarded a dominion away from Egypt (and henceforth, was deemed by the Egyptians to have become an Asiatic deity). [These guys should write soap operas!]

Osiris had begun his reign (of Lower Egypt) circa 7220 B.C.E., and continued for some 450 years (or until the old coffin party trick). Seth, who was never shown without his animal disguise (i.e. his face was never seen) had begun his reign of Upper Egypt circa 6870 B.C.E., and lasted for some 350 years. [The meaning of Seth's name, incidentally, still defies Egyptologists, despite the name being identical to Adam and Eve's third son.]

Some scholars have noted that the seven gods -- from Ptah to Ra to Horus -- reigned a total of 12,300 years.

Horus took over the reigns of kingship circa 6420 B.C.E., and held sway for 300 years. He was followed by twelve divine Rulers (gods), including Thoth and Maat, who ruled for a total of 1,570 years. They were followed in turn by thirty eight demi-gods, who ruled for some 3,650 years, beginning circa 4550 B.C.E.

The Turin Papyrus (from the time of Ramses II) lists Ra, Geb, Osiris, Seth, and Horus as the kings of Egypt, and later, Thoth, Maat and others. The papyrus also lists the 38 semi-divine rulers (split between 19 Chiefs of the White Wall and 19 Venerables of the North).

Between the semi-divine rulers and Menes, beginning circa 2900 B.C.E., human kings ruled under the patronage of Horus -- their epithet being Shamsu-Hor! For some 350 years, these human kings ruled over what was apparently a chaotic time period. Names of those who wore only the red cap of Lower Egypt, include "Scorpion", Ka, Zeser, Narmer, and Sma. This dynasty is considered by some scholars as "Dynasty O".

Circa 2550 B.C.E., another human, Mena (Menes), reunited Upper and Lower Egypt and established his capital at Memphis. He was followed by 8 or 9 other Kings, all of whom collectively formed the 1st dynasty of the Old Kingdom of Egypt. [Other scholars place Menes as beginning his reign circa 3130 B.C.E. However, when one accounts for the 580 "ghost years" proposed by Immanuel Velikovsky in his Ages in Chaos thesis, then the more likely date is 2550 B.C.E.]

With Menes began the history of ancient Egypt.

It is worth noting as an aside Rene Schwaller de Lubicz's views concerning the spiritual and cosmological insights of ancient Egypt. Clearly, ancient Egypt is the bailiwick of Ptah/Enki, and thus is likely to be far more than a mere collection of dates and reigns.

301– THE BEGINNING OF EARTH HISTORY.

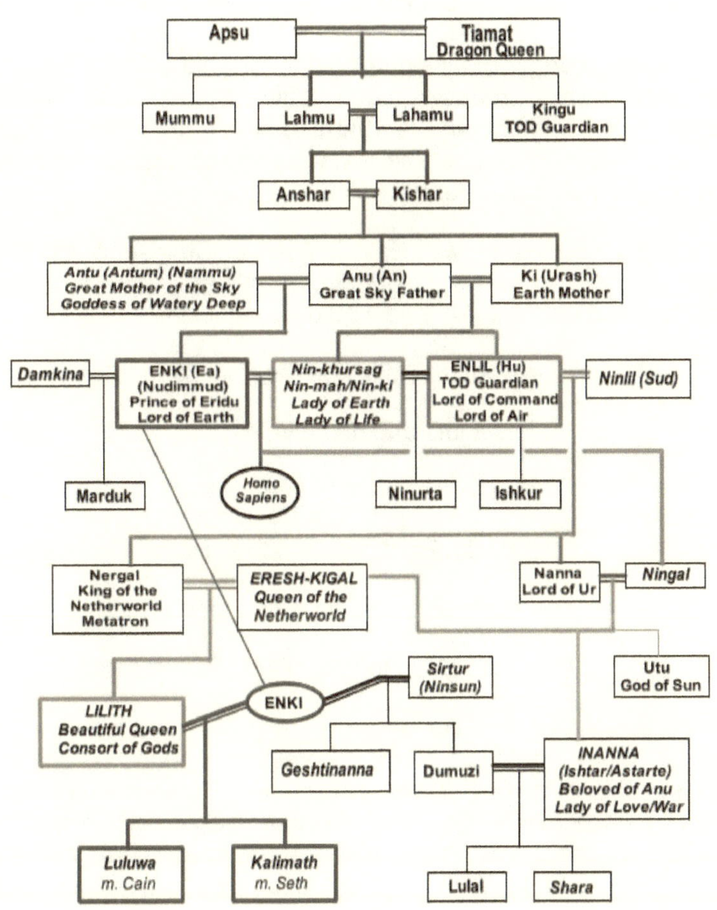

Enki and Enlil

According to the ancient Sumerian texts, the Sumerian god, Anu, the "supreme Lord of the Sky", the currently reigning titular head of the Sumerian Family Tree, had two sons. They were Enki (Ea), Lord of the Earth and Waters (whose mother was Antu), and Enlil (Ilu), Lord of the Air and Lord of the Command (whose mother was Ki). These two half-brothers -- surprise, surprise -- did not get along.

Critical to their rivalry -- particularly from Earth's viewpoint -- was the fact that Enki had been the first of the Anunnaki to hazard a trip to Earth to begin a mining operation for gold. When this effort did not apparently produce gold in sufficient quantities, Enlil was brought in, given command, and armed with a new plan (an early version of the New Deal). The revised program was to mine gold from deep mines in the Earth.

According to Sumerian texts (as detailed in *Genesis of the Grail Kings* [1]), during a visit by their father, Anu (the archetypal absentee landlord), the Anunnaki made a decision:

> "The gods had clasped their hands together,
>
> Had cast lots and had divided.
>
> Anu then went up to heaven.
>
> To Enlil the Earth was made subject.
>
> The seas, enclosed as with a loop,
>
> They had given to Enki, the Prince of Earth."

Sounds fair. However. As Laurence Gardner points out: "Enki was not happy about his brother's promotion because, although Enlil was the elder of the two, his mother (Ki) was Anu's junior sister, whereas Enki's mother (Antu) was the senior sister. True kingship, claimed Enki, progressed as a matrilineal institution through the female line, and by this right of descent Enki maintained that he was the first born of the royal succession."

> "I am the great brother of the gods.
>
> I am he who has been born as the first son of the divine Anu."

If there is a philosophy of Enki, it manifests and explains itself in early Mesopotamian and Egyptian thought, where the true creator of the universe was manifest within nature, and that nature enveloped both the Anunnaki, and the humans. Nature, as the Great Mother, was still supreme, despite any patriarchal scheme to the contrary. Admittedly, Enki's claim of his birthright, the one being based on a matrilineal succession -- essentially the mitochondria DNA link, which is wholly passed through the female line -- was in Enki's best interests. But Enki was also the *maternal* grandfather who came to the aid of Inanna when things went badly during her Descent into the Underworld.

With the arrival of Enlil, however, who in *his* best interests must demean the matriarchal line of succession, and thus nature itself -- everything changed. The Great Mother was dethroned and replaced by a supreme male (as opposed to a male consort for the Queen). The idea of cooperation -- as exemplified by the council of Anunnaki making cooperative decisions -- was quickly replaced by competition, and harmony was forsaken in favor of subservience. The supreme god became abstract, and any physical connection with human or nature was lost -- and thus the link between nature and human also destroyed. When Enlil hit town, there was a whole new deal put into effect.

According to Laurence Gardner **[1], "The dominant tenet of the new thought was based wholly on the utmost fear of Enlil,** who was known to have instigated the great Flood [or else acquiesced in not warning the humans, or making any attempt to save them], and to have facilitated the invasion and destruction of civilized Sumer. **Here was a deity who spared no mercy for those who did not comply with his dictatorial authority.**

"Abraham had experienced the vengeful Enlil first hand at the fall of Ur, and he was not about to take any chances with his own survival. He was even prepared to sacrifice the life of his young son, Isaac, to appease the implacable God (Genesis 32:9)." "The oriental scholar Henri Frankfort summarized the situation by making the point that... 'Those who served Jehovah must forego the richness, the fulfillment, and the consolation of a life which moves in tune with the great rhythms of the earth and sky."

Bramley [3] has noted that, "We therefore find Ea [Enki] as the reputed culprit who tried to teach early man (Adam) the way to spiritual freedom. This suggests that Ea intended his creation, *Homo sapiens*, to be suited for Earth labor, but at some point he changed his mind about using spiritual enslavement as a means."

From a Biblical perspective, it was Enki who (with the critical assistance of his half-sister, Ninki, aka Nin-khursag) created **Adam and Eve**. It was Enlil, on the other hand, who created "Edin". Enki was the serpent in the garden, who urged Adam and Eve to eat of the fruit of the Tree of Knowledge of Good and Evil (which was infinitely beneficial to their spiritual growth). It was Enlil, who drove them out of Edin, while Enki was there to clothe them. It is worth noting that **Zecharia Sitchin** [2] claims that the biblical word for "snake" is *nahash*, which comes from the root word NHSH, and which means "to decipher, to find out." In other words, Enki, the God of Wisdom.

In the time of Noah, it was Enlil who either created the Great **Deluge**/Flood as a means of wiping out mankind (because they supposedly made too much noise), or else refused to warn the humans or do anything to save them or help them to save themselves. Enki, on the other hand, apparently against orders of the **Anunnaki** (who Enlil now controlled), provided the boat plans for Noah to build his Ark, and thus save him, his family (and likely a fair number of helpful artisans and their families). Enki included as well the seed of other living things (a "natural" kind of thing to do).

In the Sumerian texts, we have the stories of Enki and Enlil, and for the most part there is portions devoted to each. But in **Genesis**, Enlil seemingly reigns supreme. Enlil knew early on, that a pound of good Public Relations effort is worth a ton of truth.

Abraham and his descendants served Enlil, and followed his precepts. The Egyptians, on the other hand, were Enki's protégés, and based on food management practices during the devastating droughts around the time of Jacob and Joseph, were doing a lot better than Enlil's followers. Obviously Noah backed the right horse in that Enki shared boat plans with the righteous fellow -- whom Enlil later claimed as his own.

But at one point, circa 2000 B.C.E., all hell broke loose. In an all out war of Enki's humans against Enlil's humans -- complete with all manner of diplomatic subterfuge in the mix -- Sodom and Gomorrah took the brunt of the action and were destroyed. By nuclear weaponry! They were A-bombed. The decision for this, however, was not, as you might have expected, due to Enlil's instigation. Instead, it was due to the actions of his sons, Ninurta and Ningal. The (radioactive) fallout of their actions then resulted in the final destruction of the Sumerian civilization (circa 2000 B.C.E.). Curiously, this event in the Annals of Earth turned out to be something of a Waterloo for Enlil. Not that the guy (dba "God") fled the scene, but thereafter, the idea of unilateral actions was a bit more constrained. Enlil was no longer the undisputed Lord of the Command among his peers.

Which might be just as well. As Laurence Gardner [1] phrased it: **"This muddled and unparalleled concept of Jehovah being right when he was wrong, honest when he was dishonest, was born out of an inherent fear of his vengeful power and unbounded wrath.** Whether as Jehovah (in Genesis) or as Enlil (in Mesopotamian record) it was he who had instigated the Semitic invasions which led to the 'confusion of tongues' and the fall of Sumer. It was he who had brought about the devastating Flood, and it was he who had leveled the cities of Sodom and Gomorrah -- not because of their wickedness, as related in Genesis (18-19), but because of the wisdom and insight of their inhabitants, as depicted in the Coptic *Paraphrase of Shem*. It was Jehovah who had removed the Israelites from their homeland and sending them into seventy years of captivity by King Nebuchadnezzar II and his five Babylonian successors down to King Belshazzer (545-539 BC)."

This latter event is critical as another turning point in the Enki and Enlil warfare, as it reflects a time, **circa 600 B.C.E., when Enlil was stepping back from the overt control of Earth. (A fact which does not necessarily imply stepping back from covert control!)**

Zecharia Sitchin [2] has taken a different, decidedly pro-Jehovah, pro-Enlil approach in his writings. While admitting to the complicity of Enlil's sons in the destruction of Sodom and Gomorrah, Sitchin tends to blame the female (i.e. Inanna) for seducing King Shulgi of Ur (and thus destroying a once thriving civilization). Sitchin also charges Enki's son, Marduk, who became the Babylonian god, with being perhaps the prime culprit of all the bad news that was extant in what Sitchin refers to as "The Fateful Century" (2123 - 2023). At one point in his book, *The Wars of God and Men*, Sitchin writes: "There was great jubilation in the land when the great temple was rededicated to Enlil and Ninlil [Enlil's wife], in the year 1953 B.C.E.; it was only then that the cities of Sumer and Akkad were officially declared habitable again."

And just guess who was responsible for their being uninhabitable in the first place!?

To appreciate the *continuing* sage of Enki versus Enlil, it is instructive to note their place in the **Sumerian Family Tree**, aka the "Grand Assembly of the Anunnaki".

Marduk (who would become the god of the Babylonians) was Enki's first born, and that of Enki's wife, the goddess Damkina. Enki's other wife was his half-sister, Nin-khursag (meaning "Mountain Queen"), the Lady of Life, also known as Nin-mah, the Great Lady.

Enlil was also espoused to Nin-khursag and their son was Ninurta (Ningirsu), the Mighty Hunter. By another wife, Ninlil (Sud), Enlil had a second son, Nanna (Suen), known as the Bright One. Nanna and his wife, Ningal, were the parents of Inanna (who was called Ishtar by the Babylonians), and who married the Shepherd King Dumu-zi (the latter given in the Semitic Old Testament book of Ezekiel 8:14 as Tammuz).

Another son of Enlil and Ninlil was Nergal (Meslamtaea), King of the Underworld. He married Eresh-kigal, the Queen of the Netherworld, the daughter of Nanna and Ningal (i.e. Inanna's sister), and the mother of **Lilith** (who became handmaiden to Inanna, her maternal aunt). Lilith is also notorious as the first wife of Adam, but it was Lilith who rejected him (and thus incurred the wrath of every reject-worthy male on the planet).

By some accounts, Inanna was also the granddaughter of Enki (as well as Enlil). This strange situation was critical in Inanna's classic tale of her **Descent into the Underworld.** (I.e. Enlil turned a blind eye, while Enki saved Inanna's lovely little fanny.) Even more crucial to the plot was the fact that Inanna was also a favorite of the supreme Anu. Thus she was never, never at a loss as to what she thought she could do and get away with. Her story has been well told two books by Susan Ferguson: *Inanna Returns* and *Inanna, Hyperluminal*. (Ms. Ferguson does include Enki's son, Marduk, as the bad guy, but on the other hand, keeps Enki as a favorite. She can do that. It's her books.)

Speaking of Marduk, not only was he the arch-enemy of Inanna (thus explaining Susan's plotting), but Marduk thoroughly angered just about everyone about him. Even his father, Enki, must have wondered where he went wrong in raising his first son -- a question not uncommon to *any* father. At the same time, it must be admitted, Marduk was without question a serious pain in the rear (and elsewhere) for Enlil, and thus Enki might have had moments of genuine pride.

Just as Enki may have been given temporary, overt control over the Earth during the **Age of Pisces**, Marduk, who was identified with the planet Mars, and thus the astrological sign of *Aries*, had assumed he would be in charge during the *Age of Aries*. Depending on the time allotted to each sign -- whether it is 1/12th, or more likely the actual time spent in the sign -- Marduk's Age of Aries likely ran from roughly 2,000 B.C.E. to about **600 B.C.E.** This was his time, therefore, and *The Wars of Gods and Men* told by Sitchin was in large part Marduk's attempts to wrest control from Enlil, and the Anunnaki who supported the latter. The fact that it became a very messy war was not necessarily Marduk's fault.

For the fact remains that, circa 1950 B.C.E., after Enlil's son, Ninurta, had failed to rally the Anunnaki troops on his own behalf -- and thoroughly *bombed* on his venture to Sodom and Gomorrah -- Marduk finally got his chance.

"Lord Anu, lord of the gods who from Heaven came to Earth,

and Enlil, lord of Heaven and Earth

who determines the destinies of the land,

Determined for Marduk, the firstborn of Enki,

the Enlil-functions over all mankind;

Made him great among the gods who watch and see,

Called Babylon by name to be exalted, made it supreme in the world;

And established for Marduk, in its midst, an everlasting kingship."

Marduk, from Babylon, ultimately took vengeance on the Enlil supporters known as the Hebrews, who had opposed Marduk's reign, and they thereafter spent seventy years in captivity. During this time, Enlil never raised a hand to assist them. In Enlil's view, they were quite expendable. Obviously, someone -- unlike their ancestral patriarch, Noah, had failed to back the right horse.

For sometime (i.e. the Age of Aries), Marduk took over Enlil's subjugation of the humans -- politics of the slavery kind made strange bedfellows. But the Age of Aries (unlike the Age of Pisces) was mercifully short. And it had the decided advantage of prepping the Anunnaki for Enki's take over about 600 B.C.E., when the Age of Pisces began.

[1] Laurence Gardner, *Genesis of the Grail Kings*, Bantam Press, New York, 1999.

[2] Zecharia Sitchin, *The 12th Planet*, 1976, *The Stairway to Heaven*, 1980, *The Wars of Gods and Men*, 1985, *The Lost Realms*, 1990, *When Time Began*, 1990, *Genesis Revisited*, 1990, *Divine Encounters*, 1995, Avon Books, New York.

[3] William Bramley, *The Gods of Eden*, Avon Books, New York, 1989, 1990.

02-ENKI, HUMAN HERITAGE.

The Serpent of Life and Wisdom

by **Estelle Nora Harwit Amrani**
November, 1998

from Vibrani Website
recovered through BibliothecaAlexandrina Website

The caduceus is one of the most ancient of symbols. You might best know this symbol as the DNA structure and healing used by the medical profession. Since ancient Mesopotamia the caduceus presented two serpents intertwined (the central nervous system) around a staff (the spinal column) with the wings (the "swan") on either side (the two hemispheres of the brain, with the circle in the center representing the pineal gland, or the central sun and psychic center within). It also symbolized the kundalini energy.

This was originally the symbol for the Anunnaki-Sirian creator god, EA, or EN.KI (who has become an Archetype), was the chief of the magicians, "the one who knows," and infamous for being the serpent of the Garden of Eden who created lifeforms in test tubes half a million years ago with his half-sister Ninharsag, at the suggestion of his son, Marduk, to create humans to be the workers for the gods. (The symbol is also based upon the winged globe for the planet Nibiru, the symbol of the royal Anunnaki family.)

Biblical writers called the healing serpent Nehushtan. The Hebrew word for serpent is "nahash." The root of the word are the Hebrew letters Nun, Het and Shin, which means "to guess." This was translated into other languages as "satan," which some say mean "enemy," or "adversary."

Enki's identity, as *Lord of Earth* or *In Earth* (EN.KI), also known as EA ("whose house is water") is reflected in other names, as well:

- Adonai
- Ptah
- Aton
- Aten
- Adom
- Adam
- Amen

(Linguistic paleontology is a marvelous and vast area for proving these connections.)

The name EARTH also comes from EA/Enki. Actually, the name "earth" can be traced to Enki (a.k.a. EA), and "human" is related to Ninharsag/Ninhursag, who was Hathor (the House of Horus): HU (Horus) is also a transliteration of the ancient Sumerian EA (Grimms' law of interchangeable letters and sounds). If we use Hebrew, HU means "she."

India, the "nagas" were the serpent gods and goddesses. In the Americas there was Quetzlcoatl (or Thoth). The entire world has worshipped the serpent for its wisdom, but ironically, it was not really about snakes at all - unless you feel you have to "guess" what a snake is up to! Why was the snake chosen? For its cleverness, ability to survive in the harshest of environments, and again, its shape resembling the flow of energy up the spine - to the crown chakra, and the third eye. The snake sheds its skin and is reborn. The snake is clever. And perhaps because it naturally instilled a bit of caution or awe in people. Was Enki really a snake? Not literally. He has many different appearances.

The serpent always represents spiritual wisdom, life and healing. The first symbols of serpents were attributed to Enki and then Ninhursag. However, the story of the serpent becoming an evil symbol began with the wars between Enki and his brother, **Enlil (later known by the name "Allah").** These conflicts began at birth and had to do with birthright to the royal throne of the Nibiruan civilization in which their father, Anu, was the leader and father to Enki and Enlil. There was a time when Anu felt Enki (due to his wisdom and magical abilities) was the only savior of the Anunnaki people ("Enuma Elish").

This story later was reflected in Cain and Abel, and all the stories throughout your times of brothers competing for power, favoritism and inheritance. Enlil's anger with Enki caused him to twist the truth around to make the serpent evil, which later became what you know as the story in *the Bible.* What you think of as being Satan is not that at all, but THE REVERSE!

Although there was love between Enki and Enlil, they often did not see eye to eye on many issues, especially when it came to supporting human beings. Enlil never had patience or compassion for people, and on several occasions, Sodom and Gomorrah as one example, he literally nuked them out of existence. He attempted this again during the time of the Great Flood but Enki (and those who supported him) took swift action to alert the Noahs around the planet of the forthcoming dangers. Some of the Anunnaki outraged with Enki for doing so but saw they had little choice in finally carrying out the rescue. Anu supported saving humanity.

In the *Garden of Eden situation*, Enlil was furious that Enki permitted humans to have access to knowledge, the mixing of the Anunnaki with human genes, thereby becoming more "godly," and *equal to the Anunnaki.* To strike back at Enki, and in the attempt to regain his power over humans, Enlil vowed to *tarnish Enki's reputation by spreading the idea that the serpent of wisdom was evil.* Enlil tried to wipe out knowledge of the DNA coding Enki gave humans, and of what the Anunnaki used in order to have *longevity* (gold).

However, Enlil was not completely successful because most of Enki's plan had worked. For centuries afterwards, **humans attempted to duplicate the concoction of gold the gods used to maintain their youth and health,** and those with the knowledge were able to manufacture substitutes for a while. But, much also had to do with the DNA content of the individual.

The more pure Anunnaki DNA, the better chance one had of having longevity, etc. Then, combine the DNA with spiritual awakening to the body, blood, and spirit with nutritional supplements, and **each human will know who they are - gardeners, and caretakers of the Earth, not owners.** Humans are here to maintain beauty, harmony and balance that was first given to us after the Earth was created. We are not to be interested only in ourselves.

Why did Adam eat from the *Tree of Knowledge* and not from the

Tree of Life? Without getting into complex detail, Enki told me simply: *(NOT THE AUTHOR BOB MADDISON)*

> "With the Tree of Knowledge humans had the chance to figure out everything on their own in time, to be equal to the Anunnaki. Had they eaten only from the Tree of Life, they would live but not have been more the wiser."

The Garden of Eden, by the way, was a literal place, but also a genetic metaphor.

Enki knew had Adam (Adapa) eaten from the other tree, it would not ensure wisdom or spiritual evolution. Instead, it would more likely result in primitive human living for eons without evolution. **The story of this translated from the ancient Babylonian texts is very interesting with Adapa's confusion over whom to believe, which to eat.** It resulted in him eating "the wrong" thing, but actually it was the right thing, in terms of DNA, which would eventually bring our spirituality back from whence it faltered, and remind people to tend to the Earth, which was not created by extraterrestrials, but by *God. (THAT OMNIFIC F.O.U. – Bob)*

The battle between the brothers continued into the time of the pyramid wars and Exodus. The staff with the caduceus was also one of Moses' tools. The serpent, for the Hebrews represented salvation and wisdom. Moses's copper serpent staff, often utilized by his brother, Aaron, was made famous for performing miracles. Another connection between the staff and the serpents occurred during the Exodus when the staff was seen to transform into snakes. Aaron was high priest and had been trained in magic. *(That all makes sense now. –Bob)*

He and Moses received *instructions* from a collective of that *main Anunnaki family* (who taught Moses the alphabet). Isn't it interesting that during the Exodus, the name Jehovah, YHWH, took over and the name Adonai began to disappear? This was at the time when Enki departed the Earth and Marduk became the leader of Enki's family. Enlil's family was given the Sinai (taken away from Ninharsag) and Enlil's son, Sin, was its new ruler. *His symbol is the crescent moon* (which became the *symbol for Islam*).

Luckily, the heritage of the caduceus lived on. In some versions the staff is capped with a solar disk or even a crescent moon.

Nisaba, one of Enki's daughters, also held a similar staff topped with an "ankh." For some, the staff symbolized Hermes/Mercury. Throughout time different civilizations in India, the Americas, Greece, Egypt, including the great mystery schools and secret societies have renamed and used it. In Christianity the archangel Michael was associated with this staff. The sirens became the staff for two serpents they held in their hands. This staff was considered to be so powerful it was able to raise up the dead.

The symbol of the caduceus was later transferred to one of Enki's counterparts, Ningishzida (Thoth), the healing god, and then to others of his family because it was a code for the bloodline of Enki's heritage. From this symbol of the serpent the power was transferred to the symbol of the dragon, who continued to hold the knowledge. The dragon could "divine." This is one reason why we say we and you are of "divine" heritage. You will see a version of the caduceus as the winged solar disk in Ancient Egypt, which **incorporated the knowledge of one's divinity and eternal soul**, in the third eye chakra, along with the traditional knowledge of what it stood for.

Therefore, you will see this symbol above arches and entrances into temples and royal structures to remind those who enter who they really are. And this means not only the DNA connection to the Anunnaki, but the divine soul-being as coming from the Source, itself. The symbol of the winged Isis represents the original female mother goddess, Ninharsag, and blood connection in birthing humans who mated with "the gods," those who fly as a free spirit.

When you see the caduceus, know you are triggering your own genetic memory and **seeing your heritage**. **Each one of you holds within you the DNA, in varying degrees, the wisdom from *the Anunnaki* and *the Source.*** Since Enki and Ninhursag, there have been other off-planet beings who added in their own DNA to the human species.

So, you are all a mixture - and all one family from the One *God*. Within you is the consciousness with which you can liberate yourselves through SELF-knowledge and return to the garden.

303-WHO WAS JEHOVAH?

God, the Extraterrestrial

So, who was Yahweh?

Was He one of them? Was He an extraterrestrial?

*The question, with its implied answer, is not so outrageous. Unless we deem Yahweh - "God" to all whose religious beliefs are founded on the Bible - to have been one of us Earth-lings, then He could only be not of this Earth - which "extraterrestrial" ("outside of, not from Terra") means. And the story of Man's Divine Encounters, the subject of this book, is so filled with parallels between the biblical experiences and those of encounters with **the Anunnaki** by other ancient peoples, that the possibility that Yahweh was one of "them" must be seriously considered.*

The question and its implied answer, indeed, arise inevitably. That the biblical creation narrative with which the Book of Genesis begins draws upon the Mesopotamian Enuma elish is beyond dispute. That the biblical Eden is a rendering of the Sumerian E.DIN is almost self-evident. That the tale of the Deluge and Noah and the ark is based on the Akkadian Atra-Hasis texts and the earlier Sumerian Deluge tale in the Epic of Gilgamesh, is certain. That the plural "us" in the creation of The Adam segments reflects the Sumerian and Akkadian record of the discussions by the leaders of the Anunnaki that led to the genetic engineering that brought Homo sapiens about, should be obvious.

*In the Mesopotamian versions it is **Enki**, the Chief Scientist, who suggests the genetic engineering to create the Earthling to serve as a Primitive Worker, and it had to be Enki whom the Bible quotes as saying " Let us make the Adam in our likeness and after our image." An Epithet of Enki was **NU. DIM.MUD**, "He who fashions;" the Egyptians likewise called Enki **Ptah** - "The Developer," "He who fashions things," and depicted him as fashioning Man out of clay, as a potter. "The Fashioner of the Adam," the Prophets repeatedly called Yahweh ("fashioner," not "creator"!); and comparing Yahweh to a potter fashioning Man of clay was a frequent biblical simile.*

As the master biologist, Enki's emblem was that of the Entwined Serpents, representing the double-helixed DNA - the genetic code that enabled Enki to perform the genetic mixing that brought about The Adam; and then (which is the story of Adam and Eve in the Garden of Eden) to again genetically manipulate the new hybrids and enable them to procreate. One of Enki's Sumerian epithets was BUZUR; it meant both "He who solves secrets" and "He of the mines," for the knowledge of mineralogy was considered knowledge of Earth's secrets, the secrets of its dark depths.

The biblical tale of Adam and Eve in the Garden of Eden - the tale of the second genetic manipulation - assigns to the serpent the role of triggering their acquisition of "knowing" (the biblical term for sexual procreation). The Hebrew term for serpent is Nahash; and interestingly, the same word also means soothsayer, "He who solves secrets" - the very same second meaning of Enki's epithet.

Moreover, the term stems from the same root as the Hebrew word for the mineral copper, Nehoshet. It was a Nahash Nehoshet, a copper serpent, that Moses fashioned and held up to stop an epidemic that was afflicting the Israelites during the Exodus; and our analysis leaves no alternative but to conclude that what he had made to summon divine intervention was an emblem of Enki. A passage in II Kings 18:4 reveals that this copper serpent, whom the people nicknamed Nehushtan (a play on the triple meaning serpent-copper-solver of secrets) had been kept in the Temple of Yahweh in Jerusalem for almost seven centuries, until the time of King Hezekiah.

Pertinent to this aspect might have been the fact that when Yahweh turned the shepherd's crook that Moses held into a magical staff, the first miracle performed with it was to turn it into a serpent. Was Yahweh, then, one and the same as Enki? The combination of biology with mineralogy and with the ability to solve secrets reflected Enki's status as the God of knowledge and sciences, of the Earth's hidden metals; he was the one who set up the mining operations in southeastern Africa.

All these aspects were attributes of Yahweh.

> *"It is Yahweh who giveth wisdom, out of His mouth cometh knowledge and understanding," Proverbs asserted (2:6), and it was He who granted wisdom beyond comparison to Solomon, as Enki had given the Wise Adapa.*
> *"The gold is mine and the silver is mine," Yahweh announced*
> ***(Haggai 2:8);***
> *"I shall give thee the treasures of the darkness and the hidden riches of the secret places," Yahweh promised Cyrus*
> ***(Isaiah 45:3).***

The clearest congruence between the Mesopotamian and biblical narratives is found in the story of the Deluge. In the Mesopotamian versions it is ***Enki*** *who goes out of his way to warn his faithful follower Ziusudra/Utnapishtim of the coming catastrophe, instructs him to build the watertight ark, gives him its specifications and dimensions, and directs him to save the seed of animal life. In the Bible, all that is done by Yahweh.*

The case for identifying Yahweh with Enki can be bolstered by examining the references to Enki's domains. After Earth was divided between the Enlilites and the Enki'ites (according to the Mesopotamian texts), Enki was granted dominion over Africa. Its regions included the Apsu (stemming from AB.ZU in Sumerian), the gold-mining region, where Enki had his principal abode (in addition to his "cult center" Eridu in Sumer).

The term Apsu, we believe, explains the biblical term Apsei-eretz., usually translated "the ends of earth," the land at the continent's edge - southern Africa, as we understand it. In the Bible, this distant place, Apsei-eretz, is where "Yahweh shall judge" (I Samuel 2:10), where He shall rule when Israel is restored (Micah 5:3). Yahweh has thus been equated with Enki in his role as ruler of the Apsu.

This aspect of the similarities between Enki and Yahweh becomes more emphatic - and in one respect perhaps even embarrassingly so for the monotheistic Bible - when we reach a passage in the Book of Proverbs in which the unsurpassed greatness of Yahweh is brought out by rhetorical questions:

> *Who hath ascended up to*
> *Heaven,*

and descended too?
Who hath cupped the wind in his hands,
and hound the waters as in a cloak?
Who hath established the Apsei-eretz -
What is his name,
and what is his son's name -
if thou can tell?

According to the Mesopotamian sources, when ***Enki*** *divided the African continent among his sons, he granted the Apsu to his son Nergal. The polytheistic gloss (of asking the name of the Apsu's ruler and that of his son) can be explained only by an editorial inadvertent retention of a passage from the Sumerian original texts - the same gloss as had occurred in the use of "us" in "let us make the Adam" and in "let us come down" in the story of the Tower of Babel. The gloss in Proverbs (30:4) obviously substitutes "Yahweh" for Enki.*

Was Yahweh, then, Enki in a biblical-Hebrew garb?

Were it so simple ... If we examine closely the tale of Adam and Eve in the Garden of Eden, we will find that while it is the Nahash - Enki's serpent guise as knower of biological secrets - who triggers the acquisition by Adam and Eve of the sexual "knowing" that enables them to have offspring, he is not Yahweh but an antagonist of Yahweh (as Enki was of Enlil). In the Sumerian texts it was Enlil who forced Enki to transfer some of the newly fashioned Primitive Workers (created to work in the gold mines of the Apsu) to the E.DIN in Mesopotamia, to engage in farming and shepherding.

In the Bible, it is Yahweh who "took the Adam and placed him in the garden of Eden to tend it and to maintain it." It is Yahweh, not the serpent, who is depicted as the master of Eden who talks to Adam and Eve, discovers what they had done, and expels them. In all this, the Bible equates Yahweh not with Enki but with Enlil.

Indeed, in the very tale - the tale of the Deluge - where the identification of Yahweh with Enki appears the clearest, confusion in fact shows up. The roles are switched, and all of a sudden Yahweh plays the role not of Enki but of his rival Enlil. In the

Mesopotamian original texts, it is Enlil who is unhappy with the way Mankind has turned out, who seeks its destruction by the approaching calamity, and who makes the other Anunnaki leaders swear to keep all that a secret from Mankind.

In the biblical version (chapter 6 of Genesis) it is Yahweh who voices his unhappiness with Mankind and makes the decision to wipe Mankind off the face of the Earth. In the tale's conclusion, as Ziusudra/Utnapishtim offers sacrifices on Mount Ararat, it is Enlil who is attracted by the pleasant smell of roasting meat and (with some persuasion) accepts the survival of Mankind, forgives Enki, and blesses Ziusudra and his wife. In Genesis, it is to Yahweh that Noah builds an altar and sacrifices animals on it, and it was Yahweh "who smelted the pleasant aroma."

So was Yahweh Enlil, after all?

A strong case can be made for such an identification. If there had been a "first among equals" as far as the two half brothers, sons of Anu, were concerned, the first was Enlil. Though it was Enki who was first to come to Earth, it was ***EN.LIL*** *("Lord of the Command") who took over as chief of the Anunnaki on Earth. It was a situation that corresponds to the statement in Psalms 97:9: "For thou, O Yahweh, art supreme over the whole Earth; most supreme art Thou over all the Elohim."*

The elevation of <u>Enlil</u> to this status is described in the Atra-Hasis Epic in the introductory verses, prior to the mutiny of the gold-mining Anunnaki:

> *Anu, their father, was the ruler;*
> *Their commander was the hero Enlil.*
> *Their warrior was Ninurta;*
> *Their provider was Marduk.*
> *They all clasped hands together,*
> *cast lots and divided:*
> *Anu ascended to heaven;*
> *The Earth to Enlil was made subject.*
> *The bounded realm of the sea*
> *to princely Enki they had given.*
> *After Anu had gone up to heaven,*
> *Enki went down to the Apsu.*

(Enki, interchangeably called in the Mesopotamian texts E.A. - "Whose home is water" - was thus the prototype of the sea God

Poseidon of Greek mythology, the brother of Zeus who was head of the pantheon).

After ***Anu****, the ruler on Nibiru, returned to Nibiru after visiting Earth, it was* ***Enlil*** *who summoned and presided over the council of the Great Anunnaki whenever major decisions had to be made. At various times of crucial decisions - such as to create The Adam, to divide the Earth into four regions, to institute Kingship as both buffer and liaison between the Anunnaki Gods and Mankind, as well as in times of crisis between the Anunnaki themselves, when their rivalries erupted into wars and even use of nuclear weapons - "The Anunnaki who decree the fates sat exchanging their counsels."*

Typical was the manner in which one discussion is described in part: "Enki addressed to Enlil words of lauding: 'O one who is foremost among the brothers, Bull of Heaven, who the fate of Mankind holds.' " Except for the times when the debate got too heated and became a shouting match, the procedure was orderly, with Enlil turning to each member of the Council to let him or her have a say.

The monotheistic Bible <u>lapses several times</u> into describing Yahweh in like manner, chairing an assembly of lesser deities, usually called Bnei-elim - "sons of Gods." The Book of Job begins its tale of the suffering of a righteous man by describing how the test of his faith in God was the result of a suggestion made by Satan,

> *"one day, when the sons of the Elohim came to present themselves before Yahweh."*
> *"the Lord stands in the assembly of the Gods, among the Elohim He judges," we read in Psalms 82:1.*
> *"Give unto Yahweh, o sons of Gods, give unto Yahweh glory and might," Psalms 29:1 stated, "bow to Yahweh, majestic in holiness."*

The requirement that even the "sons of the Gods" bow to the Lord paralleled the Sumerian description of the status of Enlil as the Commander in Chief: "The Anunnaki humble themselves before him, the Igigi bow down willingly before him; they stand by faithfully for the instructions."

It is an image of Enlil that matches the exaltation in the Song of Miriam after the miraculous crossing of the Sea of Reeds:

> *"Who is like thee among the Gods, Yahweh? Who is like thee mighty in holiness, awesome in praises, the maker of miracles?"*
> ***(Exodus 15:11).***

As far as personal characters were concerned, Enki, the fashioner of Mankind, was more forebearing, less stringent with both Gods and mortals. Enlil was stricter, a "law and order" type, uncompromising, unhesitant to mete out punishments when punishment was due. Perhaps it was because while Enki managed to get away with sexual promiscuities, Enlil, transgressing just once (when he date-raped a young nurse, in what turned out to be his seduction by her), was sentenced to exile (his banishment was lifted when he married her as his consort Ninlil).

He viewed adversely the intermarriage between Nefilim and the "daughters of Man." When the evils of Mankind became overbearing, he was willing to see it perish by the Deluge. His strictness with other Anunnaki, even his own offspring, was illustrated when his son Nannar (the Moon God Sin) lamented the imminent desolation of his city Ur by the deathly nuclear cloud wafting from the Sinai. Harshly Enlil told him: "Ur was indeed granted Kingship; but an everlasting reign it was not granted."

Enlil's character had at the same time another side, a rewarding one. When the people carried out their tasks, when they were forthright and God-fearing, Enlil on his part saw to the needs of all, assured the land's and the people's well-being and prosperity. The Sumerians lovingly called him "Father Enlil" and "Shepherd of the teeming multitudes." A Hymn to Enlil, the All-Beneficent stated that without him "no cities would be built, no settlements founded; no stalls would be built, no sheepfolds erected; no king would be raised, no high priest born." The last statement recalled the fact that it was Enlil who had to approve the choice of kings, and by whom the line of Priesthood extended from the sacred precinct of the "cult center" Nippur.

These two characteristics of ***Enlil*** *- strictness and punishment for transgressions, benevolence and protection when merited - are similar to how Yahweh has been pictured in the Bible. Yahweh can bless and Yahweh can accurse, the Book of Deuteronomy explicitly states (11:26). If the divine commandments shall be followed, the people and their offspring shall be blessed, their* *crops shall be plentiful, their livestock shall multiply, their enemies shall be defeated, they shall be successful in whatever trade they choose; but if they forsake Yahweh and his commandments, they, their homes and their fields shall be accursed and shall suffer afflictions, losses, deprivations, and famines (Deuteronomy 28).*

> *"Yahweh thy Elohim is a merciful God," Deuteronomy 4:31 stated; He is a vengeful God, the same Deuteronomy stated a chapter later (5:9) . . .*

It was Yahweh who determined who shall be the priests; it was He who stated the rules for Kingship (Deuteronomy 17:16) and made clear that it will be He who chooses the king - as indeed was the case centuries after the Exodus, beginning with the selection of Saul and David. In all that, Yahweh and Enlil emulated each other.

NUMBERS 7 AND 50.

Significant, too, for such a comparison was the importance of the numbers seven and fifty. They are not physiologically obvious numbers (we do not have seven fingers on a hand), nor does their combination fit natural phenomena (7 x 50 is 350, not the 365.25 days of a solar year). The "week" of seven days approximates the length of a lunar month (about 28.5 days) when multiplied by four, but where does the four come from? Yet the Bible introduced the count of seven, and the sanctity of the seventh day as the sacred Sabbath, from the very beginning of divine activity.

The accursation of Cain was to last through seven times seven generations; Jericho was to be circled seven times so that its walls would fall down; many of the priestly rites were required to be repeated seven times, or to last seven days. Of a more lasting commandment, the New Year Festival was deliberately shifted from the first month Nisan to the seventh month Tishrei and the principal holidays were to last seven days. The number fifty was the principal numerical feature in the construction and equipping of the Ark of the Covenant and the Tabernacle and an important element in the future Temple envisioned by Ezekiel.

It was a calendrical count of days in priestly rites; Abraham persuaded the Lord to spare Sodom if fifty just men would be found there. More important, a major social and economic concept of a Jubilee Year in which slaves would be set free, real property would revert to its sellers and so on, was instituted. It was to be the fiftieth year: "Ye shall hallow the fiftieth year and proclaim freedom throughout the land," was the commandment in Leviticus chapter 25.

Both numbers, seven and fifty, were associated in Mesopotamia with ***Enlil****. He was "the God who is seven" because, as the highest-ranking Anunnaki leader on Earth, he was in command of the planet which was the seventh planet. And in the numerical hierarchy of the Anunnaki, in which Anu held the highest numeral 60, Enlil (as his intended successor on Nibiru) held the numerical rank of fifty (Enki's numerical rank was forty). Significantly, when Marduk took over the supremacy on Earth circa 2000 B.C., one of the measures taken to signify his ascendancy was to grant him fifty names, signifying his assumption of the Rank of Fifty.*

The similarities between Yahweh and Enlil extend to other aspects. Though he might have been depicted on cylinder seals (which is not certain, since the representation might have been of his son Ninurta), he was by and large an unseen God, ensconced in the innermost chambers of his ziggurat or altogether away from Sumer.

In a telltale passage in the Hymn to Enlil, the All-Beneficent it is thus said of him:

> *When in his awesomeness he decrees the*
> *fates,*
> *no God dares look at him;*
> *Only to his exalted emissary, Nusku,*
> *the command, the word that is in his heart,*
> *does he make known.*
> *No man can see me and live,*
> *Yahweh told Moses in a similar vein;*
> *and His words and commandments were*
> *known*
> *through Emissaries and Prophets.*
> *While all these reasons for equating*
> *Yahweh with Enlil*
> *are fresh in the reader's mind,*
> *let us hasten to offer the contrary evidence*
> *that points to other, different*
> *identifications.*

One of the most powerful biblical epithets for ***Yahweh is El Shaddai.*** *Of an uncertain etymology, it assumed an aura of mystery and by medieval times became a code word for kabbalistic mysticism. Early Greek and Latin translators of the Hebrew Bible rendered Shaddai as "omnipotent," leading to the rendering of El Shaddai in the King James translation as "God Almighty" when the epithet appears in the tales of the Patriarchs (e.g. "And Yahweh appeared unto Abram and said to him: 'I am El Shaddai; walk before me and be thou perfect'," in Genesis 17:1), or in Ezekiel, in Psalms, or several times in other books of the Bible.*

Advances in the study of Akkadian in recent years suggest that the Hebrew word is related to shaddu, which means "mountain" in Akkadian; so that El Shaddai simply means "God of mountains." That this is a correct understanding of the biblical term is indicated by an incident reported in I Kings chapter 20. The Arameans, who were defeated in an attempt to invade Israel (Samaria), recouped their losses and a year later planned a second attack. To win this time, the Aramaean king's generals suggested that a ruse be used to lure the Israelites out of their mountain strongholds to a battlefield in the coastal plains.

> *"Their God is a God of mountains," the generals told the king, "and that is why they prevailed over us; but if we shall fight them in a plain, we shall be the stronger ones."*

Now, there is no way that Enlil could have been called, or reputed to be, a "God of mountains," for there are no mountains in the great plain that was (and still is) Mesopotamia. In the Enlilite domains the land that was called "Mountainland" was Asia Minor to the north, beginning with the Taurus ("Bull") mountains; and that was the region of ***Adad****, Enlil's youngest son. His Sumerian name was ISH.KUR (and his "cult animal" was the bull), which meant "He of the mountainland." The Sumerian ISH was rendered shaddu in Akkadian; so that Il Shaddu became the biblical El Shaddai.*

Scholars speak of Adad, whom the Hittites called Teshub (see Fig. 80) as a "storm God," always depicted with a lightning, thundering, and windblowing, and thus the God of rains. The Bible credited Yahweh with similar attributes. "When Yahweh uttereth His voice," Jeremiah said (10:13), "there is a rumbling of waters in the skies and storms come from the ends of the earth;

He maketh lightnings with the rain, and blows a wind from its sources." The Psalms (135:7), the Book of Job, and other Prophets reaffirmed Yahweh's role as giver or withholder of rains, a role initially expounded to the Children of Israel during the Exodus.

While these attributes tarnish the similarities between Yahweh and Enlil, they should not carry us away to assume that, if so, Yahweh was the mirror image of Adad. The Bible recognized the existence of Hadad (as his name was spelled in Hebrew) as one of the "other Gods" of other nations, not of Israel, and mentions various kings and princes (in the Aramean Damascus and other neighboring capitals) who were called Ben-Hadad ("Son of Adad"). In Palmyra (the biblical Tadmor), capital of eastern Syria, Adad's epithet was Ba'al Shamin, "Lord of Heaven," causing the Prophets to count him as just one of the Ba'al Gods of neighboring nations who were an abomination in the eyes of Yahweh. There is no way, therefore, that Yahweh could have been one and the same as Adad.

The comparability between Yahweh and Enlil is further diminished by another important attribute of Yahweh, that of a warrior. "Yahweh goes forth like a warrior, like a hero He whips up His rage; He shall roar and cry out and over His enemies He shall prevail," Isaiah (42:13) stated, echoing the verse in the Song of Miriam that stated, "A Warrior is Yahweh" (Numbers chapter 15). Continuously, the Bible refers to and describes Yahweh as the "Lord of hosts," "Yahweh, the Lord of hosts, a warring army commands," Isaiah (13:4) declared. And Numbers 21:14 refers to a Book of the Wars of Yahweh in which the divine wars were recorded.

There is nothing in the Mesopotamian records that would suggest such an image for Enlil. The warrior par excellence was his son, ***Ninurta****, who fought and defeated Zu, engaged in the Pyramid Wars with the Enki'ites, and fought and imprisoned Marduk in the Great Pyramid. His frequent epithets were "the warrior" and "the hero" and hymns to him hailed him as "Ninurta, Foremost Son, possessor of divine powers . .. Hero who in his hand the divine brilliant weapon carries."*

His feats as a warrior were described in an epic text whose Sumerian title was Lugal-e Ud Melam-bi that scholars have called

The Book of The Feats and Exploits of Ninurta. Was it, one wonders, the enigmatic Book of the Wars of Yahweh of which the Bible spoke?

In other words, could Yahweh have been Ninurta?

As Foremost Son and heir apparent of Enlil, ***Ninurta*** *too bore the numerical rank of fifty, and could thus qualify no less than Enlil to have been the Lord who decreed the fifty-year Jubilee and other fifty-related aspects mentioned in the Bible. He possessed a notorious Divine Black Bird that he used both for combat and on humanitarian missions; it could have been the Kahod flying vehicle that Yahweh possessed. He was active in the Zagros Mountains to the east of Mesopotamia, the lands of Elam, and was revered there as Ninshushi-nak, "Lord of Shushan city" (the Elamite capital).*

At one time he performed great dyking works in the Zagros mountains; at another, he diked and diverted mountain rain channels in the Sinai peninsula to make its mountainous part cultivable for his mother Ninharsag; in a way he, too, was "God of mountains." His association with the Sinai peninsula and the channeling of its rainwaters, that come in winter bursts only, into an irrigation system is still recalled to this day: the largest Wadi (a river that fills up in winter and dries up in summer) in the peninsula is still called Wadi El-Arish, the wadi of the Urash - the Ploughman - a nickname of Ninurta from way back. An association with the Sinai peninsula, through his waterworks and his mother's residence there, also offers links to a Yahweh identification.

Another interesting aspect of Ninurta that invokes a similarity to the Biblical Lord comes to light in an inscription by the Assyrian king Ashurbanipal, who at one time invaded Elam. In it the king called him, "The mysterious God who lingers in a secret place where no one can see what his divine being is about." An unseen God!

But ***Ninurta****, as far as the earlier Sumerians were concerned, was not a God in hiding, and graphic depictions of him, as we have shown, were not even rare. Then, in conflict with a Yahweh-Ninurta identification, we come across a major ancient text,* *dealing with a major and unforgettable event, whose specifics seem to tell us that Ninurta was not Yahweh.*

One of the most decisive actions attributed in the Bible to Yahweh, with lasting effects and indelible memories, was the upheavaling of Sodom and Gomorrah. The event, as we have shown in great detail in The Wars of Gods and Men, was described and recalled in Mesopotamian texts, making possible a comparison of the deities involved.

In the biblical version Sodom (where Abram's nephew and his family lived) and Gomorrah, cities in the verdant plain south of the Sea of Salt, were sinful. Yahweh "comes down" and, accompanied by two Angels, visits Abram and his wife Sarai in their encampment near Hebron. After Yahweh predicts that the aged couple would have a son, the two Angels depart for Sodom to verify the extent of the cities' "sinning." Yahweh then reveals to Abram that if the sins would be confirmed, the cities and their residents would be destroyed. Abram pleads with Yahweh to spare Sodom if fifty just men be found there, and Yahweh agrees (the number was bargained by Abram down to ten) and departs.

The Angels, having verified the cities' evil, warn Lot to take his family and escape. He asks for time to reach the mountains, and they agree to delay the destruction. Finally, the cities' doom begins as,

> *"Yahweh rained upon Sodom and Gomorrah sulfurous fire, coining from Yahweh from the skies; and He upheavaled those cities and the whole plain and all the inhabitants thereof, and all that which grew upon the ground... And Abraham went early in the morning to the place where he had stood before Yahweh, and gazed in the direction of Sodom and Gomorrah, toward the land of the Plain, and he beheld vapor arising from the earth as the smoke of a furnace"*
> ***(Genesis chapter 19).***

*The same event is well documented in Mesopotamian annals as the culmination of Marduk's struggle to attain supremacy on Earth. Living in exile, **Marduk** gave his son **Nabu** the assignment of converting people in western Asia to become followers of Marduk. After a series of skirmishes, Nabu's forces were strong enough to invade Mesopotamia and enable Marduk to return to Babylon, where he declared his intention to make it the Gateway* *of the Gods (what its name, Bab-Ili, implied). Alarmed, the Council of the Anunnaki met in emergency sessions chaired by Enlil. Ninurta, and an alienated son of Enki called Nergal (from the south African domain), recommended drastic action to stop Marduk. Enki vehemently objected.*

Ishtar pointed out that while they were debating, Marduk was seizing city after city. "Sheriffs" were sent to seize Nabu, but he escaped and was hiding among his followers in one of the "sinning cities." Finally, Ninurta and Nergal were authorized to retrieve from a hiding place awesome nuclear weapons, and to use them to destroy the Spaceport in the Sinai (lest it fall into Mardukian hands) as well as the area where Nabu was hiding.

The unfolding drama, the heated discussions, the accusations, and the final drastic action - the use of nuclear weapons in 2024 B.C. - are described in great detail in a text that scholars call the Erra Epic.

In this document ***Nergal*** *is referred to as Erra ("Howler") and* ***Ninurta*** *is called Ishum ("Scorcher"). Once they were given the go-ahead they retrieved "the awesome seven weapons, without parallel" and went to the Spaceport near the "Mount Most Supreme." The destruction of the Spaceport was carried out by Ninurta/Ishum:*

> *"He raised his hand; the Mount was smashed; the plain by the Mount Most Supreme he then obliterated; in its forests not a tree-stem was left standing."*

Now it was the turn of the sinning cities to be upheavaled, and the task was carried out by Nergal/Erra.

He went there by following the King's Highway that connected the Sinai and the Red Sea with Mesopotamia:

> *Then, emulating Ishum,*
> *Erra the King's Highway*
> *followed.*
> *The cities he finished off,*
> *to desolation he overturned*
> *them.*

The use of nuclear weapons there broke open the sand barrier that still partly exists in the shape of a tongue (called El Lissan), and the waters of the Salt Sea poured south, inundating the low-lying plain. The ancient text records that Erra/Nergal "dug through the sea, its wholeness he divided." And the nuclear weapons turned the Salt Sea to the body of water now called the Dead Sea:

> *"That which lives in it he made wither," and what used to be a thriving and verdant plain, "as with fire he scorched the animals, burned its grains to become as dust."*

As was the clear-cut case of the divine actors in the Deluge tale, so we find in this one concerning the upheavaling of Sodom, Gomorrah, and the other cities of that plain astride the Sinai peninsula, whom does and whom does not Yahweh match when the biblical and Sumerian texts are compared. The Mesopotamian text clearly associates Nergal and not Ninurta as the one who had upheavaled the sinning cities.

Since the Bible asserts that it was not the two Angels who had gone to verify the situation, but Yahweh himself who had rained destruction on the cities, Yahweh could not have been Ninurta.

(The reference in Genesis chapter 10 to Nimrod as the one credited with starting Kingship in Mesopotamia, which we have discussed earlier, is interpreted by some as a reference not to a human king but to a God, and thus to Ninurta to whom the task of setting up the first Kingships was assigned. If so, the biblical statement that Nimrod "was a mighty hunter before Yahweh" also nullifies the possibility that Ni-nurta/Nimrod could have been Yahweh).

But Nergal too was not Yahweh. He is mentioned by name as the deity of the Cutheans who were among the foreigners brought over by the Assyrians to replace the Israelites who were exiled. He is listed among the "other Gods" that the newcomers worshiped and for whom they set up idols. He could not have been "Yahweh" and Yahweh's abomination at one and the same time.

*If **Enlil** and two of his sons, Adad and Ninurta, are not finalists in the lineup to identify Yahweh, what about Enlil's third son, Nannar/Sin (the "Moon God")?*

His "cult center" (as scholars call it) in Sumer was Ur, the very city from which the migration of Terah and his family began. From Ur, where Terah performed priestly services, they went to Harran on *the Upper Euphrates - a city that was a duplicate (even if on a smaller scale) of Ur as a cult center of **Nannar**. The migration at that particular time was connected, we believe, with religious and royal changes that might have affected the worship of Nannar. Was he then the deity who had instructed Abram the Sumerian to pick up and leave?*

Having brought peace and prosperity to Sumer when Ur was its capital, he was venerated in Ur's great ziggurat (whose remains rise awesomely to this day) with his beloved wife NIN.GAL ("Great Lady"). At the time of the new moon, the hymns sung to this divine couple expressed the people's gratitude to them; and the dark of the moon was considered a time of,

> *"the mystery of the great Gods, a time of Nannar's oracle," when he would send "Zaqar, the God of dreams during the night" to give commands as well as to forgive sins. He was described in the hymns as "decider of destinies in Heaven and on Earth, leader of living creatures ... who causes truth and justice to be."*

It all sounds not unlike some of the praises of Yahweh sung by the Psalmist...

*The Akkadian/Semitic name for Nannar was **Sin**, and there can be no doubt that it was in honor of Nannar as Sin that the part of the Sinai peninsula called in the Bible the "Wilderness of Sin" and, for that matter, the whole peninsula, were so named. It was in that part of the world that Yahweh appeared to Moses for the first time, where the "Mount of the Gods" was located, where the greatest Theophany ever had taken place. Furthermore, the principal habitat in the Sinai's central plain, in the vicinity of what we believe is the true Mount Sinai, is still called Nakhl in Arabic after the Goddess Ningal whose Semitic name was pronounced Nikal.*

Was it all indicative of a Yahweh = Nannar/Sin identification?

*The discovery several decades ago of extensive Canaanite literature ("myths" to scholars) dealing with their pantheon revealed that while a God they called Ba'al (the generic word for "Lord" used as a personal name) was running things, he was in fact not entirely independent of his father **El** (a generic term meaning "God" used as a personal name). In these texts El is* *depicted as a retired God, living with his spouse Asherah away from the populated areas, at a quiet place where "the two waters meet" - a place that we have identified in The Stairway To Heaven as the southern tip of the Sinai peninsula, where the two gulfs extending from the Red Sea meet.*

ISLAM, MOON CRESCENT ETC.

*This fact and other considerations have led us to the conclusion that the Canaanite **El** was the retired **Nannar/Sin**; included in the reasons upon which we had expounded is the fact that a "cult center" to Nannar/Sin has existed at a vital crossroads in the ancient Near East and even nowadays, the city known to us as Jericho but whose biblical/Semitic name is Yeriho, meaning "City of the Moon God"; and the adoption by tribes to the south thereof of **Allah** - "**El**" in Arabic - as the God of Islam represented by the Moon's crescent.*

Described in the Canaanite texts as a retired deity, El as Nannar/Sin would indeed have been forced into retirement: Sumerian texts dealing with the effects of the nuclear cloud as it wafted eastward and reached Sumer and its capital Ur, reveal that Nannar/Sin - refusing to leave his beloved city - was afflicted by the deathly cloud and was partly paralyzed.

The image of Yahweh, especially in the period of the Exodus and the settlement of Canaan, i.e. after - not prior to - the demise of Ur, does not sound right for a retired, afflicted, and tired deity as Nannar/Sin had become by then. The Bible paints a picture of an active deity, insistent and persistent, fully in command, defying the Gods of Egypt, inflicting plagues, dispatching Angels, roaming the skies; omnipresent, performing wonders, a magical healer, a Divine Architect. We find none of that in the descriptions of Nannar/Sin.

Both his veneration and fear of him stemmed from his association with his celestial counterpart, the Moon; and this celestial aspect serves as a decisive argument against identifying him with Yahweh: In the biblical divine order, it was Yahweh who ordered the Sun and the Moon to serve as luminaries; "the Sun and the Moon praise Yahweh," the Psalmist (148:3) declared. And on Earth, the crumbling of the walls of Jericho before the trumpeters of Yahweh symbolized the supremacy of Yahweh over the Moon God Sin.

There was also the matter of Ba'al, the Canaanite deity whose worship was a constant thorn in the side of Yahweh's faithful. The discovered texts reveal that Ba'al was a son of El. His abode in the mountains of Lebanon is still known as ***Baalbek,*** *"The valley of Ba'al" - the place that was the first destination of Gilgamesh in his search for immortality. The biblical name for it was BeitShemesh - the "House/abode of Shamash;" and Shamash, we may recall, was a son of Nannar/Sin.*

The Canaanite "myths" devote much clay tablet space to the shenanigans between Ba'al and his sister Anat; the Bible lists in the area of Beit-Shemesh a place called Beit Anat; and we are as good as certain that the Semitic name Anat was a rendering of Anunitu ("Ami's beloved") - a nickname of Inanna/Ishtar, the twin sister of Utu/Shamash.

All that suggests that in the Canaanite trio El-Ba'al-Anat we see the Mesopotamian triad of Nannar/Sin-Utu/Shamash-Inanna/Ishtar - the Gods associated with the Moon, the Sun, and Venus. And none of them could have been Yahweh, for the Bible is replete with admonitions against the worship of these celestial bodies and their emblems.

If neither Enlil nor any one of his sons (or even grandchildren) fully qualify as Yahweh, the search must turn elsewhere, to the sons of Enki, where some of the qualifications also point.

The instructions given to Moses during the sojourn at Mount Sinai were, to a great extent, of a medical nature. Five whole chapters in Leviticus and many passages in Numbers are devoted to medical procedures, diagnosis and treatment. "Heal me, O Yahweh, and I shall be healed," Jeremiah (17:14) cried out:

> *"My soul blesses Yahweh . . . who heals all my ailments," the Psalmist sang (103:1-3).*

Because of his piety, King Hezekiah was not only cured on Yahweh's say-so of a fatal disease, but was also granted by Yahweh fifteen more years to live (II Kings chapter 19). ***Yahweh could not only heal and extend life, he could also (through his Angels and Prophets) revive the dead; an extreme example was provided by Ezekiel's vision of the scattered***

dry bones that came back alive, their dead resurrected by Yahweh's will.

The biological-medical knowledge underlying such capabilities was possessed by Enki, and he passed such knowledge to two of his sons: ***Marduk*** *(known as Ra in Egypt), and* ***Thoth*** *(whom the Egyptians called Tehuti and the Sumerians NIN.GISH.ZIDDA - "Lord of the Tree of Life"). As for Marduk, many Babylonian texts refer to his healing abilities; but - as his own complaint to his father reveals - he was given knowledge of healing but not that of reviving the dead.*

On the other hand, Thoth did possess such knowledge, employing it on one occasion to revive Horus, the son of the God Osiris and his sister-wife Isis. According to the hieroglyphic text dealing with this incident, Horus was bitten by a poisonous scorpion and died. As his mother appealed to the "God of magical things," Thoth, for help, he came down to Earth from the heavens in a sky boat, and restored the boy back to life.

When it came to the construction and equipping of the Tabernacle in the Sinai wilderness and later on of the Temple in Jerusalem, Yahweh displayed an impressive knowledge of architecture, sacred alignments, decorative details, use of materials, and construction procedures - even to the point of showing the Earthlings involved scale models of what He had designed or wanted.

Marduk has not been credited with such an all-embracing knowledge; but Thoth/Ningishzidda was. In Egypt he was deemed the keeper of the secrets of pyramid building, and as Ningishzidda he was invited to Lagash to help orientate, design, and choose materials for the temple that was built for Ninurta.

Another point of major congruence between Yahweh and Thoth was the matter of the calendar. It is to ***Thoth*** *that the first Egyptian calendar was attributed, and when he was expelled from Egypt by Ra/Marduk and went (according to our findings) to Mesoamerica, where he was called "The Winged Serpent" (Quetzalcoatl), he devised the Aztec and Mayan calendars there. As the biblical books of Exodus, Leviticus, and Numbers make clear, Yahweh not only shifted the New Year to the "seventh* *month," but also instituted the week, the Sabbath, and a series of holidays.*

Healer; reviver of the dead who came down in a sky boat; a Divine Architect; a great astronomer and designer of calendars. The attributes common to Thoth and Yahweh seem overwhelming.

So was Thoth Yahweh?

Though known in Sumer, he was not considered there one of the Great Gods, and thus not fitting at all the epithet "the God Most High" that both Abraham and Melchizedek, priest of Jerusalem, used at their encounter. Above all, he was a God of Egypt, and (unless excluded by the argument that he was Yahweh), he was one of those upon whom Yahweh set out to make judgments.

Renowned in ancient Egypt, there could be no Pharaoh ignorant of this deity. Yet, when Moses and Aaron came before Pharaoh and told him, "So sayelh Yahweh, the God of Israel: Let My people go that they may worship Me in the desert," Pharaoh said: "Who is this Yahweh that I should obey his words? I know not Yahweh, and the Israelites I shall not let go."

If Yahweh where Thoth, not only would the Pharaoh not answer thus, but the task of Moses and Aaron would have been made easy and attainable were they just to say, Why - "Yahweh" is just another name for Thoth .. . And Moses, having been raised in the Egyptian court, would have had no difficulty knowing that - if that were so.

If Thoth was not Yahweh, the process of elimination alone appears to leave one more candidate: ***Marduk****.*

That he was a "God most high" is well established; the Firstborn of Enki who believed that his father was unjustly deprived of the supremacy on Earth - a supremacy to which he, Marduk, rather than Enlil's son Ninurta, was the rightful successor. His attributes included a great many - almost all - the attributes of Yahweh. He possessed a Shem, a sky-chamber, as Yahweh did; when the Babylonian king Nebuchadnezzar II rebuilt the sacred precinct of Babylon, he built there an especially strengthened enclosure for the "chariot of Marduk, the Supreme Traveler between Heaven and Earth."

When Marduk finally attained the supremacy on Earth, he did not discredit the other Gods. On the contrary, he invited them all to reside in individual pavilions within the sacred precinct of Babylon. There was only one catch: their specific powers and attributes were to pass to him - just as the "Fifty Names" (i.e. rank) of Enlil had to.

A Babylonian text, in its legible portion, listed thus the functions of other great Gods that were transferred to Marduk:

Ninurta = Marduk of the hoe
Nergal = Marduk of the attack
Zababa = Marduk of the combat
Enlil = Marduk of lordship and counsel
Nabu = Marduk of numbers and counting
Sin = Marduk the illuminator of the night
Shamash = Marduk of justice
Adad = Marduk of rains

This was not the monotheism of the Prophets and the Psalms; it was what scholars term henotheism - a religion wherein the supreme power passes from one of several deities to another in succession. Even so, ***Marduk*** *did not reign supreme for long; soon after the institution of Marduk as national God by the Babylonians, it was matched by their Assyrian rivals by the institution of Ashur as "lord of all the Gods."*

Apart from the arguments that we have mentioned in the cases of Thoth that negate an identification with any major Egyptian deity (and Marduk was the great Egyptian God Ra after all), the Bible itself specifically rules out any equating of Yahweh with Marduk. Not only is Yahweh, in sections dealing with Babylon, portrayed as greater, mightier, and supreme over the Gods of the Babylonians - it explicitly foretells their demise by naming them. Both Isaiah (46:1) and Jeremiah (50:2) foresaw Marduk (also known as Bel by his Babylonian epitheht) and his son Nabu fallen and collapsed before Yahweh on the Day of Judgment.

Those prophetic words depict the two Babylonian Gods as antagonists and enemies of Yahweh; Marduk (and for that matter, Nabu) could not have been Yahweh.

(As far as Ashur is concerned, the God Lists and other evidence suggest that he was a resurgent Enlil renamed by the Assyrians "The All Seeing;" and as such, he could not have been Yahweh).

As we find so many similarities, and on the other hand crucial differences and contradicting aspects, in our search for a matching "Yahweh" in the ancient Near Eastern pantheons, we can continue only by doing what Yahweh had told Abraham: Lift thine eyes toward the heavens . . .

The Babylonian king Hammurabi recorded thus the legitimization of Marduk's supremacy on Earth:

> *Lofty Anu,*
> *Lord of the Anunnaki,*
> *and Enlil,*
> *Lord of Heaven and Earth*
> *who determines the destinies*
> *of the land,*
> *Determined for Marduk, the*
> *firstborn of Enki,*
> *the Enlil-functions over alt*
> *mankind*
> *and made him great among*
> *the Igigi.*

As this makes clear, even Marduk as he assumed supremacy on Earth recognized that it was ***Anu****, and not he, who was "Lord of the Anunnaki." Was he the "God Most High" by whom Abraham and Melchizedek greeted each other?*

The cuneiform sign for Anu (AN in Sumerian) was a star; it had the multiple meanings of "God, divine," "heaven," and this God's personal name. Anu, as we know from the Mesopotamian texts, stayed in "heaven"; and numerous biblical verses also described Yahweh as the One Who Is in Heaven.

> *It was "Yahweh, the God of Heaven," who commanded him to go to Canaan, Abraham stated (Genesis 24:7).*

> *"I am a Hebrew and it is Yahweh, the God of Heaven that I venerate," the Prophet Jonah said (1:9).*
> *"Yahweh, the God of Heaven commanded me to build for Him a House in Jerusalem, in Judaea," Cyrus stated in his edict regarding the rebuilding of the Temple in Jerusalem (Ezra 1:2).*

When Solomon completed the construction of the (first) Temple in Jerusalem, he prayed to Yahweh to hear him from the heavens to bless the Temple as His House, although, Solomon admitted, it was hardly possible that "Yahweh Elohim" would come to dwell on Earth, in this House, "when the heaven and the heaven-of-heavens cannot contain Thee" (I Kings 8:27); and the Psalms repeatedly stated,

> *"From the heaven did Yahweh look down upon the Children of Adam" (14:2)*
> *"From Heaven did Yahweh behold the Earth" (102:20)*
> *"In Heaven did Yahweh establish His throne" (103:19)*

Though Anu did visit Earth several times, he was residing on Nibiru; and as the God whose abode was in Heaven, he was truly an unseen God: among the countless depictions of deities on cylinder seals, statues and statuettes, carvings, wall paintings, amulets - his image does not appear even once!

Since Yahweh, too, was unseen and unrepresented pictorially, residing in "Heaven," the inevitable question that arises is, Where was the abode of Yahweh? With so many parallels between Yahweh and Anu, did Yahweh, too, have a "Nibiru" to dwell on?

*The question, and its relevance to Yahweh's invisibility, does not originate with us. It was sarcastically posed by a heretic to a Jewish savant, **Rabbi Gamliel**, almost two thousand years ago; and the answer that was given is truly amazing!*

*The report of the conversation, as rendered into English by **S.M. Lehrman** in The World of the Midrash, goes thus:*

> *When Rabbi Gamliel was asked by a heretic to cite the exact location of God, seeing that the world is so vast and there are seven oceans, his reply was simply, "This I cannot tell you."*
>
> *Whereupon the other tauntingly retorted:*

"And this you call Wisdom, praying to a God, daily, whose whereabouts you do not know?"
The Rabbi smiled:

"You ask me to put my finger on the exact spot of His Presence, albeit that tradition avers that the distance between heaven and earth would take a journey of 3,500 years to cover. So, may I ask you the exact whereabouts of something which is always with you, and without which you cannot live a moment?" The pagan was intrigued. "What is this?" he eagerly queried.

The Rabbi replied: "The soul which God had planted within you; pray tell me where exactly is it?" It was a chastened man that shook his head negatively.

It was now the Rabbi's turn to be amazed and amused. "If you do not know where your own soul is located, how can you expect to know the precise habitation of One who fills the whole world with His glory?"

Let us note carefully what Rabbi Gamliel's answer was: according to Jewish tradition, he said, the exact spot in the heavens where God has a dwelling is so distant that it would require a journey of 3,500 years ...

How much closer can one get to the 3,600 years that it takes Nibiru to complete one orbit around the Sun?

Although there are no specific texts dealing with or describing Anu's abode on Nibiru, some idea thereof can be gained indirectly from such texts as the tale of Adapa, occasional references in various texts, and even from Assyrian depictions. It was a place - let us think of it as a royal palace - that was entered through imposing gates, flanked by towers. A pair of Gods (Ningishzidda and Dumuzi are mentioned in one instance) stood guard at the gates.

Inside, Anu was seated on a throne; when Enlil and Enki were on Nibiru, or when Anu had visited Earth, they flanked the throne, holding up celestial emblems.

(The Pyramid Texts of ancient Egypt, describing the Afterlife ascent of the Pharaoh to the celestial abode, carried aloft by an "Ascender," announced for the departing king: "The double gates of heaven are opened for thee, the double gates of the sky are opened for thee" and envisioned four scepter-holding Gods announcing his arrival on the "Imperishable Star").

In the Bible, too, Yahweh was described as seated on a throne, flanked by Angels. While Ezekiel described seeing the Lord's image, shimmering like electrum, seated on a throne inside a Flying Vehicle, "the throne of Yahweh is in Heaven," the Psalms (11:4) asserted; and the Prophets described seeing Yahweh seated on a throne in the Heavens. The Prophet Michaiah ("Who is like Yahweh?"), a contemporary of Elijah, told the king of Judaea who had sought a divine oracle (I Kings chapter 22):

> *I saw Yahweh sitting on his throne,*
> *and the host of heaven were standing by Him,*
> *on His right and on His left.*

The Prophet Isaiah recorded (chapter 6) a vision seen by him "in the year in which king Uzziah died" in which he saw God seated on His throne, attended by fiery Angels:

> *I beheld my Lord seated on a high and lofty throne,*
> *and the train of His robe filled the great hall.*
> *Seraphs stood in attendance on Him,*
> *each one of them having six wings:*
> *with twain each covered his face,*
> *with twain each covered his legs,*
> *and with twain each one would fly.*
> *And one would call out to the other:*
> *Holy, holy, holy is the Lord of Hosts!*

Biblical references to Yahweh's throne went farther: they actually stated its location, in a place called <u>Olam</u>. "Thy throne is established forever, from Olam art Thou," the Psalms (93:2) declared; "Thou, Yahweh, are enthroned in Olam, enduring through the ages," states the Book of Lamentations (5:19).

Now, this is not the way these verses, and others like them, have been usually translated. In the King James Version, for example, the quoted verse from Psalms is translated "Thy throne is established of old, thou art from everlasting," and the verse in Lamentations is rendered "Thou, O Lord, remainest for ever: thy throne from generation to generation." Modern translations likewise render Olam as "everlasting" and "forever" (The New American Bible) or as "eternity" and "for ever" (The New English Bible), revealing an indecision whether to treat the term as an adjective or as a noun.

Recognizing, however, that Olam is clearly a noun, the most recent translation by the Jewish Publication Society adopted "eternity," an abstract noun, as a solution.

The Hebrew Bible, strict in the precision of its terminology, has other terms for stating the state of "lasting forever." One is Netzah, as in Psalm 89:47 that asked, "How long, Yahweh, wilt Thou hide Thyself - forever?" Another term that means more precisely "perpetuity" is Ad, which is also usually translated "for ever," as in "his seed I will make endure for ever" in Psalm 89:30. There was no need for a third term to express the same thing.

Olam, often accompanied by the adjective Ad to denote its everlasting nature, was itself not an adjective but a noun derived from the root that means "disappearing, mysteriously hidden." The numerous biblical verses in which Olam appears indicate that it was deemed a physical place, not an abstraction. "Thou art from Olam," the Psalmist declared - God is from a place which is a hidden place (and therefore God has been unseen).

It was a place that was conceived as physically existing: Deuteronomy (33:15) and the Prophet Habakkuk (3:6) spoke of the "hills of Olam." Isaiah (33:14) referred to the "heat sources of Olam. " Jeremiah (6:16) mentioned the "pathways of Olam" and (18:5) "the lanes of Olam," and called Yah-weh "king of Olam" (10:10) as did Psalms 10:16.

The Psalms, in statements reminiscent of the references to the gates of Anu's abode (in Sumerian texts) and to the Gates of Heaven (in ancient Egyptian texts), also spoke of the "Gates of

Olam" that should open and welcome the Lord Yahweh as He arrives there upon His Kabod, His Celestial Boat (24:7-10):

Lift up your heads, O gates of Olam
so that the King of Kabod may come in!
Who is the King of Kabod?
Yahweh, strong and valiant, a mighty
warrior/
Lift up your heads, O gates of Olam,
and the King of Kabod shall come in!
Who is the King of Kabod?
Yahweh lord of hosts is the King of Kabod.

"Yahweh is the God of Olam," declared Isaiah (40:28), echoing the biblical record in Genesis (21:33) of Abraham's "calling in the name of Yahweh, the God of Olam." No wonder, then, that the Covenant symbolized by circumcision, "the celestial sign," was called by the Lord when he had imposed it on Abraham and his descendants "the Covenant of Olam:"

And my Covenant shall be in your flesh, the Covenant of Olam.
(Genesis 17:13)

In post-biblical rabbinic discussions, and so in modern Hebrew, Olam is the term that stands for "world." Indeed, the answer that ***Rabbi Gamliel*** *gave to the question regarding the Divine Abode was based on rabbinic assertions that it is separated from Earth by seven heavens, in each of which there is a different world; and that the journey from one to the other requires five hundred years, so that the complete journey through seven heavens from the world called Earth to the world that is the Divine Abode lasts 3,500 years.*

This, as we have pointed out, comes as close to the 3,600 (Earth) years' orbit of Nibiru as one could expect; and while Earth to someone arriving from space would have been the seventh planet, Nibiru to someone on Earth would indeed be seven celestial spaces away when it disappears to its apogee.

Such a disappearing - the root meaning of Olam - creates of course the "year" of Nibiru - an awesomely long time in human terms. The Prophets similarly, in numerous passages, spoke of the "Years of Olam" as a measure of a very long time. A clear sense of periodicity, as would result from the periodic appearance and disappearance of a planet, was conveyed by the frequent use of "from Olam to Olam" as a definite (though extremely long) measure of time: "I had given you this land from Olam to Olam," the Lord was quoted as saying by Jeremiah (7:7 and 25:5).

And a possible clincher for identifying Olam with Nibiru was the statement in Genesis 6:4 that the Nefilim, the young Anunnaki who had come to Earth from Nibiru, were the "people of the Shem" (the people of the rocketships), "those who were from Olam."

With the obvious familiarity of the Bible's editors, Prophets, and Psalmists with Mesopotamian "myths" and astronomy, it would have been peculiar not to find knowledge of the important planet Nibiru in the Bible. It is our suggestion that yes, the Bible was keenly aware of Nibiru - and called it Olam, the "disappearing planet."

Does all that mean that therefore Anu was Yahweh? Not necessarily ...

Though the Bible depicted Yahweh as reigning in His celestial abode, as Anu did, it also considered Him "king" over the Earth and all upon it - whereas Anu clearly gave the command on Earth to Enlil. Anu did visit Earth, but extant texts describe the occasions mostly as ceremonial state and inspection visits; there is nothing in them comparable to the active involvement of Yahweh in the affairs of nations and individuals.

Moreover, the Bible recognized a God, other than Yahweh, a "God of other nations" called ***An****; his worship is noted in the listing (II Kings 17:31) of Gods of the foreigners whom the Assyrians had resettled in Samaria, where he is referred to as An-melekh ("Anu the king"). A personal name Anani, honoring Anu, and a place called Ana-tot, are also listed in the Bible. And the Bible had nothing for Yahweh that paralleled the genealogy of Anu (parents, spouse, children), his lifestyle (scores of concubines) or his fondness for his granddaughter Inanna (whose worship as the "Queen of Heaven"-Venus was deemed an abomination in the eyes of Yahweh).*

And so, in spite of the similarities, there are also too many essential differences between Anu and Yahweh for the two to have been one and the same.

Moreover, in the biblical view Yahweh was more than "king, lord" of Olam, as Anu was king on Nibiru. He was more than once hailed as El Olam, the God of Olam (Genesis 21:33) and El Elohim, the God of the Elohim (Joshua 22:22, Psalms 50:1 and Psalms 136:2).

The biblical suggestion that the Elohim - the "Gods," the Anunnaki - had a God, seems totally incredible at first, but quite logical on reflection.

At the very conclusion of our first book in The Earth Chronicles series (The 12th Planet), having told the story of the planet Nibiru and how ***the Anunnaki*** *(the biblical Nefilim) who had come to Earth from it "created" Mankind, we posed the following question:*

> *And if the Nefilim were the "Gods" who "created" Man on Earth, did evolution alone, on the Twelfth Planet, create the Nefilim?*

Technologically advanced, capable hundreds of thousands of years before us to travel in space, arriving at a cosmological explanation for the creation of the Solar System and, as we begin to do, to contemplate and understand the universe - the Anunnaki must have pondered their origins, and arrived at what we call Religion - their religion, their concept of God.

Who created the Nefilim, the Anunnaki, on their planet?

The Bible itself provides the answer. Yahweh, it states, was not just "a great God, a great king over all of the Elohim" (Psalms 95:3); He was there, on Nibiru, before they had come to be on it: "Before the Elohim upon Olam He sat," Psalm 61:8 explained. Just as the Anunnaki had been on Earth before The Adam, so was Yahweh on Nibiru/Olam before the Anunnaki. The creator preceded the created.

We have already explained that the seeming immortality of the Anunnaki "Gods" was merely their extreme longevity, resulting from the fact that one Nibiru-year equaled 3,600 Earth-years; and that in fact they were born, grew old, and could (and did) die. A time measure applicable to Olam ("days of Olam" and "years of Olam") was recognized by the Prophets and Psalmist; what is more impressive is their realization that the various Elohim (the Sumerian DIN.GIR, the Akkadian Ilu) were in fact not immortal - but Yahweh, God, was.

Thus, Psalm 82 envisions God passing judgment on the Elohim and reminding them that they - the Elohim! - are also mortal: "God stands in the divine assembly, among the Elohim He judges," and tells them thus:

> *I have said, ye are Elohim, all of you sons of the Most High; But ye shall the as men do, like any prince ye shall fall.*

We believe that such statements, suggesting that the Lord Yahweh created not only the Heaven and the Earth but also the Elohim, the Anunnaki "Gods," have a bearing on a puzzle that has baffled generations of biblical scholars. It is the question why the Bible's very first verse that deals with the very Beginning, does not begin with the first letter of the alphabet, but rather with the second one. The significance and symbolism of beginning the Beginning with the proper beginning must have been obvious to the Bible's compilers; yet, this is what they chose to transmit to us:

> *Breshit bara Elohim*
> *et Ha'Shamaim v'et Ha'Aretz*
> *...which is commonly translated, "In the beginning God created the Heaven and the Earth."*

Since the Hebrew letters have numerical values, the first letter, Aleph (from which the Greek alpha comes) has the numerical value "one, the first" - the beginning. Why then, scholars and theologians have wondered, does the Creation start with the second letter, Beth, whose value is "two, second"?

While the reason remains unknown, the result of starting the first verse in the first book of the Bible with an Aleph would be astounding, for it would make the sentence read thus:

> *Ab-reshit bara Elohim,*
> *et Ha'Shamaim v'et Ha'Aretz*
> *The Father-of-Beginning created the Elohim, the Heavens, and the Earth.*

By this slight change, by just starting the beginning with the letter that begins it all, an omnipotent, omnipresent Creator of All emerges from the primeval chaos: Ab-Reshit, "the Father of Beginning." The best modern scientific minds have come up with the Big Bang theory of the beginning of the universe - but have yet to explain who caused the Big Bang to happen. Were Genesis to begin as it should have, the Bible - which offers a precise tale of Evolution and adheres to the most sensible cosmogony - would have also given us the answer: the Creator who was there to create it all.

And all at once Science and Religion, Physics and Metaphysics, converge into one single answer that conforms to the credo of Jewish monotheism: "I am Yahweh, there is none beside me!" It is a credo that carried the Prophets, and us with them, from the arena of Gods to the God who embraces the universe.

One can only speculate why the Bible's editors, who scholars believe canonized the Torah (the first five books of the Bible) during the Babylonian exile, omitted the Aleph. Was it in order to avoid offending their Babylonian exilers (because a claim that Yahweh had created the Anunnaki-Gods would have not excluded Marduk)? But what is, we believe, not to be doubted is that at one time the first word in the first verse in the Bible did begin with the first letter of the alphabet. This certainty is based on the statements in the Book of Revelation ("The Apocalypse of St. John" in the New Testament), in which God announces thus:

> *I am Alpha and Omega, the Beginning and*
> *the End, the First and the Last.*

The statement, repeated three times (1:8, 21:6, 22:13), applies the first letter of the alphabet (by its Greek name) to the Beginning, to the divine First, and the last letter of the (Greek) alphabet to the End, to God being the Last of all as He has been the First of All.

That this had been the case at the beginning of Genesis is confirmed, we believe, by the certainty that the statements in Revelation harken back to the Hebrew scriptures from which the parallel verses in Isaiah (41:6, 42:8, 44:6) were taken, the verses in which Yahweh proclaims His absoluteness and uniqueness:

> *I, Yahweh, was the First And the Last I will*
> *also be!*
> *I am the First*
> *and I am the Last;*
> *There are no Elohim without Me!*
> *I am He,*

I am the First,
I am the Last as well.

It is these statements that help identify the biblical God by the answer that He himself gave when asked: Who, O God, are you? It was when He called Moses out of the Burning Bush, identifying Himself only as "the God of thy father, the God of Abraham, the God of Isaac and the God of Jacob."

Having been given his mission, Moses pointed out that when he would come to the Children of Israel and say, "the God of your forefathers has sent me to you, and they will say to me: What is His name? - what shall I tell them?"

And God said to Moses:
Ehyeh-Asher-Ehyeh *-*
this is what thou shall say
unto the Children of Israel:
Ehyeh sent me.
And God said further to Moses:
Thus shalt say unto the Children of Israel:
Yah wen, the God of your fathers,
the God of Abraham, the God of Isaac,
and the God of Jacob,
hath sent me unto you;
This is my name unto Olam,
this is my appellation unto all generations.
(Exodus 3:13-15)

The statement, Ehyeh-Asher-Ehyeh, has been the subject of discussion, analysis, and interpretation by generations of theologians, biblical scholars, and linguists. The King James Version translates it "I am that I am ... I am hath sent me to you." Other more modern translations adopt "I am, that is who I am ... I am has sent you."

The most recent translation by the Jewish Publication Society prefers to leave the Hebrew intact, providing the footnote, "meaning of the Hebrew uncertain."

The key to understanding the answer given during this Divine Encounter are the grammatical tenses employed here. Ehyeh-Asher-Ehyeh is not given in the present but in the future tense. In simple parlance it states: "Whoever I shall be, I shall be."

And the Divine Name that is revealed to a mortal for the first time (in the conversation Moses is told that the sacred name, the Tetragrammaton YHWH, had not been revealed even to Abraham) combines the three tenses from the root meaning "To Be" - the One who was, who is, and who shall be. It is an answer and a name that befit the biblical concept of Yahweh as eternally existing - One who was, who is, and who shall continue to be.

A frequent form of stating this everlasting nature of the biblical God is the expression "Thou art from Olam to Olam." It is usually translated, "Thou art everlasting;" this conveys undoubtedly the sense of the statement, but not its precise meaning. Literally taken it suggests that the existence and reign of Yahweh extended from one Olam to another - that He was "king, lord" not only of the one Olam that was the equivalent of the Mesopotamian Nibiru - but of other Olams, of other worlds!

No less than eleven times, the Bible refers to Yahweh's abode, domain, and "kingdom" using the term Olamim, the plural of Olam - a domain, an abode, a kingdom that encompasses many worlds. It is an expansion of Yahweh's Lordship beyond the notion of a "national God" to that of a Judge of all the nations; beyond the Earth and beyond Nibiru, to the "Heavens of Heaven" (Deuteronomy 10:14, I Kings 8:27, II Chronicles 2:5 and 6:18) that encompass not only the Solar System but even the distant stars (Deuteronomy 4:19, Ecclesi-astes 12:2).

THIS IS THE IMAGE OF A COSMIC VOYAGER.

All else - the celestial planetary "Gods," ***Nibiru*** *that remade our Solar System and remakes the Earth on its near passages, the Anunnaki "Elohim," Mankind, nations, kings - all are His manifestations and His instruments, carrying out a divine and universal everlasting plan. In a way we are all His Angels, and when the time comes for Earthlings to travel in space and emulate the Anunnaki, on some other world, we too shall only be carrying out a destined future.*

It is an image of a universal Lord that is best summed up in the hymnal prayer Adon Olam that is recited as a majestic song in Jewish synagogue services on festivals, on the Sabbath, and on each and every day of the year:

> *Lord of the universe, who has reigned Ere all that exists had yet been created. When by His will all things were wrought, "Sovereign" was His name was then pronounced.*
>
> *And when, in time, all things shall cease, He shall still reign in majesty. He was, He is, He shall remain, He shall continue gloriously.*

Incomparable, unique He is, No other can His Oneness share.
Without beginning, without end. Dominion's might is His to bear.

....

Bob's note: (re; Jehovah, Yahweh etc.)

I think that "The Name" is more likely a "title", and used by whoever was the leading or acting chief or general "god" in this particular area (geographical) at any time.

IF THERE WAS A SUPERIOR OR SUPREME "GOD" (not as in the case of this "family" patriarch, Anu) but a "God" to Anu, Enlil, Enki, et al, then such would have been known and recorded. For example Thoth was the designated chronicler of the Anunnaki etc.

It seems that no such record of a "God" (to them) exists from any source whatsoever, except for the record of the Hebrew, who so designated a supreme "god" for and by them alone.

Thus the god as later recorded by the Hebrew, was and obviously is NOT a universal "God" or even known or recognized by anyone apart from them. As I mention, if such a one existed as a fact, it would surely be known and recorded by the Anunnaki. As there is no such recording, I am left to conclude that "The Fabric Of the Universe" is the sole Omnific source of all things. (As discussed in my essays etc herein.)

304-Are the Extraterrestrials Who First Came to Earth Still Here?

With respect to "Prime Directive" Violations, and descriptions in ancient Sumerian texts of the Anunnaki ("those who from heaven to earth came"), there are three possibilities:

1) The extraterrestrials who first arrived on Earth, transformed Homo erectus into Homo sapiens (via DNA manipulations), and ruled the Earth for generations, have left the Earth and are not coming back.

2) They're still here (and have been), but they're staying out of the limelight. [In this connection, they just might have gone underground!]

Or

3) They left, but they're coming back (and they may not be too happy about what has happened since taking their "sabbatical"), i.e. get ready for Future Interventions!

These possibilities -- and even a fourth or fifth possibility -- are discussed in a paper first presented at the 1995 conference of the International Forum on New Science. Inasmuch as the material is still relevant, the paper is presented here without further editing.

Are the Extraterrestrials Who First Came to Earth Still Here?

Dan Sewell Ward

Abstract

Zecharia Sitchin has provided a fascinating documentation on the existence of extraterrestrials in the history of mankind. These extraterrestrials are described in ancient Sumerian texts (circa 1500 to 3500 BCE -- "Before Current Era") as being the Anunnaki ("those who from heaven to earth came"). According to Sitchin (1985, 1990-1) and any open-minded interpretation of the various ancient texts available to us today, the Anunnaki were instrumental in creating man some 250,000 years ago, and were thereafter extensively involved in mankind's history during the pre-Christian *eras. The Book of **Genesis**, as just one example, can best be interpreted in a consistent manner by incorporating the ancient Sumerian description of the Anunnaki.*

*[**Laurence Gardner** and others have provided ever more detail on the subject, including the recent publications of Genesis of the Grail Kings and Lost Secrets of the Sacred Ark -- both written by Gardner, and includes comprehensive, extensively researched notes and justifications for the Anunnaki legacy.]*

Despite Zecharia Sitchin's [et al] comprehensive research and scholarly interpretation of the ancient texts, two major questions concerning the Anunnaki have not been fully explored. The first question, which relates to the second, is why did they come to the Earth to begin with? The second question is whether or not the Anunnaki, after having been so completely immersed in the history of mankind, are still on the planet earth today.

This paper provides a brief summary on the background of the origins, history, and involvement in human activities of the Anunnaki, as well as the issue of why the Anunnaki came to earth in the first place and how man became a part of the Anunnaki history. The paper will then consider a variety of theories, ideas, and pieces of evidence which bear on the essential issue of whether or not the Anunnaki are still here. In developing the thesis of this paper, facts from a wide range of scholarly fields will be utilized, including ancient and recent histories, quantum physics, astronomy, mythology and philosophy.

The Anunnaki

According to Zecharia Sitchin (1985) and his interpretation of ancient Sumerian texts, extraterrestrials first arrived on the planet Earth some 485,000 years ago. These extraterrestrials were referred to by the Sumerians as the Anunnaki ("those who from heaven to earth came"). They are believed to have been from the planet Nibiru, a planet allegedly orbiting about our sun in a strongly elliptical orbit every 3,600 years. The planet's perihelion (closest point of approach to the Sun) is thought to be within the main asteroid belt between Mars and Jupiter, at a distance from the Sun of approximately 2.75 A.U. (an A.U. being the distance from the Sun to the Earth). The fact Nibiru is not known to modern science is likely due to the planet's aphelion (furthermost point in the planet's orbit from the Sun) being more than eight times the distance from the Sun to the planet Pluto (roughly 40 A.U.). Furthermore, Nibiru is now far out in deep space and unlikely to be easily detected.

The Anunnaki are an extremely long-lived race, potentially living as long as 500,000 Earth years, or about 140 "Nibiru years". Accordingly, Anunnaki events move along very slowly compared to the hectic pace of humans. After the first Anunnaki arrived on Earth, for example, the first notable event was the discovery of gold 36,000 years later. This led to an Anunnaki mission to Earth to recover the gold from sea water roughly 445,000 years ago. This initial effort was led by an Anunnaki named EA, who assumed the title EN.KI ("Lord of Earth").

After the initial operation failed to produce the expected quantities of gold, an enlarged operation was commenced, this time under the command of EN.LIL (translated as "Lord of the Command"). Enlil was the half-brother of Enki. But while Enki was the older of the two brothers, Enlil was the ***Heir Apparent*** *of Nibiru in the same manner as Isaac was Abraham's favored son over Ishmael. The resulting* ***Enki and Enlil*** *rivalry turned out to be very important in the history of mankind and in understanding such historical documents as the biblical Book of Genesis.*

The original gold extraction operation had been located in Mesopotamia at the head of the Persian Gulf. The new plan was to mine gold from the ground in southern Africa (referred to in the Sumerian texts as the AB.ZU), ship it back to Mesopotamia, and then lift it off the planet for trans-shipment to Nibiru. Enki was given authority over Africa (including Egypt), while Enlil took control of Mesopotamia and the overall Earth mission.

Roughly 272,000 years ago, the Anunnaki laboriously mining the gold from the South African mines mutinied! The mutiny and the difficulties the Anunnaki had encountered in working the mines were resolved when Enki proposed to create a "primitive worker" to do the work in lieu of the Anunnaki. In effect, Enki proposed, "Let us make man in our image, after our likeness." Given the go-ahead, Enki and his half-sister, Ninki, created man, ***Homo sapiens sapiens****, through genetic engineering for the sole purpose of providing workers to mine the gold! Man was created by cross-breeding Homo erectus with an extraterrestrial!*

Ninki herself carried the first "mixed worker" to term and gave birth to a being she called the "Lulu". Later, fourteen "birth goddesses" (female Anunnaki) were utilized to produce additional workers. This solution was only a moderate success in that the Lulu (or "mixed worker") was a hybrid and incapable of procreation (just as the mule, the cross between a horse and a donkey, cannot reproduce itself). And as the "birth goddesses" became weary of continually being pregnant, Enki and Ninki were forced to modify the genetic structure of the Lulu in order for it to be able to reproduce itself. They created the "Adama". This latter step, however, was taken without Enlil's knowledge!

Eventually, many of the Lulus were taken to Mesopotamia to work in agriculture and aid the Anunnaki efforts to raise food for themselves and their new workers. According to the Sumerian texts, Enlil created E.DIN, a special place where new strains of edible crops could be developed and later implemented. Lulus were placed inside E.DIN, but without Enlil having learned of the Lulu's recently acquired ability to procreate.

When the obvious result became obvious even to Enlil, when the Lulu's had effectively eaten from the Tree of Knowledge of Good and Evil (i.e. sexual knowing), the result was the Adama's expulsion from E.DIN. At the same time, Enki was the god who clothed the now homeless Adam and Eve. In fact, the duality of the God of Genesis can be explained by the often opposing actions of Enlil and Enki. Even the Jesuits of the Catholic Church have begun to acknowledge the reality of at least two gods in the story of Genesis. The existence of the Anunnaki as "gods" also explains Genesis 1:26, 3:22, 6:2, and 6:4.

Scientific confirmation for the creation of mankind from Homo erectus roughly 250,000 years ago comes from two sources. The first is from Cann, et al (1987) wherein mitochondrial DNA (a form of DNA transmitted only maternally) had been used to postulate a single woman living in Africa roughly 250,000 years ago who became the mother of every human being now living on the planet. More recently, Dorit, et al (1995) have found no intraspecific polymorphism whatsoever in a gene paternally inherited, and concluded a date of the last common male ancestor of all humans currently on the planet to be 270,000 BCE.

*The ancient Sumerian texts also describe in great detail the events leading up to the Flood and **Deluge** of biblical renown (circa 11,600 BCE) and the histories of mankind down through the age of Abraham and his grandchild, Jacob. The Sumerian version has been found to corroborate the Book of Genesis at every turn, and in fact has provided a much more detailed description of the biblical events. In many respects, the Book of Genesis is an executive summary of the far more ancient Sumerian texts.*

Sitchin (1990-1) has given extensive documentation for the corroboration of the Sumerian texts and their interpretation by modern scientific discoveries. These Sumerian texts are readily available (e.g. Dalley, 1989) and any open-minded interpretation tends to confirm Sitchin's hypothesis. This is particularly true of the Sumerian Enuma Elish which describes the creation of our solar system in a scientifically consistent manner.

*The Sumerian texts in describing the details of the ancient history of mankind place a strong emphasis on the fact that gold belonged to the gods (the human term for the Anunnaki). In this regard, it is worth noting **Genesis 2:11-12, where after God had created the heaven and the earth and placed man in the garden of Eden (but before creating Eve), the narrative informs us of the existence of gold in Havilah and the fact "the gold of that land is good." This supports the ancient Sumerian texts in their assigning an overriding importance of gold to the Anunnaki.***

*This emphasis, in turn, leads us to the primary mission of the Anunnaki in coming to Earth -- the why of their mission. But before describing this theory, it is necessary to digress momentarily and discuss the theories of **David Radius Hudson** (1994).*

Orbitally Rearranged Mono-atomic Elements

*The **ORME** (i.e. Orbitally Rearranged Mono-atomic Elements) is a term coined by David Hudson, and which, coincidentally, is the same as the Hebrew word meaning "**The Tree of Life**". According to Hudson, a group of eight elements, known as "the precious metals" and including ruthenium, **Rhodium and Iridium**, palladium, osmium, platinum, silver, and gold, are to be found in nature in a mono-atomic form. In this mono-atomic state the precious metals no longer display the characteristics of metals, but lose their chemical reactivity and change the configuration of their* *nuclei. Within the new configuration, the atoms interact in two dimensions, with the superdeformed nuclei reaching high spin, low energy states. In this state the elements become perfect superconductors, with their electrons combining in "Cooper Pairs" and thus becoming photons of light. In addition, the mono-atomic elements exhibit a **Meisner Field**, a non-polar magnetic field which repels all other magnetic fields and is a primary indicator of achieving superconductivity.*

*The physical characteristics of these mono-atomic elements have been described in some detail in the scientific literature (see, for example, Duncan, et al 1989, Greiner, et al, 1990, Macchaiavelli, et al, 1988, Dejbakhsh, et al, 1988, Lim, et al, 1989, Shimizu, et al, 1990, Randeria, et al, 1989, Adrian, et al, 1992, Boyer, 1975, Puthoff, 1989, and Haisch, 1994). The connection of the Orme phenomena with **Zero-Point-Energy** is particularly profound. For example Puthoff (1989) in describing a zero-point Lorentz force has predicted a weight loss of 4/9th when nuclei interact in only two dimensions. Hudson (1994) has experimentally observed this weight loss, as well as the mono-atomic element's Meisner field levitating on the earth's geomagnetic field! The latter is based on the Meisner field excluding all external magnetic fields, including the earth's.*

Of particular importance in David Hudson's work is the fact that the precious metals in their mono-atomic form can not be identified as such by standard methods of analysis such as emission spectroscopy -- unless special procedures are followed. In the case of emission spectroscopy, for example, the test must be conducted for some twenty times the standard length of test.

The inability of standard procedures to identify elements such as rhodium and iridium in their mono-atomic state has made these elements appear to be extremely rare, and thus expensive! One source of the precious metals in South Africa, for example, produces one-third of one ounce of all of the precious elements per ton of ore. David Hudson has found widespread ore samples in the United States and elsewhere which are, in reality, producing in excess of 2,200 ounces per ton of ore! In the case of rhodium, which currently sells in the neighborhood of $3,000/ounce, Hudson is obtaining 1200 ounces per ton of ore!

The inability of standard scientific methods in detecting the precious elements in their mono-atomic state is extremely important, as is the fact these elements are almost always found in their natural state in combination with gold. (Hudson, for example, made his discoveries of the mono-atomic state, while attempting to separate the gold from the refined ore.) Thus the search and overriding importance of gold to the Anunnaki may be, in fact, an attempt on the part of these extraterrestrials to acquire the precious elements in their mono-atomic form, and in large quantities. Why this would be important strikes at the very heart of why the Anunnaki came to Earth in the first place.

The Tree of Life

Based on a grand synthesis of historical, philosophical, mythological and scientific evidence, David Hudson (1994) has determined that the Orme he has identified is truly The Tree of Life. Hudson has noted, for example, the scientific basis for human cells to exhibit superconductivity (see e.g. Giudice, et al, 1981), and the extensive amount of research currently being conducted on treating cancer and other diseases with precious elements. These precious elements appear to be correcting the DNA, literally "flowing the light of life" within the body.

Books on ***Alchemy*** *have talked about* **"the white powder of gold"** *-- the same form in which Hudson found all eight of the mono-atomic precious elements. In addition, the alchemists of old, in their quest to accomplish "The Master Work" (in other words, to prepare the Elixir of Life and* ***The Philosopher's Stone****; or in more anecdotal, traditional terms, to transmute lead into gold), taught that the key to success was to "divide, divide, divide..." The implication of this advice was that only in freeing the atoms of an element from the confines of its crystalline-like structure of many atoms of the same or diverse elements, could one hope to achieve the alchemist's esoteric goal. In effect, the alchemists were attempting to reach the mono-atomic form of gold and the other precious elements!*

Hudson has also connected the Orme with the ***Hebrew tradition of the manna*** *-- the white powder of gold -- as well as noting the use of* ***the term, manna -- which literally means "What is it?"*** *-- in The Egyptian Book of the Dead (Budge 1967). The historical and philosophical implications also include the* ***Orme as part of***

the Melchizedek priesthood *and the metallurgical foundry at Qumrun, where the Essenes were located.*

The Essenes are described in The Dead Sea Scrolls Uncovered (Eisenman and Wise, 1993), while other references include the mixture of the white powder of gold in water as being "that which issues from the mouth of the creator," ***The Golden Tear from the Eye of Horus****, and "the semen of the father in heaven." In the Essene tradition, the white powder of gold is referred to as the "teacher of righteousness" which is swallowed and taken internally.* ***Hudson also believes Moses knew the secrets of making the Orme, and that the Ark of the Covenant was merely the container of the Orme. As a volatile superconducting electrical device, the Orme could thus explain the incredible properties of the Ark -- from levitation to "blasts of heavenly displeasure".***

The implications of the existence of the Orme and taking the White Powder of Gold (and the other precious elements) is astounding. Based on the immense philosophical, mythological, historical, and scientific literature, Hudson believes a human ingesting the Orme in the correct manner can fulfill all the dreams of the esoteric alchemist: To have perfect telepathy, be able to levitate and/or bilocate, know good and evil when it's in the room with you (i.e. eat of the fruit of the Tree of Knowledge of Good and Evil), project one's thoughts into someone else's mind, heal by the laying on of hands, and cleanse or resurrect the dead within two or three days after they have died. The latter is evidenced by the traditions of the Anunnaki raising their own from the dead, as well as the gospels of Jesus Christ. In effect, the evidence suggests that an individual taking the Orme can become a fifth dimensional being!

Implications

The astounding implication of the juxtaposition of Sitchin, [Gardner] and Hudson's work is the almost inescapable conclusion that the Anunnaki's purpose on arriving on Earth was to acquire the mono-atomic elements for the purpose of ingesting them and thus being able to have the abilities to accomplish all of the incredible feats described in six centuries of literature and philosophy!

In effect, extraterrestrials from the planet Nibiru came to Earth to mine gold and precious elements, created mankind in order to have someone work the mines, provided the knowledge to a very limited, select number of human beings on the importance of the Orme and the method by which it could be made and taken internally by the individual, ***and then rather lost control of mankind as the human population increased exponentially!***

It is noteworthy that the first attempt by the Anunnaki to extract gold was from sea water, specifically the head waters of the Persian Gulf. According to Hudson, the primary form which gold occurs in sea water is in the mono-atomic state. The precious elements also occur in herbs and numerous vegetables. Grapes, for example, can be a primary source. Volcanoes are also a source in that the interior of the Earth is the primary manufacturer of the elements. Hudson has also discovered in the brain tissue of pigs and cows, over 5% of the brain tissue by dry matter weight consists of rhodium and iridium.

It is worth recalling Genesis 2:11-12, previously cited, wherein there was not only gold in Havilah, but "the gold of that land is good." The emphasis on the gold being good -- as opposed to simply being there -- may imply the "good" gold is the mono-atomic gold and/or the gold which includes in its deposits the other precious elements in their mono-atomic state. I.e. the mono-atomic state is what makes the gold "good"!

Early indications from the continuing research on the Orme and its effect on human beings suggest rhodium to be the primary element in the health of the individual, while iridium appears to provide the incredible talents promised by the Orme. The key to benefiting from the Orme, however, appears to be the specific procedure by which the "teacher of righteousness" is swallowed. (Hudson (1994) ***has suggested the need for a forty day fast****, in which after ten days of water only, the Orme is ingested in doses of some 500 milligrams for the subsequent thirty days.)*

In addition to all of the other possible talents the Orme might provide mankind, it is worth noting the extremely long life spans of the Anunnaki may be due in part to the Orme and its proper ingestation. One might also assume the ***long life times of The Adam's Family (from Adam to Noah, and to a lesser extent from Shem through Abraham)*** *might also have been due to the Anunnaki sharing the Orme with the patriarchs and other carefully selected individuals.*

Given the background provided thus far, it is possible to now consider the thesis of this paper: Whether or not the extraterrestrials (the Anunnaki) who came to Earth nearly a half million years ago are still here. The answer to this question, of course, depends upon the reason the Anunnaki came to Earth. But, if in fact the "why" of their arrival and long-term stay was in order to acquire the mono-atomic truly precious elements for purposes of ingesting them and acquiring the Orme talents and abilities (including possibly the extremely long life spans of the Anunnaki), then the question becomes whether or not the Anunnaki are still here and still mining the precious elements of Earth.

Points of Evidence

There are three possibilities: 1) The Anunnaki left and are not coming back. 2) They're still here (and have been), but they're staying out of the limelight. 3) They left, but they're coming back (and they may not be too happy about what has happened since taking their sabbatical).

I. *Currently, there is a lack of evidence for their overt presence. This is particularly true when one considers their overwhelming presence and complete involvement in the affairs of mankind from the creation of man until as late as the time of Alexander the Great. Sitchin (1985) has stated, that from the time of Alexander's death in 323 BCE, "the Wars of Men have been the wars of men alone." However, Sitchin has also stated (1995) that "I have never written that they ever left." [There is also the pivotal date of* ***600 B.C.E.****]*

In ancient times, the population of mankind was extremely small compared to modern times. Thus mankind may have been much easier to control by a few hundred gods, or a mere dozen of the Anunnaki who were highly influential in the large scale interactions between the Anunnaki and mankind. The exponential increase in mankind's population, however, may have taken from the Anunnaki the ability to easily and overtly control mankind.

An analogy is perhaps in order here. To step on an ant or keep it under control is considered trivial for a human. But if this same human is faced with the onslaught of army ants on the move, a man is virtually helpless. His only recourse may be to flat get out of the way! Stopping army ants is not an option.

The Anunnaki may feel the same way about the billions of human beings currently dominating the Earth. A few might be easy to control, but large numbers of independent souls are chaotic. And ***one aspect is abundantly clear: the so-called gods are very definitely mortal, even if they are long lived.*** *This point is emphasized in the story of Isis and Ra as related in The Egyptian Book of the Dead (Budge, 1967, pg Lxxxix), and by numerous examples in the Sumerian texts (Sitchin, 1985).*

II. *There are some very distinct and contrary views of the Anunnaki -- roughly as many different philosophies and reasons to act as there are major figures. Not only does one encounter the strong rivalry between* ***Enki and Enlil****, but also between Marduk and Ninurta (the two eldest sons, respectively), between various factions in their several major wars (from "The Battle of the Titans" to the first and second "Pyramid Wars"), and the fact of the goddess* ***Inanna****'s attempts to take power on several occasions. Peace among the various factions has been difficult at best, with such goddesses as Ninki being the primary peace facilitator.*

In this connection there are also two "camps", so to speak, divided by those Anunnaki born on Earth and those born on their home planet, Nibiru. The latter include Enki, Enlil, Ninki, Marduk, and Ninurta. While Enki and Ninki may have a special relationship with mankind by virtue of having created us (and because there may be little motivation for them to return to Nibiru and relative obscurity), ***the others may have little or no interest in the welfare of Man and/or Earth.*** *On the other hand, those Anunnaki such as Inanna, her twin brother, Utu, as well as Nanna, Ningal, and Ishkur, were born on Earth* ***and may feel this planet is where their destiny lies. At the same time, however, they may not feel any concerns for the fate of mankind.***

The divergent opinions and motivations for the Anunnaki may imply the need for the various factions to cease their overt warfare and go undercover. ***There are references in the ancient literature where mankind occasionally began to object to being used as cannon fodder in the Anunnaki wars, and thus the gods may have of necessity been obliged to control the affairs of mankind without our knowledge. There is also the distinct possibility of the continuing wars on the planet today being instigated or inspired by hidden forces and agendas, including potentially those of the differing Anunnaki factions.***

III. *There is also a possibility the Anunnaki recognized the incredible rate of man's progress (which relatively speaking was enormously faster than theirs). They may have taken action accordingly. If a typical Anunnaki, for example, is capable of living in excess of 400,000 years, and man is (currently) living roughly 80 years, this amounts to a multiple of 5000. Thus, if a human sleeps 8 hours, does this imply an Anunnaki sleeps roughly four and a half years? Furthermore, if the Anunnaki rate of progress in terms of their personal growth is measured by their history, one might have to conclude the Anunnaki time scales are extremely different from that of Mankind's.*

It has been suggested by authors such as Ferguson (1995) and Frissel (1994) that the speed at which mankind is evolving is an extreme rarity in the universe, and probably the result of a wildly successful interplanetary experiment in genetic manipulation (when Enki and Ninki cross bred Homo erectus and the Anunnaki). As such, mankind might pose a serious threat to a race who could conceivably sleep through World War II! There is also the implication of our spiritual growth being extremely rapid and the resultant motivation of the Anunnaki to do everything in their power to keep us down on the farm.

IV. *Lao Tzu*, ***Confucius, Guatama Budda, and Zoroaster were all born in the one hundred years from 550 to 650 BCE., i.e. circa*** *600 B.C.E.*. ***With them were born philosophies and religious teachings, which for the first time, departed from the worship of flawed*** *Gods and Goddesses*, ***and became more involved in profound, universal-God philosophies. From this viewpoint, the god(s) of Genesis is not the God of the***

universe, but is instead, a description of Enki and Enlil, and to a lesser extent, the other Anunnaki.

Were the more profound philosophies which arose around the turn of the Sixth Century BCE a coincidence, a universal-God inspired event, or perhaps, a gift from Enki and/or Ninki? ***Keep in mind Ninki is very likely the original focus of the matriarchal, goddess religions which flourished for tens of thousands of years up until the final destruction of Crete circa 1250 BCE when the last "official" version of the goddess religions was forced to go underground.*** *The goddess religions may, in fact, have been derived separately from the mainstream of the Enlil-dominated, Sumerian-civilized version of his-story, and probably populated by ancient men and women who grew weary of working the mines and lit off on their own for the uncivilized (and infinitely more appealing) wilderness life.*

The question which still begs for an answer is whether or not Ninki, or some combination of other Anunnaki gods, arranged to have true philosophies introduced six hundred years before the Current Era. It is a distinct possibility.

V. *Numerous races in the history of mankind have abruptly vanished from the scene after having established a thriving and often boisterous culture. Several MesoAmerican cultures such as the Olmecs, Toltecs, and Mayans, comparatively isolated from other influences appear to have simply left one day, leaving their cities and worldly goods intact. Teotihuacan, for example appears to have been abandoned circa 200 BCE.*

An explanation of why such cultures could simply drop out of sight may derive in part from the ancient primary god of Mexico and Central America, Quetzalcoatl. According to Sitchin (1990-2), the plumed serpent god, Quetzalcoatl, was none other than the Sumerian deity, Ningishzida, one of the younger sons of Enki and easily one of the least belligerent of the Anunnaki. Quetzalcoatl is also identified as the Egyptian god, Thoth.

(In this regard Marduk is believed to be Ra, Enki is Ptah, and Inanna is Isis. This cross identification is then used to connect the Isis/Ra conflict and myths with those of Inanna and Marduk. It also suggests Marduk may no longer be on the scene!)

As the God of Wisdom, Enki was proficient in genetics, engineering, architecture (including sacred geometry), life sciences (including how to raise the dead), and other such esoteric disciplines. Much of this wisdom and knowledge was apparently passed onto his son, Quetzalcoatl/Ningishzida/Thoth, quite possibly from Enki's viewpoint, the only worthy recipient.

Quetzalcoatl, therefore, may be one Anunnaki on the side of mankind, and may have assisted in the growth and enlightenment (and ascension?) of those people under his sway -- the cultures in Mexico and Central America, in addition to ancient Egypt (specifically the Old and Middle ***Kingdoms****) where Thoth may have provided similar esoteric knowledge. In particular, Moses was compared to Thoth in a second century BCE work by the Judaeo-Greek philosopher Artapanus, who credited the prophet with a range of "scientific"' inventions (Hancock 1992). Thus,* **Moses'** *ability to make the Orme -- from his time as a prince of Egypt -- may have come directly or indirectly from Thoth!*

Husdon (1994) believes, for example, **Moses** *used his time on Mount Sinai to build a foundry and create the Orme. Clearly, his having to wear a veil thereafter because of major burns on his face make this a viable consideration. In addition, Moses was to bring down from the Mount, "tables of stone, and a law, and commandments which I have written; that thou mayest teach them." (Exodus 24:12) Not commandments on tables of stone, but tables of stone, and the law, and commandments. The tables of stone were undoubtedly the Orme, which Moses had made on the Mount which was covered by clouds for six days, and which was seen a "devouring fire on the top of the mount". (Exodus 24:17) Also, you don't teach laws written on tablets kept hidden!*

VI. During the time of the Spanish Conquistadors conquest of Mexico and Peru, an absolutely phenomenal amount of gold and silver was taken and shipped back to Spain. Estimates of the amount of gold and silver "extracted" from the Inca Empire alone amount to over 60 million ounces of gold and 300 million ounces of silver! And the Spaniards never really believed they had managed to find El Dorado, the mother-lode of all the gold.

If gold was so incredibly precious to the Anunnaki, why did this overwhelming surplus of gold reside in Mexico and Peru in the mid 1500s AD? Had the Anunnaki not made a pickup in recent centuries, allowing for the accumulation? Were the Indian miners still actively recovering the gold, even when there was no god hanging around to demand its continuing production?

Keep in mind in the case of the Inca Empire, that the South American territory was under the control of Ishkur, identified as the Inca Great God, Viracocha. There was a clear demarkation between Quetzalcoatl's territory and Viracocha's. And while, there is some indication Viracocha/Ishkur, the "God of War", may have shoved Quetzalcoatl/Thoth/ Ningishzida aside at one point -- thus accounting for the legends of Quetzalcoatl's departure from Central America and his promise to return (possibly in 1987?) -- ***the fact remains that the methods of operation in South America were not for the benefit of the human population. Ishkur was not the sort of Anunnaki to free the slaves and cease demanding the gold which "belonged to the gods"!***

On the other hand, and this may be critically important, the Anunnaki may ultimately have had little interest in yellow gold, essentially the gold in its metallic state. If the mono- atomic gold (along with mono-atomic rhodium, iridium, etcetera) was what the gods were truly interested in acquiring, then the yellow gold might have been thought of as a waste or by-product of mining the mono-atomic elements, the latter the all-important Orme. Thus, ***the surplus of gold in the ancient Mexican and Inca Empires does not necessarily imply the absence of the Anunnaki. I.e., the yellow gold of the Incas was not gold that was "good".***

VII. *Hoagland (1994) has presented a fascinating concept in his analysis of the so-called "Face on Mars" located near the area of* ***Cydonia****. The region includes what appear to be pyramids situated at precise angles in relation to one another and to the "Face" itself. The evidence for artificial constructions by intelligent beings at one time on Mars is impressive. And the similarity between these structures and those of in Sumeria, Egypt and Mexico (e.g. Teotihuacan) suggest a connection between them, and by implication with the space-faring Anunnaki. More recent disclosures have also suggested structures on the Moon.*

*The Martian and Lunar structures may or may not be active at the present time, but with the limited amount of information available, this can not be clarified one way of the other. It is likely, however, such structures may have been a base or small-scale colonization effort in the past, and there is some evidence (Sitchin, 1990-1) of recent activity near Mars. Such evidence tends, however, to come under the same category as reports of **UFOs**. It is extremely difficult to make definitive judgements as to their accuracy. On the other hand, UFO activity around volcanoes might tie in with the fact the primary source of mono-atomic elements are volcanoes. Such considerations, unfortunately, are highly speculative, and at the present time, are unlikely to provide clear answers.*

Nevertheless, there is at least one UFO style event which might suggest a continuing presence by the Anunnaki in this century. In their book, The Fire Came By, Baxter and Atkins (1976) describe an incredible event which occurred on June 30, 1908 AD. Witnesses described a cylindrical object, shining very brightly (apparently much too bright for the naked eye) and with a bluish-white light. According to the eyewitnesses, the object was moving vertically for about ten minutes.

The object appears to have initially streaked across the nighttime sky in a direction about 10 degrees from true north. Whereupon it turned to move almost due east, and then just as suddenly changed to proceed just slightly north of due west. The object then caused an immense explosion in the region of Stony Tunguska, some 40 miles NNE of Vanavara, in the frozen wasteland of Siberia.

The Tunguska Explosion*, as it came to be called, exceeded in sheer force the enormous size, destructive capability and long-term effects of the volcanic eruptions at Thera (Santorini) circa 1500 BCE, Vesuvius (August 24, 79AD), and Krakatoa (August 26, 1883 AD) between Java and Sumatra. The explosion at Tunguska was comparable only with the explosive capabilities of the heaviest hydrogen bombs. Furthermore, the Tunguska explosion had all the ear marks of a mid-air explosion, and none of the characteristics of an object slamming into the Earth.*

The trajectory of the object could be likened to an attempt by an intelligently controlled air or space craft to evade a pursuer, only to fail at the attempt, resulting in its eventual massive destruction. In effect, a mid-air "dog fight" or aerial combat between two opposing forces. Inasmuch as in 1908, the United States War Department still had letters from Orville Wright in their "crank" file; such alleged aerial maneuvers would have to be between extraterrestrials -- either of the same or different species. And because the Anunnaki have amply demonstrated their ability to war among themselves, it is not necessary to imagine other extraterrestrials in order to account for an apparent continual state of war between opposing forces.

However, there were no reports of a second object in the skies above Siberia. Another interpretation may be needed.

The exact time and location of the explosion is well known. An astrological natal chart of the event may be able to provide an alternative explanation. This chart is sufficiently striking, particularly when the ingredients are analyzed in some detail, the overall impression of the chart becomes ones of a clear signal of some kind, as in a warning. It is noteworthy the explosion did not kill a single human being, but was inescapable in terms of being witnessed by a variety of methods. Thus one is forced to evaluate the event as being of profound importance, perhaps in the same category as the phenomena of crop circles.

VIII. *The interspecies warfare among the Anunnaki has manifested itself in a variety of forms. This includes items as relatively mundane as different philosophical viewpoints with respect to law codes. For example, while the original Sumerian so-called laws were really codes of behavior ("take care of widows", "do not cheat laborers", and so forth), a later code instigated by Marduk detailed specific punishments for different crimes. Known as the Hammurabi code, the latter utilized fear of reprisal as the motivating force, instead of the earlier concept of doing what is right, or adhering to* ***The Golden Rule****.*

The different Anunnaki factions definitely had different modes of operating. This becomes relevant to our thesis if we question who precisely might be in charge. *The myth of Ra and Isis (or Marduk and Inanna), details a time when Ra was old and senile, and Isis proceeded to take away his powers. The end result was Isis "knew Ra by his own name", and Ra "hid himself from the gods, and his place in the boat of millions of years was empty." The implication* *is that Ra was vanquished and lost his position in the Anunnaki hierarchy. It is important to realize the myth derives from the XVIIIth Dynasty, roughly [according to* ***Immanuel Velikovsky****] the time of King Solomon and Queen Hatshepsut of Egypt (circa 1000 BCE). Thus Marduk may very well have inspired much of the Babylonian history (from 2000 to 1000 BCE), but then lost it all in Egypt at the hands of the "great goddess, Isis."*

In addition, the Anunnaki attached considerable importance to astronomical alignments. They dated the beginning of the Nippurian calendar (the same as the Jewish calendar) from 3,760 BCE in the Age of Taurus, based on a mass conjunction of planets. Two millennia later, they changed the calendar such that instead of Taurus, it became the Age of Aries. [The ages proceed backward, due to the Precession of the Equinoxes of the Earth.] At that time, astrology was changed to place Aries on the Ascendant instead of Taurus. (It was also a time when the female was demoted in relation to the male -- but that's another story.) *The beginning of the new age of Aries was marked by an astronomical alignment of all of the visible planets (including the Sun and Moon) in March 1993 BCE -- which is also the beginning of the Chinese Calendar.*

The significance of the astrological ages is important in that Marduk was temporarily held off from his bid for power in Babylon by the claims of his brother, Nergal, that it was not yet time, not yet the Age of Aries -- Aries being identified as Marduk's planet Mars. The fact Marduk may have survived in power for only half his tenure does not eliminate the idea. We, in fact, have our own abruptly ended Presidential examples.

By the same token, the Age of Pisces,, beginning roughly 200 to 600 years BCE is identified with the planet Neptune, which in turn is identified with Enki. This may explain why the Astrological Age of Pisces began with the philosophers (Lao Tze, Budda, etcetera -- *perhaps even including Jesus Christ). It may also explain why the Anunnaki are not making themselves obvious and painfully present -- Enki may currently be in charge. For our purposes, the relevant issue may be the forthcoming Age of Aquarius -- traditionally associated with the planet Uranus, and identified with Anu, the father of Enki and Enlil.*

The question of who may be in charge in the new millennia may appear somewhat in doubt. However, it might be pointed out the name Inanna means "beloved of Anu". And Anu, the quintessential dirty old man, was notorious for providing all manner of benefits to his great grand-daughter, Inanna!

Thus the temporary apparent absence of the Anunnaki may be due to the Age of Pisces in which mankind is being allowed something of a breathing space to grow and develop. It may also include the presentation and subsequent availability of profound philosophies the basis of which would aid in our growth. At the same time, the Age of Pisces may be ending in the very near future (if not already) with the most likely date being ***2012 A.D.****!* ***[However, there is another spectacular alignment of planets on September 8, 2040 A.D., and thus this later date may be the actual beginning of the Age of Aquarius.]***

Conclusions and Future Implications

*The accumulated evidence of several millennia strongly supports the contention initially proposed by Zecharia Sitchin, [****Laurence Gardner*** *and others] of extraterrestrials having arrived on Earth for the purpose of mining gold. These Anunnaki from the planet Nibiru created and subsequently interacted in overt and substantial ways with mankind over thousands of years. Based on the work of David Hudson, it becomes clear the Anunnaki came to Earth to mine gold and acquire supplies of the precious elements (such as rhodium, iridium, silver, and gold) in their mono-atomic state in order to ingest them as a means of acquiring and maintaining their powers and potentially their longevity.* ***The preponderance of the evidence makes this conclusion virtually inescapable.***

In addition, there is no evidence (other than the apparent lack of their interaction in our day-to-day lives) the Anunnaki ever left the Earth. There is also no known motivation which might have encouraged them to leave. On the other hand, there is reasonable and convincing arguments to suggest the Anunnaki are still on the planet (albeit they may be commuting from either Mars or the Earth's Moon, and/or simply staying out of sight). They may be, however, maintaining whatever control they are exercising from behind closed doors. Furthermore, it is very likely there are several factions of the Anunnaki with disparate goals and agendas for mankind, and that these disagreements may very well be spilling over into the affairs of humanity.

These conclusions suggest two notable implications. The first is one of awareness, in which each of us begins to view our world and society from a different perspective. ***This is the process of evaluating our Paradigms -- our view of looking at our world that has developed from birth and which includes possibly serious misinterpretations -- and considering if we wish to revise our paradigm.*** *It is a process which demands we "pay attention" to events in our lives, and* ***recognize the fundamental difference between "the gods" and the creator of the universe.***

The other implication is whether or not we as a race can expect to see a renewed overt involvement by the Anunnaki in our lives. The Anunnaki's demonstrated appreciation of astronomical alignments as a means of determining when certain events should proceed has already been mentioned. We know, for example, that an alignment circa 4000 BCE apparently initiated the Sumerian Civilization, an alignment circa 3100 BCE may have initiated the Egyptian and Indus Valley Civilizations (as well as the ***Mayan Calendar****), a 1993 BCE alignment initiated the Chinese Calendar and effectively began the reign of Marduk, the Temple of Solomon (circa 1000 BCE) was delayed from the time of David ("because it was not yet time"), and the Christian Era arrived approximately 4 BCE, shortly after the profound philosophies of the East began and after the Anunnaki dropped out of sight.*

References:

Adrian, F.J. and Cowan, D.O., 1992. "The New Superconductors", Chemical and Engineering News, December 21.

Baxter, J. & Atkins, T., 1976. The Fire Came By, Doubleday, NY

Boyer, T.H., 1975. "Random electrodynamics: The theory of classical electrodynamics with classical electromagnetic zero-point radiation", Physical Review D, Vol 11, No 4, 15 February.

Budge, E.A.W., 1967. The Egyptian Book of the Dead, The Papyrus of Ani, Dover Publications, New York.

Cann, R.L., Stoneking, M, and Wilson, A.C. 1987. "Mitochondrial DNA and human evolution", Nature, Vol 325, 1 January.

Dalley, S., 1989. Myths from Mesopotamia, Oxford Univ Press, NY.

Dejbakhsh, H., Schmitt, R.P., and Mouchaty, G., 1988. "Collective and single particle structure in 103Rh", Physical Review C, Vol 37, No 2, February .

Dorit, R.L., Akashi, H., Gilbert, W. 1995. "Absence of Polymorphism at the ZFY Locus on the Human Y Chromosome," Science, Vol 268, 26 May.

Duncan, M.A. & Rouvray, D.H., 1989. "Microclusters", Scientific American, December.

Eisenman, R. and Wise, M., 1993. The Dead Sea Scrolls Uncovered, Penquin Books, New York.

Frissel, B., 1994. Nothing in This Book Is True, But It's Exactly How Things Are, Frog, Ltd., Berkeley.

Ferguson, V.S., 1995. Inanna Returns, Thel Dar Pub., Seattle.

Gardner, Laurence, 1999. Genesis of the Grail Kings, Bantam Press, London.

Gardner, Laurence, 2003. Lost Secrets of the Sacred Ark, HarperCollins, London.

Giudice, E.D., Doglia, S., Milani, M., Smith, C.W., and Vitiello, G., 1981. "Magnetic Flux Quantization and Josephson Behavior in Living Systems", Physica Scripta, Vol 40. See also: Giudice, E.D., Doglia, S., and Milani, M., 1981. "Nonlinear Properties of Coherent Electric Vibrations in Living Cells", Physics Letters, Vol 85A, No 6.7, 12 October.

Greiner, W. and Sandulescu, A., 1990. "New Radioactivities", Scientific American, March.

Haisch, B., Rueda, A., and Puthoff, H.E., 1994. "Inertia as a zero-point-field Lorentz force" Physical Review A, Vol 49, No 2, February.

Hancock, G., 1992. The Sign and the Seal, Touchstone Books, NY.

Hoagland, R.C., 1994. Lecture presented at the International Forum on New Science, Fort Collins, September.

Hudson, D., 1994. "Orbitally Rearranged Mono-atomic Elements", Lecture at the International Forum on New Science, September.

Lim, C.S., Spear, R.H., Vermeer, W.J., and Fewell, M.P., 1989. "Possible discontinuity in octupole behavior in the Pt-Hg region", Physical Review C, Vol 39, No. 3, March .

*Macchiavelli, A.O., Burde, J., Diamond, R.M., Beausang, C.W., Deleplanque, M.A., McDonald, R.J. , Stephens, F.S., and Draper, J.E., 1988. "**Superdeformation** in 104,105Pd", Physical Review C, Vol 38 No 2, August.*

Puthoff, H.E., 1989. "Gravity as a zero-point-fluctuation force", Physical Review A, Vol 39, No 5, March 1.

Randeria, M., Duan, J.M., and Shieh, L.Y., 1989. "Bound States, Cooper Pairing, and Bose Condensation in Two Dimensions", Physical Review Letters, Vol 62, No 9, 27 February.

Shimizu, Y.R., Vigezzi, E., and Broglia, R.A., 1990. "Inertias of superdeformed bands", Physical Review C, Vol 41, No 4, April. Also, Shimizu, Y.R. and Broglia, R.A., 1990. "Quantum size effects in rapidly rotating nuclei", Physical Review C, Vol 41, No. 4, April.

Sitchin, Z., 1985. The Wars of Gods and Men, Avon Books, NY.

Sitchin, Z., 1990-1. Genesis Revisited, Avon Books, New York.

Sitchin, Z., 1990-2. The Lost Realms, Avon Books, New York.

Sitchin, Z., 1995. Private Communication.

305-MORE INFORMATION

Although a lot of the information in this section is included or duplicated in other sections included in this book, there are some elements hereunder that are totally unique and most insightful to this entire paradigm. I am including this writing as nothing that is within it seems in any way contradictory to the burden of what else I have included in this writing. – Bob.

Who Did What When

(Jupiter) (aka: Zeus, Thor, Kishar)

Given the basics from the "personal planets" (Mercury, Venus, Mars...), we can now begin to get into the meat of it. With a foundation of how things work, we can now investigate why things were worked in quite the manner that they were. In other words: History!

The primary advantage of knowing history is that one might be able to avoid repeating the same old dramas, traumas, tragedies, and unrecognized comedies. Everything proceeds in cycles, and in order to make it a spiral toward the heavens, it's good to remember what came down the first time and thus avoid the painful part. It's like resisting reincarnation if only to skip the necessity of being a hormone-plagued teenager all over again.

*There is, for example, the age old control technique of tyrants who claim imaginary enemies and/or threats in order to go to war and thereby distract the population from the serious defects, deficiencies, and incompetence of those same national leaders. Everyone from Napoleon to Hitler has applied this **con-game** in order to maintain order under the guise of **security and safety**. In all cases, of course, the greater threat is from the leader and his co-conspirators -- sort of the **ABC's** of any "Axis of Evil".*

For example, one such perpetrator said, "Why of course the people don't want war. Why should some poor slob on a farm want to risk his life in a war when the best he can get out of it is to come back to his farm in one piece? Naturally the common people don't want war: neither in Russia, nor in England, nor for that matter in

Germany. That is understood. But, after all, it is the leaders of the country who determine the policy, and it is always a simple matter to drag the people along, whether it is a democracy, a fascist dictatorship, a parliament, or a communist dictatorship. Voice or no voice, the people can always be brought to the bidding of the leaders. That is easy. All you have to do is tell them they are being attacked and denounce the peacemakers for lack of patriotism and exposing the country to danger. It works the same in any country." -- Hermann Goering

Unfortunately for the average individual more prone to burying one's head in the sand, there is the old adage, "Fool me once; shame on you. Fool me twice; shame on me!" It is therefore, no one else's fault if history repeats itself. In other words, ***Scapegoatology*** *is simply not an excuse. Being Cleopatra -- de Queen of de Nile -- no longer suffices as a justification for complacency. We are each individually responsible to knowing the truth, for knowing the* ***Nature of Law*** *and applying it, for claiming our* ***Sovereignty****, and seeing beyond the* ***Illusions****,* ***Secrets****,* ***Conspiracies****, and outright lies of inexcusable leadership.* **As Rudyard Kipling once said, with reference to war, "Ask not why our fellows died; tis because our fathers lied."** *It's now time to practice* ***discrimination****, find out the truth, and get over it.*

However. In order to fully appreciate the current dismal status quo, one must go back in history far beyond the recent "current events". To truly understand the possibilities for deceit and wilful ignorance, it is preferable to start at the beginning -- in this case, prior to the beginning of the human species, affectionately known as ***Homo sapiens sapiens*** *[the latter arrogantly translated as "wise"].*

Before the dawn of man, extraterrestrials arrived on the earth in order to mine gold. [My apologies to those who might be shocked at this revelation. See ***Laurence Gardner*** *and* <**http://www.karenlyster.com/genesis.html**> *for more details.] These extraterrestrials were known to the ancient Egyptians as "Urshu", to the early Hebrews as "Nephilim", to the Australian Aborigines as "Nurrumbunguttias", and to the Sumerians as "Anunnaki". The latter is loosely translated as "Those who from heaven to earth came".*

*The **Sumerian** Civilization was the earliest known civilization (known at least to mainstream archaeology). It flourished from roughly 8000 until 2000 B.C.E. in the Tigris-Euphrates valley -- what is now modern day **Iraq**. Inasmuch as the Sumerians were considerably more prolific in documenting the history of the extraterrestrials, we will refer henceforth to the ETs by their Sumerian name, **Anunnaki** -- goldminers extraordinaire.*

Because of the physical labor involved in mining gold (as in a mile under the South African plains), certain of these Anunnaki performed genetic manipulations on Homo erectus, i.e., primates who walked erect, and who at the time, were the reigning progenitors of Homo sapiens sapiens, i.e. modern day human beings. The Anunnaki's goal was to create mineworkers -- preferably those without a union card.

Two species evolved from these genetic experiments: Neanderthal and Cro-Magnon. One was an able worker, but learned slowly, while the latter learned entirely too quickly. To accomplish the creation of Homo sapiens from the Homo erectus stock, the Anunnaki included their own DNA in a genetic manipulation; at one point utilizing the services of fourteen female Anunnaki to serve as birth mothers. Later on, the Homo sapiens were able to take up the reproduction process themselves; a tradition, which for obvious reasons, continues to be enthusiastically practiced today.

*Critical to everything going on at the time was the factionalization of the Anunnaki, a rift caused by two half-brothers vying for lordship over earth. The eldest was EN.KI. ("Lord of Earth"), the scientist who with his half-sister, NIN.KI., was responsible for creating mankind. The younger was EN.LIL. ("Lord of the Command"), who had the advantage of his mother being a half-sister to his father and was therefore the rightful heir apparent. Enlil was thus the legal firstborn. (This is the same tradition practiced by **Abraham**, his half-sister Sarah, Abraham's eldest son, Ishmael, and Abraham's son by Sarah, Jacob. In fact, the early Hebrews undoubtedly developed this incestuous tradition by virtue of their contact with the Anunnaki.)*

*The two brothers, **Enki and Enlil**, along with their descendants, established two competing **factions** for control of the earth -- a competition which may have continued to the present day. Attempts to mediate between them were never wholly successful.*

The initial reason for the Anunnaki to take up residence on Earth was their continuing need for gold. More accurately, it was for the equivalent of the **ORME** *-- a phrase which includes the "manna of heaven" in its definition. The ORME can also be thought of as an acronym for "orbitally rearranged monoatomic elements". This latter term refers to the eight precious elements, i.e., gold, silver, platinum,* **Rhodium and Iridium***, palladium, osmium, and ruthenium. Esoterically, the ORME is the "***white powder of gold***" in Hebrew tradition, or more scientifically, certain of the eight precious elements in their monoatomic, non-metallic state.*

There are competing theories with respect to the Anunnaki's preoccupation with gold, but the prevailing intelligence suggests that it is the ORME which allows the Anunnaki to live exceedingly long lives (potentially thousands upon thousands of years), and by extension, humans, who have some of the same DNA as their creators. The very long lives attributed in the Bible to Adam, Noah, and the other early patriarchs ***(The Adam's Family)*** **appears therefore to be based on fact, and not a delusion of ancient historians. The patriarchs' longevity is likely due to both their genetic heritage from the Anunnaki and to their access to the ORME** *-- an access which may have been significantly reduced by the time of Abraham, who lived a mere 70 years. From the Anunnaki viewpoint, an ORME source was fundamentally important to their continued existence as very long lived beings.*

The Sumerian records provide a rich source of materials which corroborates the above statements, and which provide hints as to the current situation. For example, the Anunnaki had a profound and all-encompassing respect for **Sacred Geometry***,* **Astrology***, and what might be termed the "esoteric sciences". This respect is shown by humankind's respect for the* **Ha Qabala** *(Jewish Kaballah, Christian Caballah or ecumenical Qaballah), all of which incorporate* **Astrology***,* **Numerology***, and* **Tarot** *into the* **Tree of Life** *teachings. The preoccupation by the Anunnaki with astrology and related sacred geometries was also reflected in the Sumerian writings in terms of astrological "ages" and an Anunnaki "god" or "goddess" associated with each Age.*

For example, the current age, the **Age of Pisces***, is associated with the planet Neptune, which, in turn, is associated with the god, Enki. Significantly, the Anunnaki gave the god or goddess to* *which the current age was associated unprecedented power or authority over the affairs of humans. Inasmuch as these ages lasted on average some 2,160 years (and perhaps longer for constellations which span a greater than 1/12th portion of the Zodiac, e.g., Pisces), the Age of Pisces may be reckoned from roughly the Sixth Century B.C.E. (Before Current Era) until approximately the current day. Significantly, it is likely that Enki has ostensibly been in charge of the affairs of Earth since as far back as* ***600 B.C.E.****, and may continue to be in overt control until the end of the* ***Mayan Calendar*** *-- currently believed to end on 21 December* ***2012 A.D.***

Despite the fact that Enki is associated with the Serpent in the Garden of Eden -- and may be likened in some ways to Lucifer ("Light Bringer") -- Enki, by virtue of his being the father of mankind, may have a much greater investment in the future of humans than Enlil (the vengeful god of the Book of ***Genesis****).*

Considerable detail on Enki as the ***archetype*** *for the serpent of life and wisdom is provided in* ***http://www.vibrani.com/serpent.htm***>*, where even the derivation of the modern day caduceus symbol of healing and* ***DNA*** *structure is discussed. Enki, or EA, is literally the Lord of Earth -- even the name of the planet being derived from this God of* ***Wisdom*** *and chief scientist / creator god of the* ***Anunnaki****.*

In counterpoint to Enki, his half brother, Enlil, was instrumental in allowing humankind to perish in Noah's flood, while Enki was busily providing the patriarch with boat plans! Enlil, aka Jehovah, was in fact the God of Storm, the god of wrath, who demonstrated a distinct distaste for human beings (they disturbed his sleep!), while Enki was countering most of Enlil's actions by helping and assisting the fledgling mankind. (There is also the possibility that Enki would delight in defying Enlil by frustrating Enlil's decisions and plans -- just in order to get his goat. Or his sheep -- the latter a disguised, covert, albeit very clever, pardonable pun.)

It is particularly noteworthy that Gnosticism is founded on the premise of a creator god which is unknowable (see, for example, the first 34 verses of Genesis), and that after that, being derived from Sophia, the Goddess of Wisdom, a "half-maker" so-called god arrived on the scene to take the glory and responsibility for creating -- the most blatant and

significant act of plagiarism in the history, herstory, and annals of anything remotely human. This god is even identified unabashedly as Jehovah, while Jesus Christ is honored for the teachings of a loving and caring deity (someone more on the order of Enki.) [The fact that both Sophia and Enki are the deities of Wisdom, tells us a lot about what counts and what doesn't. Enlil, for example, has never been described as "wise". And yet the very nature of our modern society is Enlil-based!]

*Gnosticism goes a step further and includes an appreciation for sexuality as a path to wisdom and enlightenment, the ingesting semen and menses for the nourishment of the soul, and the absolute need for a sovereignty style self-responsibility of the individual -- all of which correlates with the **Sumerian** and other ancient descriptions of everything from the priestess aided sexual rituals of ancient Sumerian, Babylonian and Egyptian temples to the **Star Fire**, **White Powder of Gold** and **ORME**, to the anti-Enlil, authority-based, anal retentive religious **straightjackets** now prevalent on the planet.*

*The original Gnosticism, deriving in the time period of Jesus Christ and possibly as early as **600 B.C.E**., seems to have been on the scene and in the know of what life and society were all about during this period. Thus Gnosticism represents an independent version of the Sumerian annals and literature (and vice versa).* ***The fact that Gnosticism was very nearly wiped out by a Catholic (same ending as alcoholic) early version of an inquisition -- including the unforgivable burning of the Library of Alexandria -- is testament to both the importance of Gnosticism in its myriad forms of interpretation, and the fact that Gnosticism is so diametrically opposed to the control freaks of many of the modern day religions.***

*Very importantly, Enki appears to have decided to use his **Age of Pisces** (which began circa **600 B.C.E.**) to provide human civilization with a religion of philosophies, the latter being considerably more than merely a structure of obedience to the whims of any member of a dysfunctional family of extraterrestrials (aka **Gods and Goddesses**). Such an introduction of philosophies clearly includes Gnosticism.*

To this end, we have, beginning about 600 B.C.E., Zarathustra (born circa 630 B.C.E.), ***Lao Tzu*** *(b. circa 600 B.C.E.), Confucius (b. 551 B.C.E.), and Gautama Buddha (b. 563 B.C.E.), and others, who* ***reject the pantheon of gods and goddesses, and simultaneously, proclaim non-exclusive human philosophies that are not dependent upon the will of gods and/or goddesses, but upon human free will to make their own choices. The Greeks, for example, went from the "heroic age" of which Homer wrote eloquently about the conflict between gods and goddesses (and who used mankind as just so much cannon fodder), to the age of Phythagoras, Aischylos, Pindar, Sophokles -- followed in kind by Herodotos, Euripides, Sokrates, Hippokrates, Thucydides,*** *Aristophanes, Plato, and Aristotle.*

Hinduism*, which predates 600 B.C.E., was amended in a very significant manner by the addition of the Upanishads at this same period of time. This latter teaching, for the first time, began to speculate on the nature of the universe and man's relation to it. Similarly,* ***Judaism****, which traditionally began during the time of Abraham (circa 2000 B.C.E.) was in its early beginnings primarily concerned with the god, Yahweh, his chosen people, and the same chosen people doing precisely what the one God, Yahweh, wanted them to do.*

According to one source: "The prophets were first and foremost teachers of religion, not of ethics. Their supreme concern was the will of God, rather than the rule of righteousness." However, in the 6th century, B.C.E. Jeremiah and Ezekiel switched to an emphasis on individual responsibility and sought to restore to the people a sense of personal relationship with God. "This was a time of general national disintegration, when religious and social organizations were rapidly breaking up." [1]

It is likely that the "sense of personal relationship with God" may have originated in the earliest Sumerian ethics, where doing right was valued above obeying the dictates of any god or goddess. Said ethics likely came down from Enki, but the needs and dictates of Enlil probably contributed to very early Judaism's departure from this ideal, and instead pointed early Judaism in the direction of a religion of discipline to the god in charge. Such a situation was obvious from the time of Abraham, when Yahweh was intimately involved with his chosen people in the day-to-day administration of earthly affairs.

Around 600 B.C.E. there appears to be a fundamental change, and with prophets such as Jeremiah and Ezekiel, there was a new emphasis. There is also the very real suggestion that the vengeful God of Judaism is none other than Enlil, while Enki is more likely associated with the god who helped Adam, Eve, Cain, Noah, et al, and the loving God of Judaism. The biblical evidence for this, incidentally, is, as it turns out, quite massive!

The end result is a sea change in religion and philosophy incipient upon the time frame of the sixth century, B.C.E. *and thereafter, and which may be construed to be the beginning of the* ***Age of Pisces****. We might assume, therefore, that Enki's age was destined by the "Lord of Earth" as a time for humankind to attain* ***Sovereignty*** *in its philosophy and sense of personal responsibility.*

Since then, in the interim, the tale has become ever more twisted and complicated. ***History 009*** *describes, for example, the intervening years in which the knowledge of the ORME was transferred via* ***the royal blood lines. These "blue bloods" were actually thought of as carrying a greater percentage of the Anunnaki DNA. Unfortunately for modern royals' dreams of superiority, the last two and a half millennia of begetting outside the family has likely diluted any previous distinction. There is a fair chance, for example, that after sixty generations almost anyone on the planet can claim King David as an ancestor.*** *In addition, the withdrawal of the ORME from the patriarchs since the flood was sufficient to leave Abraham pretty much at the level of the general mass of mankind, even before all of the seriously rampant begetting.*

The potential difficulty with this scenario today is that the Age of Pisces is rapidly being overtaken by the Age of Aquarius, when another Anunnaki will be taking over from Enki. Inasmuch as Aquarius is related to Uranus, and Uranus is considered the planet of Anu, it is likely that Enki's and Enlil's father, Anu, will be the new god-in-charge. The degree to which Anu will revise the lordship of earth is anyone's guess, but we can note the fact that the effective transfer date of responsibility is still December 21, ***2012 A.D.****!*

It should be noted that traditionally, Anu was not exactly a "hands-on" manager, and did a lot of delegating of his supreme authority. In addition, there is the story of **Cronus and Zeus**, *which may gives us a clue to what might be the coming new state of affairs -- i.e., it appears that Enlil (Yahweh) will not be returning with a vengeance (or otherwise).*

An alternative, albeit hopeful view is that Anu's decided favorite, his great granddaughter, the goddess Inanna, may be allowed to rule in his stead (with or without Enki's blessing) -- effectively heralding a **Return of the Goddess**. *In any case, we may be faced with a rendition of Don Meredith's favorite song:* **The Party's Over**!

Clearly, an understanding of such ancient to modern history makes for all manner of connecting the dots in an elaborate puzzle. But it gets better.

References:

[1] Encyclopaedia Britannica, vol 11, pg 507 and vol 13, pg 103, 1960.

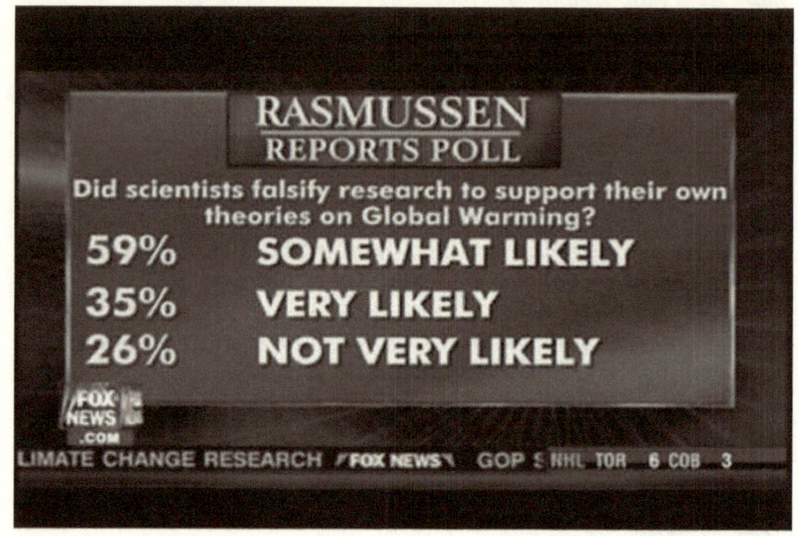

PART FOUR

AS MY FATHER ONCE TOLD ME, "YOU CANNOT BELIEVE MOST OF WHAT YOU HEAR, READ, OR ARE TOLD".

(HE WAS THERE AT GALLOPOLI 1915, AND TOLD ME STORIES THAT ARE MOST DISTURBING.)

In this section we shall look at some of the troubling aspects and facets of life on planet earth. This will include a lot of materials taken directly from the Internet that will be representative of the vast amount of materials available on any of the subjects covered.

As it is beyond the scope or ability of this book to cover exhaustively any one of the many topics concerned, accept that these small offerings are to merely indicate that indeed claims made, "facts" presented, reported on, and commonly accepted are really not alright or fully acceptable in as much as many of the claims made etc.

Some of the matters covered here have been commonly denigrated and mockingly called "conspiracy theory". Well that of course is the prerogative of those who use such dismissal techniques without seriously questioning their acceptance of status quo paradigm. Mockers and dismissers have rarely ended up being proven right. How few listened to tales of concentration camps in German occupied Europe before the truth of such were forced upon them Too few will forgo what to them is a comfortable acceptance, rather that try to look behind the curtains to establish truth. Even as liberty is blatantly eroded, many prefer to say that all is OK, our government would not lie to us or fail to have our best interests and welfare at heart. "Death by Government" is common.

So let's take a sniff around and see how many, and why so many things are the subjects of what are dismissed as "conspiracy theory". It is only when you look at them in "total" that one can identify not many, but one massive frightening mother of all conspiracy theories, and that is indeed probably not theory.

MARS ANOMALY RESEARC

Home | Comments | Get Involved | Contact Us | Book | Master Lis

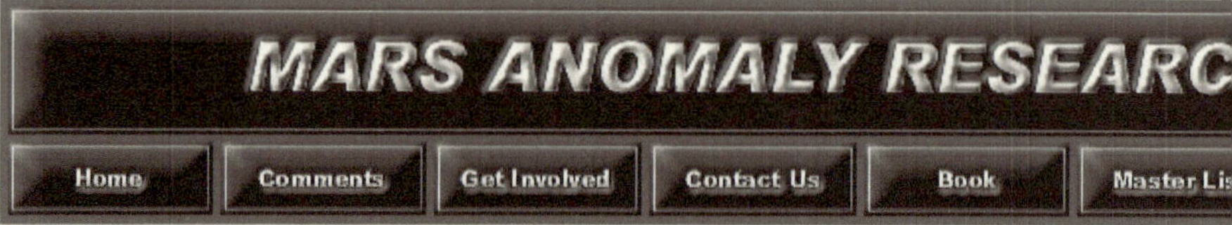

MOON & MARS TRUTH WHEN THE DUST SETTLES

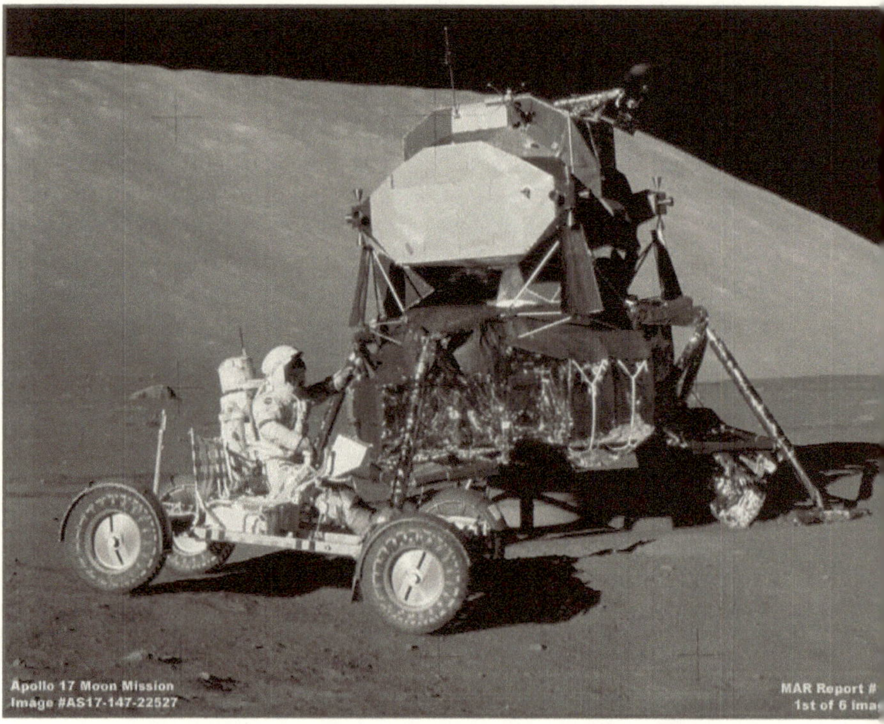

Apollo 17 Moon Mission
Image #AS17-147-22527
MAR Report #
1st of 6 ima

The above first image is just a context image to familiarize some of you with the appearance of the equipment that will be under examination here in this report. On the left in the foreground, you see a Moon rover (LRV) with an astronaut in it and on the right just behind the LRV you see the Moon Lunar or Landing Module (LM) with it shiny foil showing. Note the wide stance of the LM legs and the footpad on the ground at the end of the leg on the right.

The equipment we're looking at here is from the Apollo 17 mission which was the last manned mission to the Moon to date as far as anything publicly admitted to. The last three manned Apollo 15, 16, and 17 missions all had rovers. Rovers are folded up and on a pallet that is stored against the side of the LM in one of the areas where you see the shiny foil. Upon landing on the Moon, the astronauts get out and drop down the pallet with the folded two person rover. They unfold it and assemble it. All of these missions used the same type of equipment both as to the LM and the rover.

The anomalous Moon evidence we'll be examining below in this report as just samples will be drawn from two Moon missions, Apollo 11 and Apollo 17 and then we'll move on to some Mars Opportunity rover evidence. The reason I'm doing this is to demonstrate that the type of evidence you will see here spans decades of planetary exploration from the early 1970's to the 2000's and forms a consistent pattern of secrecy behavior and are obviously not isolated activities.

The above second image is of the Apollo 17 rover's right rear wheel/tire/fender area. This image is from one side edge of the original image to the other. One of the astronauts damaged the right rear fender and it was temporarily repaired with duct tape and that is why this image was taken and what it is suppose to be all about. However, note that the soil surface immediately in front of the tire and immediately behind it has not been disturbed by rover tracks.

Since lifting and carefully placing the rover in a spot on the Moon is impossible and since driving up to this position would leave tracks, then we have a problem with the evidence in this image. Okay, I thought to myself, I gotcha. However, as I expanded out my investigation of this, I found out that I was not the first to find this after all. To save me reporting time on this, you'll find some in depth interesting reporting on this at the following link http://apolloanomalies.com/rover_tracks_rebuttal.htm.

So is it a shut out for me or can I add anything to this discovery? As it turns out, yes I believe that I can in the above third image which is just a closer shot of the Apollo 17 rover right rear tire and wheel. Note that the tire is covered by a what appears to be a metal mesh and that crossing angled bands forming increased traction cleats have been added to the bottom of the tire that makes contact with the ground and of course should leave matching tread marks in that ground.

Now note the inner wheel rim is covered with what appears to be very fine texture gray soil. On the Moon that soil is called regolith. One reason is that Moon regolith, it is theorized by the science communities, is primarily formed by objects impacting the Moon's surface. Over millions of years this impact and blasting effect grinds the soil up into very fine particles that would be considered dust or powder here on Earth. Therefore, in theory, the Moon's dusty powdery surface combined with a gravity only a 17th as much as Earth and no moisture equates to the soil often acting like a dry powder that when disturbed from the surface tends to hang in suspension in the air for a while.

Note two things in the above third image. Note that the very fine grained soil is adhering to the wheel including vertical and curving overhead surfaces as though there was some moisture in it causing it to do so with sticking power. That observation is significant suggesting the presence of moisture. Yet, even as this is clearly happening on the wheel surface, note that no soil at all, not a speck of it, appears to be adhering to the tire mesh or the cleats including even their fasteners on the bottom of the tire. Since much of the Moon soil is so finely textured, some of it should at least be embedded in the mesh but obviously isn't.

This suggests that this tire may have been positioned elevated in the air where the bottom of it was inspected and cleaned for that inspection. The bottom of this tire has obviously not yet been rolling around on the Moon's surface, if it is on the Moon at all, after having been cleaned. It was almost certainly cleaned in place on the wheel because the regolith crusted on the wheel has not been disturbed by removal from the rover. It may be that something different than what we have been told (an astronaut using a tool accidentally ripped the fender) happened requiring the tire to be inspected by raising the vehicle and then lowering it back down in this spot where the image was taken before the astronaut sitting in the rover (feet in image #2) drove it off.

The question of course is how did the bottom of that mesh and cleated tire surface get so clean. The lack of any rover tire tread marks in the soil both in front and behind the tire implies that it was lowered and placed in this spot after cleaning and that is not possible on the Moon. The astronauts do not have the equipment for this and they are too limited in their movements by those bulky awkward suits.

Then there is the issue of the lack of tire tracks in the soil/regolith. The soil is plenty soft enough for tracks because the adjacent astronaut foot prints can clearly be seen in the soil. The rover of course weighs much more than an astronaut and should leave deeper impressions. In the 2nd image, you can clearly see that the soil directly behind the tire is in a natural state and undisturbed. Although foot prints have disturbed the soil immediately in front of the tire, further forward can still be seen undisturbed soil as well as the small piece of the right front tire and its shadow under the rover. Obviously, the right front tire has not rolled across any of this regolith and left an impression either.

This evidence is not alone, there is a long litany of problems with the Moon science data like this. So you can perhaps understand why the researchers at the above link might come to the conclusion that a lot, if not all, of the Moon manned landing data was staged on a set here on Earth. Further, the incentive to fabricate was certainly there. Why? Because the Russians in those days were repeating one first after another with respect to getting into space and dealing with the Moon. This embarrassment could certainly create motivation for a panic to catch up here in America.

I don't profess to know the straight of it but under these playing catch up role circumstances, the temptation to fabricate would have been very great. It is unfortunately a too often exhibited characteristic of human behavior both on an individual level and on a collective level any where in the world.

What you are looking at in the above fourth image is one of the Apollo 11 Mission LM footpads covered in foil that is suppose to be resting on the Moon's surface. As you can see, the round footpad has a turned up shape at the edges forming a depression system within the pads interior. Further, the inner surface of the shallow bowl shape consists of all this wrinkled foil forming many different angles that would serve as good dust traps.

The combination of shallow bowl shape and wrinkled foil forming a depression and its location at the lowest point right against the ground is the perfect catch system. Now consider the LM lowering down for landing on the Moon's surface with its decent thruster burn nozzle pointed straight down right at the ground coming to within inches of it and blowing up a cloud of Moon regolith dust to hang suspended in the air over the site for a while in the low Moon gravity. You know this would be so. It would be impossible for some of the disturbed settling dust to not settle back down into this large shallow bowl shape and the shiny foil with its many traps would show every bit of it.

However, as you can clearly see in the above 4th image, there is no sign of any dust or regolith in this footpad shallow bowl shape. Further, it is the same with other shots of this and other Apollo 11 LM pads at these links: AS11-40-5902, AS11-40-5917, AS11-40-5918, AS11-40-5920, and AS11-40-5925. When you use these confirmation links, be sure to access the "Hi-Res" official image because it is much larger than the "standard" image and offers a much closer view of the Apollo 11 LM footpads than you see here confirming not a sign of dust or regolith inside the footpad shallow bowl shape. Once again I thought to myself, yep I gotcha. I should have known better on such older Moon material. Again, as I expanded my investigation on this type of evidence, I found others had beaten me to the punch with some good reporting here at http://www.aulis.com/jackstudies_22.html.

Evidence like this is small but crucial and unequivocal. Once again you can perhaps see how others could have come to the conclusion that these Moon landings were not real and may have been fakes. Now let's move on to the Mars evidence below.

Mars Opportunity Rover Sols 652-663
1/4/06 Panorama On the Rim of 'Erebus'
MAR Report 158
5th of 6 images

The above fifth image shows a portion of the Mars Opportunity rover deck. This image is drawn from the official 1/4/2006 panorama titled "On the Rim of Erebus." It is a mosaic of many smaller images taken from Sols 652 through 663. Because the whole of this huge panorama image is a 360° view, the official original is presented in a cylindrical projection that causes some distortion in the rover decking you see above but this does not impact the kind of evidence presented here.

Now what must be understood is that all of the many solar cells and many of the other flat horizontal surfaces in this image including even the upright mast are covered over by smudge image tampering. That is, in fact, routine when it comes to the rover deck upper surfaces. Even so, it is not the flat surfaces that we are interested in so much as it is the nooks and crannies where dust and soil should collect and become trapped like at the screws, brackets, fittings, etc. Look closely at these spots and, try as you might, you will not find any soil or dust accumulations. In fact all this deck appears to be in clean pristine condition.

Now think of that. This rover has theoretically been in operation moving around on Mars for up to 663 Mars days and nights, the whole time in the open. That's nearly 22 months our time or almost 2 years and more importantly going through multiple warm and colder seasons and weather conditions. It has no doubt encountered frost as well as wind blown sediment many multiples of times. Yet there is no sign of any of this on the top deck surfaces that are natural catch systems. Even someone cleaning the deck with air blaster would be hard pressed to have gotten it this clean. The super clean rover looks like it just rolled out of a lab instead of plowing around on Mars continuously in the open for hundreds of days.

This strains credibility in obvious ways. How are we to have confidence in data that has these kind of credibility problems in it. Do you believe that this particular rover deck has been out in the open on Mars for nearly 2 years? How can you? Yet that is what we are suppose to believe.

DOCUMENTATION

http://www.apolloarchive.com/apg_thumbnail.php?ptr=231&imageID=AS17-147-22527. This link takes you to the ApolloArchive.com page for the official science data image from which my 1st report image of the LM and LRV together was drawn. On the page that opens, if the thumbnail image on the left or the "Hi-Res" text link in the middle will not work for you, then use the small text non java script "Standard" and "Hi-Res" links on the right side of the page.

http://www.apolloarchive.com/apg_thumbnail.php?ptr=382&imageID=AS17-137-20979. This link takes you to the ApolloArchive.com page for the official science data image from which my 2nd and 3rd report images of the rover right rear tire here were drawn. On the page that opens, if the thumbnail image on the left or the "Hi-Res" text link in the middle area will not work for you, use the small non java script text "Standard" and "Hi-Res" links on the right side.

http://www.apolloarchive.com/apg_thumbnail.php?ptr=622&imageID=AS11-40-5926. This link takes you to the ApolloArchive.com page for the official science data image from which my 4th report image of the LM footpad was drawn. On the page that opens, if the thumbnail image on the left or the "Hi-Res" text link in the middle area will not work for you, then use the small text non java script "Standard" and "Hi-Res" links on the right side of the page. You are urged to access the "Hi-Res" image because it will display an even larger image with a much closer view of the footpad and the absence of regolith than you see here in this report.

http://marsrovers.jpl.nasa.gov/gallery/panoramas/opportunity/2006.html. This link will take you to the page where the year 2006 Opportunity panorama images are located from which the too clean rover deck evidence was drawn. There are a large number of panoramas there listed in order from the most recent down to the oldest in that year. This report evidence is the last listed and so you must scroll down to the very bottom of the page to access the 1/4/2006 "On the Rim of EreBus" image and accompanying narrative. Note that the largest image does not display but the "Browse" and "Medium" images do display.

Joseph P. Skipper, Investigator
www.marsanomalyresearch.com©

March 1, 2009

J. P. Skipper can be contacted at: jskipper@marsanomalyresearch.com

This two part report is about Moon evidence that may or may not be real. The first part is about general image tampering in the Moon science data. I go to this trouble because you should at least be aware that the credibility of the data in general is often very questionable and so caution is advised in approaching anomalous information within it. The second part will be about some interesting anomalies and then it will be up to you to determine what you think about it all.

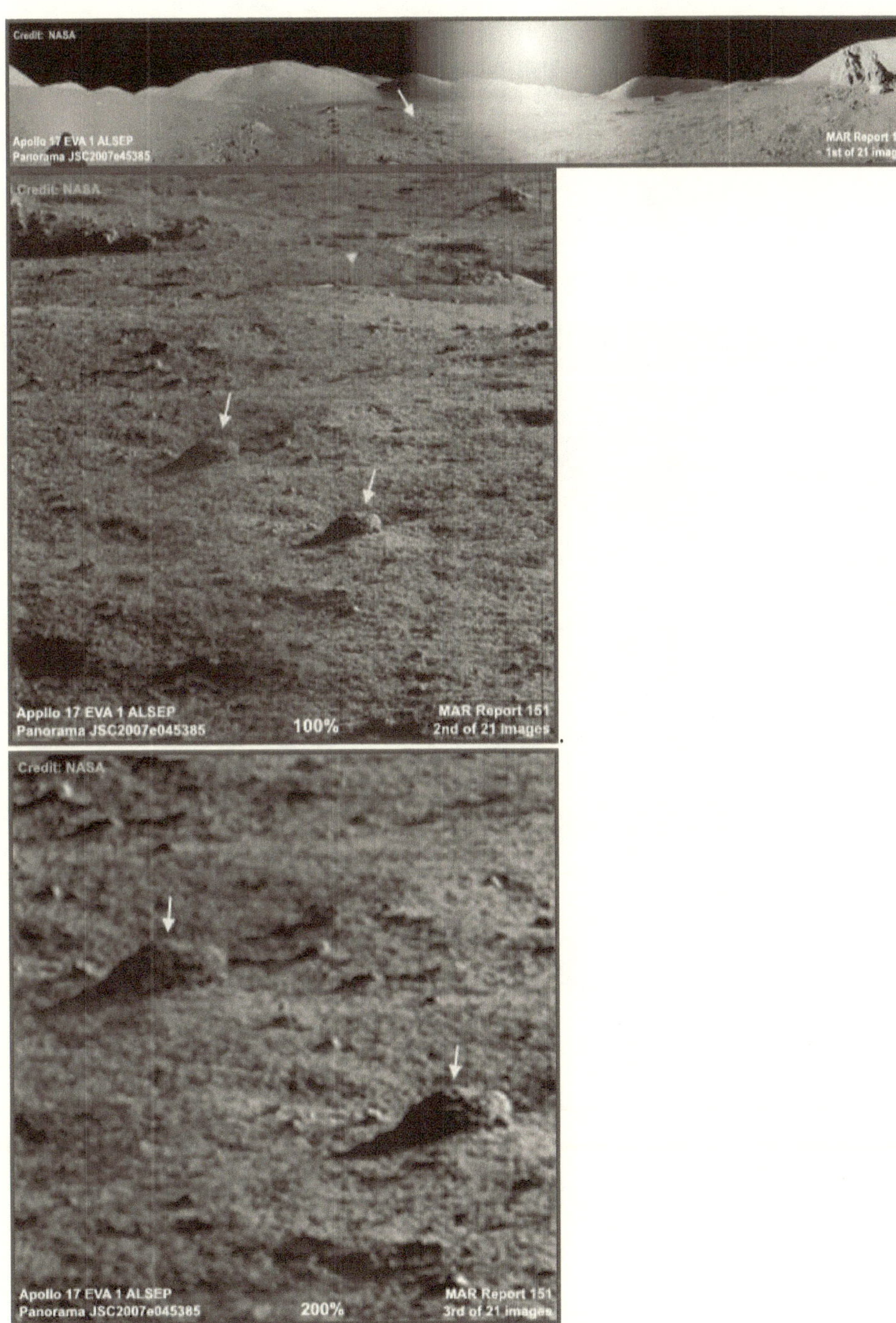
Credit: NASA
Apollo 17 EVA 1 ALSEP
Panorama JSC2007e45385
MAR Report 151
1st of 21 images
Credit: NASA
Apollo 17 EVA 1 ALSEP
Panorama JSC2007e045385
100%
MAR Report 151
2nd of 21 images
Credit: NASA
Apollo 17 EVA 1 ALSEP
Panorama JSC2007e045385
200%
MAR Report 151
3rd of 21 images

The above first image is a wide Apollo 17 EVA 1 ALSEP panorama image on the Moon demonstrating the context scene with the yellow arrow pointing out the location of the evidence spots in question for any who wish to do a follow-up search behind me on this. The evidence consists of two duplicated identical rocks in the terrain. The left second image shows a 100% full resolution view of the two rocks and the third right image shows a 200% zoom view of the two rocks so that there can be no mistaking that they are identical except for some slight light highlighting differences between them.

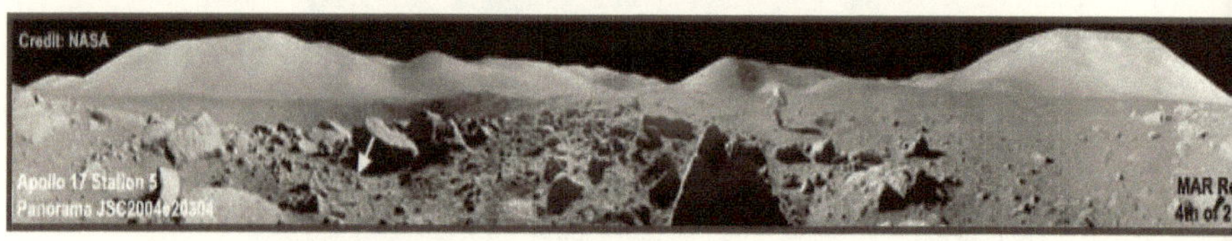

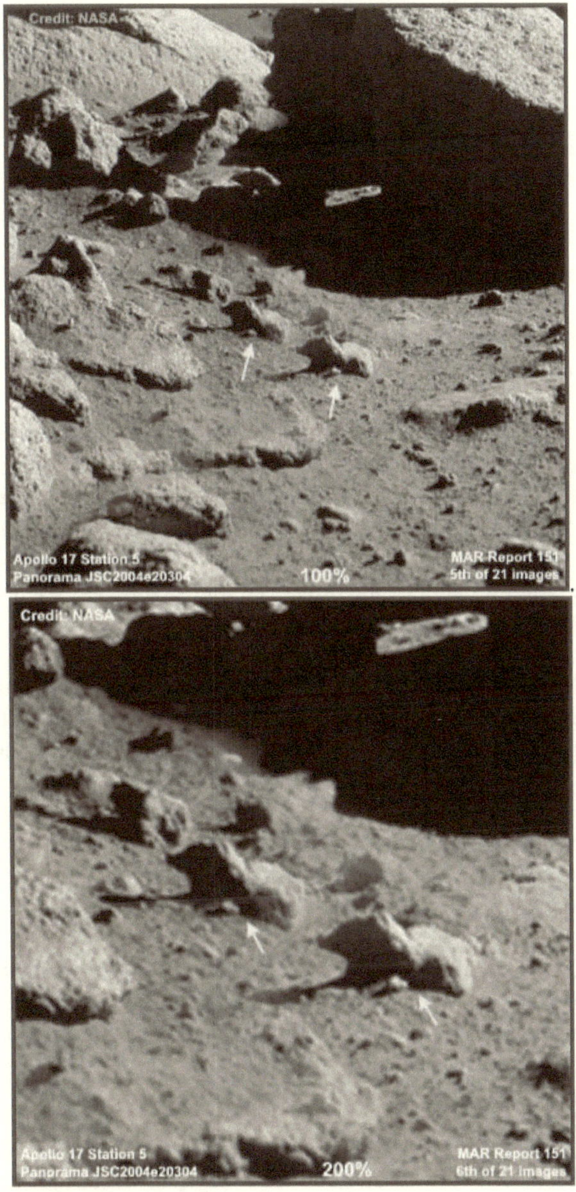

The above fourth image is a wide Apollo 17 Station 5 panorama image on the Moon again demonstrating the context scene with the yellow arrow pointing out the location of the evidence spot in question. The evidence once again consists of two duplicated identical rocks in the terrain when only one is real. Again the fifth left image shows a 100% full resolution view of the two rocks and the sixth right image shows the 200% zoom view of the two rocks so that there can be no mistaking that they are identical and duplicates of each other.

As you can see, these last two objects look a little like rodents don't they. Note the identical small rocks in the foreground right beside the two "rodents." Note also the two identical rocks in the background immediately behind the "rodents." One of these latter rocks may not be immediately apparent without a closer look because the one of the right is cleverly washed with sunlight highlights. Yes I'm also aware of the tiny flying sauser looking object to the immediate left of the "rodent" on the left and make no issue of it except to acknowledge its presence.

The reason that I'm showing you this twice over is to conclusively demonstrate that someone at official level has clearly been messing with and in this science data creating information there that isn't real and that it goes in the official record in this manner. Further, you should know that this was some very expert graphics work very difficult to detect except for the recognition factor of the obvious duplicated rocks. We are just lucky to be able to detect this at all.

You should know that these panorama images are mosaics of multiple small images attached together to form the whole. You should also know that the smaller images producing these particular evidence scenes found here at AS17-147-22552 for the Apollo 17 EVA 1 ALSEP panorama and at AS17-145-22159 for the Apollo 17 Station 5 panorama do not have these rock duplications in them. In other words, the duplications and false data are limited to the wide panorama images and not the source images for these evidence spots.

Upon reflecftion there is no telling how much of this goes undetected that does not involve easy recognition rocks. As for myself, I checked only a very few of the panorama images, quickly ran across this evidence, and did not look any further because I had the sample evidence I needed to prove my point. So there are likely more of them to be found.

J. P. Skipper can be contacted at:
jskipper@marsanomalyresearch.com

Here we go with more Moon evidence again. The problem with evidence from this source hanging so close right over our heads every night is that the bulk of the imaging evidence dates back into the 1960s. Those 1960s efforts were in turn the outgrowth of World War II leadership loss of innocence in the 1940s when it came to the question of alien presence along with the development of the atomic age and the polarization of sides in the Cold War right after the World War II.

Competition between East and West was fierce during the Cold War while national pride and confidence in leadership issues were at stake all around. As the space race began and while school children were being trained to take cover under desks and tables in the event of a nuclear blast in their territory, let's face it, initially the Russians were first and leading the way time after time beating the USA to the technology punch in spite of the fact that the West got the chief German scientists after World War II and Russia got the leftovers. Here in the USA the old USSR leadership was perceived as some of the most ruthless people on the planet killing many millions and comparable to Hitler himself. I personally lived through this time as a kid developing into a young man.

However, there is a difference in the matter of perception as well as who is doing the perceiving in those more simple times. Here in the USA populations, academics, scientists, the media, and the great bulk of congressional leaders saw this space race as a matter of pride and achievement in technology. However, while some more hidden inside leaders saw it this way as well, they also saw it primarily as a race to achieve alien contact and advantage in a ruthless global power struggle where national survival itself hung in the balance. It was this environment and perception that bred and initiated the secrecy that we now live with to this day.

Why? Because emotional populations uncontrolled and undirected are the worst possible scenario from a leadership point of view. For example, if Americans get really angry at the "sneak" attack on Pearl Harbor and flood into the military to "make things right" as they saw it and the doves in Congress quickly convert to hawks freeing money for war, that is perceived as a good thing first directed at Germany and then Japan. In fact it might even have been a "directed" thing.

On the other hand, alien presence of unknown agenda that can't be denied scaring the crap out of entire populations who then demand of their leaders that this be "fixed" in some way when no one has a clue what to do about it is understandably not perceived by them as a good thing. That's because the next step is the hunt for those that may be better able and that means current leadership heads rolling and loss of too much control at minimum.

All of this means that anxiety is very high and that of course leads to extremes, compromises and secrecy. As a boy I remember when the Sputnik satellite first went up. Immediately on the heels of this, there was a newspaper story with a "leaked" photo showing a secret warehouse full of many satellites and the theme of the story was that we Americans were already there development wise but that we wanted to take a slower more careful approach. As a older boy, I thought yea careful is good! In other words, I bought right into it. Later as a man exercising good old hind sight and remembering all those satellites pictured in large numbers, I thought what a dummy I was to have bought into that obvious load of crapolla!

The point is that once the deceit and secrecy starts in order to cover one's tracks and preserve one's position, then there's no end to it unless of course one is willing to pay the prices that will be forthcoming. So it becomes a never ending stream of secrecy upon secrecy and the web of it gets more and more convoluted. Worse, it corrupts everyone as it becomes entrenched into the culture and future generations who never know anything but this and tend to take it granted without question. In other words, it becomes institutionalized as a psychologically accepted norm.

This is the atmosphere the 1960s Moon science data exists in and what has influenced it. Obviously the obfuscation technology of the 1960s and 1970s would be poorer than it is in more recent times and that guarantees old mistakes. It also started out as film moving to digital media technology and that too is mistake prone and particularly trying to get all the data together scattered all over the place over the years. These "mistakes" have been dealt with many times both as film, as digital material not admitted to at the time, and then openly as digital material in more recent times. So the data winds up being a nightmare of different "fix" intrusions so that it is hard to tell what is real and what is not and that means that all of it must be approached very cautiously.

With all that said, following is some of the most obvious evidence of image tampering and incompetence in doing it in the Moon science data for anyone to see. It may be old hat for some but an eye opener for many others that may be inclined to not believe that such consistent manipulation patterns could really exist in the data well beyond the scope of mere incompetence. As for you in the secrecy crowd, rest easy as this is all just part of the process of awakening to truth and starting to deal with it.

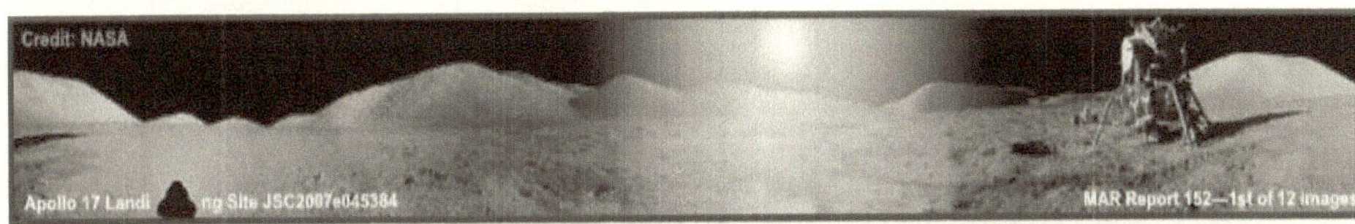

The above two panorama thumbnail images are from the Apollo 17 mission to the Moon. The above first JSC2007e045384 image as labeled is suppose to be of the landing site with the landing module seen in color on the far right. The second JSC2004e52772 image below it as labeled is suppose to be of the ALSEP Station location on this mission. First note that the background horizon lines and hills are essentially identical as are the focal width of the scenes and the backgrounds they encompass. Even though these are small narrow thumbnail size, do you see anything wrong with these two scenes or perhaps I should say do you see a whole raft of things wrong with these two comparison scenes?

First note that the landing module visible on the right in the first image is not present in the second image of the same location and some other kind of equipment is there in its place. Remember, while the equipment in the second image can no doubt be disassembled and moved, the Lander in the first image can't. The Lander should be there in the second image and yet now you see the it and now you don't. Telling isn't it.

Now in the first image note the hill and its shape in the background behind and to the right of the Lander. Note also the shadow thrown on the ground to the right by the landing module. Note its direction and its shape. Now move your attention all the way over to the left side of this same first image. Note the shape of the hill there and the shadow in the foreground below the hill. Note that, even though the hill is only partially seen on the left, they are in fact the same as on the right side of the image as is the shadow. In other words, the scene of the right side of the image is replicated on the left side minus the Lander's presence even as the Lander shadow is present. Closer looks in the 3rd and 4th images below will demonstrate this more conclusively.

However, before we leave these above panorama thumbnail images, look on the left side of the second image and compare the background hill with the background hill on the right side of the image. Once again note that they are the same as on the right and as in the first image on both sides. Now on the left side of the second image just below the hill, note the presence of a group of large rocks there. Note that these rocks are not replicated on the right nor are they present in the first image of the same left location. Once again, telling isn't it.

These are just the most gross and most obvious flaws in these two panorama images created by incompetent image tampering. If you begin to look closer in the enlarged image there are many other smaller flaws involving geological terrain features that should not be changing in the scenes but are.

Then there are the shadows that are many times all wrong and in conflict with each other. Some of this can perhaps be explained by astronauts taking these images at different times during the Moon's day with different sun angles but most definitely not all of it. Remember that the panorama images are made up of a number of different smaller images. Regardless, the most telling evidence of course is terrain fixed features and the lander fixed in place being in a spot in one image and not being in the same spot in another image and same scenes being replicated on both side of an image strip.

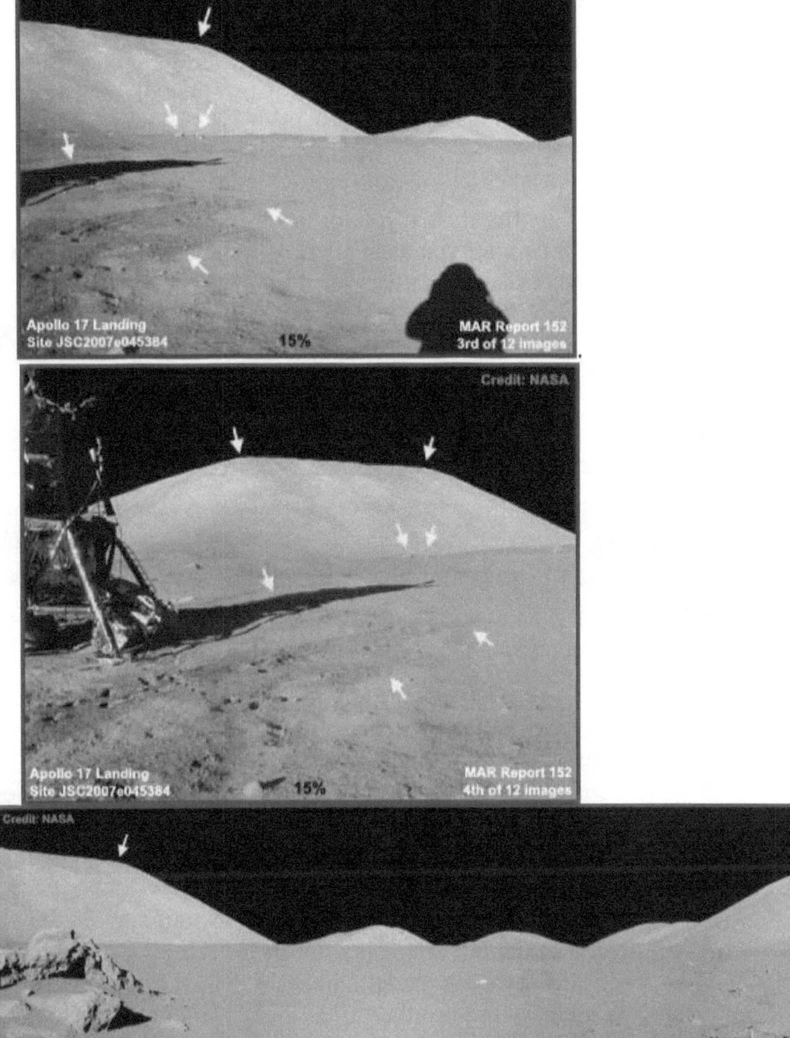

The above third, fourth, and fifth images demonstrate closer views of the far left and right scenes in the first panorama thumbnail image and the far left scene from the 2nd image. I'm including the 3rd and 4th images here to make sure you understand conclusively that these scenes on each far side of this one panorama JSC2007e045384 image are duplicates of each other with the most obvious duplicated features pointed out with yellow arrows. The 5th image from the JSC2004e52772 panorama is included to demonstrate that the scene in the more level terrain immediately below the hill on the left changes from a relatively empty terrain with the Lander shadow duplicated in it as in the upper 3rd image to a group of large rocks in it in the lower 5th image.

So does this inspire trust in you for the official Moon science data? If this doesn't quite do it for you, then below is some more of this same kind of manipulated evidence for your absorption.

The above sixth and seventh images are basically of the same same scene as confirmed by the hill outlines in the background on the horizon line. A closer look confirms that the 6th image is a ever so slightly closer view and the 7th image is a slightly pulled back more distant view but still the same. Do any flaws between these comparison images jump out at you?

As you can quickly see, by far the most obvious flaw is the very large rock that is in the 6th image but not in the 7th image. This rock should have appeared between the left and right yellow arrows in the 7th image but isn't there. A little further to the right is a large visually washed out rock pointed out by the yellow arrow. Note that it also isn't in the upper 6th image. The question then becomes is this lone rock the same one in the 6th image but just slightly moved to the right a bit?

As you can see in the above eighth closer view of the rock in the 6th panorama image comparing it to the ninth closer view of the distant rock in the 7th panorama image, they just are not the same. So this is a case of major size rocks or boulders appearing and disappearing in the same locations in these official science data scenes. There is other lesser evidence in these scenes doing this as well but the most gross and obvious evidence should be enough to tell you what you need to know. Any of this giving you any pause as yet?

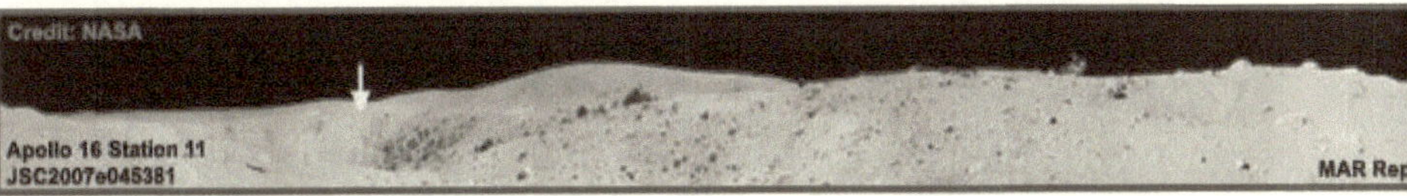

The above tenth Apollo 16 panorama JSC2007e045381 thumbnail image provides the context scene in which the evidence appears and the two evidence site locations are pointed out by the two yellow arrows. Note that this duplicated evidence is a little more subtle in that it involves smaller terrain features rather than large obvious objects. The two images below will provide a closer look at these two duplicated sites.

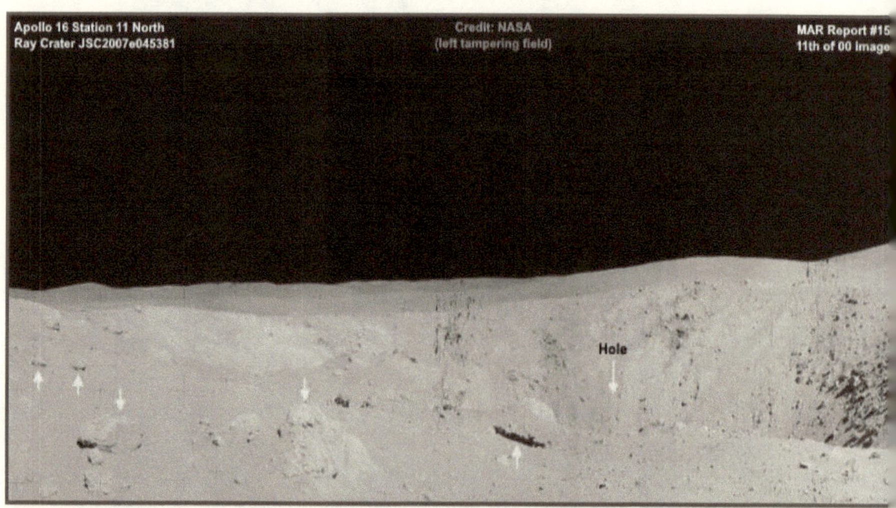

....

The two 11th and 12th images above demonstrate closer views of the duplicated image tampering areas within the JSC2007e045381 panorama strip. The 11th image demonstrates the duplicated site on the left and the 12th image the duplicated site on the right. The multiple yellow arrows point out just a few of the most obvious duplicated terrain features but not all in each site.

As you can see in this and previous reporting, there is plenty of evidence of image tampering in this old 1960s Moon data and this may barely be scratching the surface. After all, only when the tampering evidence is grossly incompetent and very obvious involving very large appearing and disappearing objects does it lend itself to discovery and more readily come to our attention. Even though it seems I've presented a fairly good amount of it here and in previous reporting, likely only a tiny portion of the tampering is so obvious with the bulk being much more subtle going without detection.

The fact is that I could go on and on with this type of reporting but, although some may think differently, I actually do restrain myself considerably relative to the massive amount of tampering evidence there is like this. I find this type of reporting both boring and unpleasant and try to avoid it until I think you should be exposed to it.

The presentation of some of it is necessary if you are to get a grasp of how convoluted and difficult it is to research this data and how untrustworthy it is to draw conclusions from. That state of affairs may be embarrassing to the secrecy crowd when it is revealed, as is happening here, but it is ultimately acceptable to them because ambiguity in the data makes truth very hard to determine and ignorance for you and I on this material is their goal ever how it is achieved.

BOB'S COMMENTS

Hereunder is the response of a typical "de-bunker", and as usual, the style is to attack the individuals, belittle, put down etc. Sad.

My greatest qualifications are a keen eye, good sense and the power of reason. When the hoax advocates make claims that are based on flimsy evidence, sloppy research, and a poor understanding of the sciences, it does not take a PhD to figure out they are wrong. In general, the main proponents of the hoax theory are people who have no special education, training or experience to qualify them to make their claims.

They are in no better position to judge the facts than you and I; so use your own sense of reason.

(Hey, as was once wisely said, I have no "qualification" but I can read. (And see.) The evidence used is all sourced from NASA official releases, so that is neither "flimsy" or "sloppy". To compare any two photographs does not require any "special education" and to infer "we have no "special education" is merely saying 'I have credentials, you have none. So shut up.' **Oh dear, I forgot. We are supposed to unquestioningly accept and believe whatever "they" tell us.**

APOLLO 11 ORIGINAL LANDING TAPES NOW MISSING!

Thursday, July 13, 2006: The original high-quality video tapes of Apollo 11, which were apparently sent by NASA to the National Archives and then were returned to the Goddard Space Flight Center, have gone missing (see the pdf by John M. Sarkissian). The quality of the video broadcast to the world on television was of much, much lower quality than the video originally received – or manufactured! - by NASA. Obviously, if you were going to fake the moon landing, you might have a motive to 'lose' the high-quality tapes, where artifacts of faking could be seen. This was by far the biggest moment in the American space program. You'd think they would care about hanging on to the evidence.

See Missing Tape Info:
http://www.apfn.org/pdf/Search_for_SSTV_Tapes.pdf

Did man really set foot on the moon?

Shocking : See what NASA has done (Long but worth reading)

Did man really walk on the Moon or was it the ultimate camera trick, asks David Milne?

In the early hours of May 16, 1990, after a week spent watching old video footage of man on the Moon, a thought was turning into an obsession in the mind of Ralph Rene.

"How can the flag be fluttering?" the 47 year old American kept asking himself when there's no wind on the atmosphere free Moon? That moment was to be the beginning of an incredible Space odyssey for the self- taught engineer from New Jersey.

He started investigating the Apollo Moon landings, scouring every NASA film, photo and report with a growing sense of wonder, until finally reaching an awesome conclusion: America had never put a man on the Moon. The giant leap for mankind was fake.

It is of course the conspiracy theory to end all conspiracy theories. But Rene has now put all his findings into a startling book entitled NASA Mooned America. Published by himself, it's being sold by mail order - and is a compelling read.

The story lifts off in 1961 with Russia firing Yuri Gagarin into space, leaving a panicked America trailing in the space race. At an emergency meeting of Congress, President Kennedy proposed the ultimate face saver, put a man on the Moon. With an impassioned speech he secured the plan an unbelievable 40 billion dollars.

And so, says Rene (and a growing number of astro-physicists are beginning to agree with him), the great Moon hoax was born. Between 1969 and 1972, seven Apollo ships headed to the Moon. Six claim to have made it, with the ill fated Apollo 13 - whose oxygen tanks apparently exploded halfway being the only casualties. But with the exception of the known rocks, which could have been easily mocked up in a lab, the photographs and film footage are the only proof that the Eagle ever landed. And Rene believes they're fake.

For a start, he says, the TV footage was hopeless. The world tuned in to watch what looked like two blurred white ghosts throw rocks and dust. Part of the reason for the low quality was that, strangely, NASA provided no direct link up. So networks actually had to film man's greatest achievement from a TV screen in Houston - a deliberate ploy, says Rene, so that nobody could properly examine it.

By contrast, the still photos were stunning. Yet that's just the problem. The astronauts took thousands of pictures, each one perfectly exposed and sharply focused. Not one was badly composed or even blurred.

As Rene points out, that's not all: The cameras had no white meters or view ponders. So the astronauts achieved this feet without being able to see what they were doing. There film stock was unaffected by the intense peaks and powerful cosmic radiation on the Moon, conditions that should have made it useless. They managed to adjust their cameras, change film and swap filters in pressurized suits. It should have been almost impossible with the gloves on their fingers.

Award winning British photographer David Persey is convinced the pictures are fake. His astonishing findings are explained alongside the pictures on these pages, but the basic points are as follows: The shadows could only have been created with multiple light sources and,in particular, powerful spotlights. But the only light source on the Moon was the sun.

The American flag and the words "United States" are always Brightly lit, even when everything around is in shadow. Not one still picture matches the film footage, yet NASA claims both were shot at the same time.

The pictures are so perfect, each one would have taken a slick advertising agency hours to put them together. But the astronauts managed it repeatedly. David Persey believes the mistakes were deliberate, left there by "whistle blowers" who were keen for the truth to one day get out.

If Persey is right and the pictures are fake, then we've only NASA's word that man ever went to the Moon. And, asks Rene, "Why would anyone fake pictures of an event that actually happened?"

The questions don't stop there. Outer space is awash with deadly radiation that emanates from solar flares firing out from the sun. Standard astronauts orbiting earth in near space, like those who recently fixed the Hubble telescope, are protected by the earth's Van Allen belt. But the Moon is to 240,000 miles distant, way outside this safe band. And, during the Apollo flights, astronomical data shows there were no less than 1,485 such flares.

John Mauldin, a physicist who works for NASA, once said shielding at least two meters thick would be needed. Yet the walls of the Lunar Landers which took astronauts from the spaceship to the moons surface were, said NASA, about the thickness of heavy duty aluminum foil.

How could that stop this deadly radiation? And if the astronauts were protected by their space suits, why didn't rescue workers use such protective gear at the Chernobyl meltdown, which released only a fraction of the dose astronauts would encounter? Not one Apollo astronaut ever contracted cancer - not even the Apollo 16 crew who were on their way to the Moon when a big flare started. "They should have been fried", says Rene.

Furthermore, every Apollo mission before number 11 (the first to the Moon) was plagued with around 20,000 defects a-piece. Yet, with the exception of Apollo 13, NASA claims there wasn't one major technical problem on any of their Moon missions. Just one effect could have blown the whole thing. "The odds against these are so unlikely that God must have been the co-pilot," says Rene.

Several years after NASA claimed its first Moon landing, Buzz Aldrin "the second man on the Moon" was asked at a banquet what it felt like to step on to the lunar surface. Aldrin staggered to his feet and left the room crying uncontrollably. It would not be the last time he did this. "It strikes me he's suffering from trying to live out a very big lie," says Rene. Aldrin may also fear for his life.

Virgil Grissom, a NASA astronaut who baited the Apollo program, was due to pilot Apollo 1 as part of the landings build up. In January 1967, he hung a lemon on his Apollo capsule (in the US, unroadworthy cars are called lemons) and told his wife Betty: "If there is ever a serious accident in the space program, it's likely to be me."

Nobody knows what fuelled his fears, but by the end of the month he and his two co-pilots were dead, burnt to death during a test run when their capsule, pumped full of high pressure pure oxygen, exploded.

Scientists couldn't believe NASA's carelessness - even a chemistry students in high school know high pressure oxygen is extremely explosive. In fact, before the first manned Apollo fight even cleared the launch pad, a total of 11 would be astronauts were dead. Apart from the three who were incinerated, seven died in plane crashes and one in a car smash. Now this is a spectacular accident rate.

"One wonders if these 'accidents' weren't NASA's way of correcting mistakes," says Rene. "Of saying that some of these men didn't have the sort of 'right stuff' they were looking."

NASA wont respond to any of these claims, their press office will only say that the Moon landings happened and the pictures are real. But a NASA public affairs officer called Julian Scheer once delighted 200 guests at a private party with footage of astronauts apparently on a landscape. It had been made on a mission film set and was identical to what NASA claimed was they real lunar landscape. "The purpose of this film," Scheer told the enthralled group, "is to indicate that you really can fake things on the ground, almost to the point of deception." He then invited his audience to "Come to your own decision about whether or not man actually did walk on the Moon."

A sudden attack of honesty? You bet, says Rene, who claims the only real thing about the Apollo missions were the lift offs. "The astronauts simply have to be on board," he says, "in case the rocket exploded. It was the easiest way to ensure NASA wasn't left with three astronauts who ought to be dead." he claims, adding that they came down a day or so later, out of the public eye (global surveillance wasn't what it is now) and into the safe hands of NASA officials, who whisked them off to prepare for the big day a week later.

And now NASA is planning another giant step - Project Outreach, a 1 trillion dollar manned mission to Mars. "Think what they'll be able to mock up with today's computer graphics," says Rene Chillingly. "Special effects was in its infancy in the 60s. This time round will have no way of determining the truth."

9 SPACE ODDITIES:

1. Apollo 14 astronaut Allen Shepard played golf on the Moon. In front of a worldwide TV audience, Mission Control teased him about slicing the ball to the right. Yet a slice is caused by uneven air flow over the ball. The Moon has no atmosphere and no air.

2. A camera panned upwards to catch Apollo 16's Lunar Landerlifting off the Moon. Who did the filming?

3. One NASA picture from Apollo 11 is looking up at Neil Armstrong about to take his giant step for mankind. The photographer must have been lying on the planet surface. If Armstrong was the first man on the Moon, then who took the shot?

4. The pressure inside a space suit was greater than inside a football. The astronauts should have been puffed out like the Michelin Man, but were seen freely bending their joints.

5. The Moon landings took place during the Cold War. Why didn't America make a signal on the moon that could be seen from earth? The PR would have been phenomenal and it could have been easily done with magnesium flares.

6. Text from pictures in the article said that only two men walked on the Moon during the Apollo 12 mission. Yet the astronaut reflected in the visor has no camera. Who took the shot?

7. The flags shadow goes behind the rock so doesn't match the dark line in the foreground, which looks like a line cord. So the shadow to the

lower right of the spaceman must be the flag. Where is his shadow? And why is the flag fluttering if there is no air or wind on the moon?

8. How can the flag be brightly lit when its side is to the light? And where, in all of these shots, are the stars?

9. The Lander weighed 17 tons yet the astronauts feet seem to have made a bigger dent in the dust. The powerful booster rocket at the base of the Lunar Lander was fired to slow descent to the moons service. Yet it has left no traces of blasting on the dust underneath. It should have created a small crater, yet the booster looks like it's never been fired.

http://www.orwelltoday.com/moonhoaxdoc.shtml

307 Deaths of key people involved with the Apollo program

In a television program about the hoax theory, Fox Entertainment Group listed the deaths of 10 astronauts and of two civilians related to the manned spaceflight program as having possibly been killings as part of a coverup.

- Ted Freeman (T-38 crash, 1964)
- Elliott See and Charlie Bassett (T-38 accident, 1966)
- Virgil "Gus" Grissom (supposedly an outspoken critic of the Space Program) (Apollo 1 fire, January 1967)
- Ed White (Apollo 1 fire, January 1967)
- Roger Chaffee (Apollo 1 fire, January 1967)
- Ed Givens (car accident, 1967)
- C. C. Williams (T-38 accident, October 1967)
- X-15 pilot Mike Adams (the only X-15 pilot killed in November 1967 during the X-15 flight test program - not a NASA astronaut, but had flown X-15 above 50 miles).
- Robert Lawrence, scheduled to be an Air Force Manned Orbiting Laboratory pilot who died in a jet crash in December 1967, shortly after reporting for duty to that (later cancelled) program.
- NASA worker Thomas Baron Train crash, 1967 shortly after making accusations before Congress about the cause of the Apollo 1 fire, after which he was fired. Ruled as suicide.

- Paul Jacobs, a private investigator from San Francisco, interviewed the head of the US Department of Geology in Washington about the 'moon rocks'. *Did you examine the Moon rocks, did they really come from the Moon?* Jacobs asked - the geologist did not respond, only laughed. Paul Jacobs and his wife died from cancer within 90 days.
- Lee Gelvani claims to have almost convinced James Irwin, an Apollo 15 astronaut whom Gelvani referred to as an "informant", to confess about a cover-up having occurred. Irwin was supposedly going to contact Kaysing about it; however he died of a heart attack in 1991, before any such telephone call occurred.

Why the Americans *NEVER* landed on the moon.

Why they would fake it

The Soviet Union had been making all the early advances and the greatest progress in the great Moon race.

The Soviet Union launched the first man and the first women in space in 1961 & 1963 and were also the first to orbit the Earth.

With the above happening the US Government had to make some kind of success with President Kennedy promising that the US would put a man on the moon by the end of the 1960's.

Many people believe that NASA had released that it was not possible to go to the moon with the technology available
(Computer chips being as powerful then as a modern washing machines chip) so they resorted to faking the landing to ensure a
victory of the Soviet Union and keep the dollars coming in for real space projects.

The Pictures

NASA have never offered any explanation whatsoever for the numerous errors in the photographs, despite repeated questioning.
These errors include:

The Apollo 11 pictures show the ground in the distance being much darker
than the ground in the foreground, as if the Astronauts were standing in a pool of light.

Several photos show evidence of extra lighting (as a professional photographer
would use fill-in lights) but no such lights were supposed to have been used.

Some photos clearly show the light coming from "impossible" angles. In one instance, Aldrin's boot is lit from below as he descends the ladder.

Some photos contradict the TV camera pictures of the same events.

Some photos of one astronaut taken by the other are clearly taken from slightly above the eye level of the subject, but in his visor, the reflection of the astronaut with the camera shows it being held at chest level.

The length of the shadows in the Apollo 12 pictures don't agree with the angle which the Sun should have been at.

Some wide area photos show shadows pointing in different directions.

In the sound recording of the lunar landing, you cannot hear the sound of the engines. As the astronaut calls out the remaining distance to the surface, he is only a few feet away from a rocket engine which should have been producing 10000 lb of thrust.

The sounds

The major point which has helped convince me that the moon landing was faked was the fact that when the control room asked a question to the Astronoughts the replies were instant with no delays. This seems strange as even with technology in the 1990's there is a delay from satellite links from the UK to the US. There is about a 0.7 second delay from London to California so how is it possible for instant replies from the Moon ?

There is also evidence that when people go into space that there voice goes tense although the Astronaughts voices have been analyzed and found to be normal, and 7/10 people said it sounded like someone reading from a script.

When Houston are talking to the module you should not be able to hear the responses at least when the module is landing and the infamous "eagle has landed" quote, this is due to the noise that should have been created by the rocket motor which generates several hundred thousand pounds of thrust 20 ft below the astronaughts. The noise would have completely drowned
the vocals out.

The Radiation

An American author has researched and found out that he believes the Apollo Spacecraft would have needed to be two meters thick to prevent cosmic radiation from cooking the Astronaughts inside.

Also in addition to the radiation protection for the astronaughts similar protection would be required for the films + cameras, NASA's official explanation of how the films were protected was that the cameras were painted with a coat of aluminium paint,
yeah right.

Over the course of a full lunar day and night, the temperature on the Moon can vary wildly, from around +200 to -200 degrees Celsius (+392 to -328 degrees Fahrenheit), so it's natural to wonder how lunar astronauts survived this huge temperature variation.
There's another problem. The moon takes 27 days to rotate once on its axis. So any place on the surface of the Moon experiences about 13 days of sunlight, followed by 13 days of darkness. So if you were standing on the surface of the Moon in sunlight, the temperature would be hot enough to boil water. And then the Sun would go down, and the temperature would drop 250 degrees in just a matter of moments.

5771
photos
taken in
4834
minutes !
One exposure
every 50 seconds !

TIME & MOTION STUDY:

Anyone with even elemental math skills and common sense can look at the facts, do the calculations, and come to their own conclusions about the alleged MASSIVE VOLUME of lunar surface photography in such a LIMITED TIME.

Here is my conclusion: IT COULD NOT BE DONE.
http://www.aulis.com/skeleton.html

It boils down not to just studying the photographs for signs of fakery, though I have examined every available Apollo photo for more than three years (and discovered many fakes). Very simply, it amounts to a study known to many businesses...A TIME AND MOTION STUDY. The elementary question is: was it possible to take the known number of photos (from NASA records) in the amount of time available (from NASA records)? But before you read my study, to understand it you need to know some basic information about the Apollo missions:

1. Of seven Apollo missions to put "men on the Moon", six were

claimed to be "successful". (Apollo 13 was "aborted".)

2. Each of the six successful missions landed two astronauts "on the Moon" in a flimsy craft NASA originally had called the Lunar Excursion Module (LEM, later shortened to LM), an unproven craft which never had an opportunity for a lunar landing test flight. But it landed and then took off six times with spectacular "success" on Apollo missions 11 and 12, and 14 through 17...once even landing within 200 feet of a pre-selected target.

3. Two astronauts rode each LEM to the Moon surface while one remained in the orbiting Command and Service Module (CSM) awaiting their return.

4. During their Extra-Vehicular Activity (lunar surface exploration) each of the two wore a bulky inflated spacesuit with clumsy gloves, greatly limiting mobility. On their backs they wore a huge and heavy Life Support System (PLSS) backpack containing an oxygen tank and circulating water air conditioning system which pumped refrigerated water throughout the suit to counteract the 200+/- degree heat (and cold) of lunar conditions. Pumps circulated both refrigerated air and water to the liquid cooling undergarment, as well as dehumidifying, removing carbon dioxide, and providing all other functions needed to survive harsh conditions in the confining suits.

5. The principal objective of all six missions was SCIENTIFIC RESEARCH projects to be carried out by the two astronauts. Most of the projects, which numbered about a half dozen each mission, were remarkably similar on all six missions. All of these science experiments involved unpacking equipment from stowage bays, assembling it, transporting it to its location, setting it up, and then doing the experiments. As you might imagine, each of these research projects would require a major portion of the TIME of the two men for each experiment.

6. Another major project besides operation of the packaged experiments was the Geological Study, which involved searching for different specimens of rocks and soils in various locations, documenting and collecting samples to return to earth. This obviously occupied much of their TIME.

7. Considerable TIME was needed for "housekeeping chores". After landing, the LEM had to be inspected to make sure it had not been damaged. Communications equipment to put them in contact with

Earth had to be set up and operated, including radio and television antennas and TV cameras. The US flag was planted in the moondust on each mission. All of this was done before any experiments were initiated. Oh, and don't forget the "ceremonial" chat with President Nixon during Apollo 11.

8. The first three missions required the astronauts to walk to each experiment location. The last three missions were supplied with a Lunar Roving Vehicle (LRV) to travel to distant locations miles away from the LEM. The partially pre-assembled LRV was attached to the outside of the LEM. The rover floor served as a pallet which was hinged to the outside of the LRV. The wheels were folded under. The "pallet" was lowered by hand to the lunar surface, and the wheels rotated into position. After the wheels were down, the vehicle had to be outfitted with all of its considerable equipment from various storage bins of the LEM. Oddly, not a single photo exists in the public domain (at least that I could find to date) of the astronauts assembling and equipping the LRVs. The battery-powered rovers had a top speed of about 8 mph, only slightly faster than walking...much like a golf cart. During the LRV travels ("traverses"), both men rode, and when moving, had no opportunity for photography. Also, the time taken in assembling the rover was not used for any photography. Though I could find no time given by NASA, surely it is reasonable to guess that it took at least an hour to unload, assemble and equip and test a rover?

9. Almost incidental to the main astronaut tasks was PHOTOGRAPHY. Each astronaut had his own camera. (Apart from the Apollo 11 EVA.) It was a square-format specially-built Hasselblad. It was mounted on a chest-plate for the astronaut to operate. The astronaut had to manually set the shutter speed and apertures while wearing bulky, pressurized gloves and without being able to see the controls. The cameras had NO VIEWFINDER, so the astronaut could only guess at what was being photographed. Each camera had a bulk film magazine holding more than a hundred exposures. The film (mainly Ektachrome color film) had a very narrow exposure range, which required PERFECT aperture and shutter settings, because according to NASA, the cameras did not have automatic exposure capability.

10. It is important to know that although each man had his own camera, they ALMOST NEVER USED THEM AT THE SAME TIME. Usually one of them was photographing the other doing some task. Therefore having two cameras DID NOT TRANSLATE TO TWICE AS MUCH TIME FOR PHOTOGRAPHY, as one might surmise. Now that you understand the missions, here is my discovery of NASA

overzealousness, which has been successfully hidden till now.

A TIME AND MOTION STUDY

For more than three years I have been collecting and analyzing nearly all the significant photos from the Apollo missions. These official photos are readily available on multiple NASA websites for downloading. Recently I noticed they were taking up many gigabytes of memory on my computer's external hard drive, so I began organizing them and deleting duplications. I did a rough estimate of the number of Apollo photos, and was amazed that I had thousands!

I visited several official NASA websites to find HOW MANY PHOTOS WERE TAKEN on the surface of the Moon. Amazingly, NASA AVOIDS THIS SUBJECT almost entirely. Two days of searching documents and text were fruitless. But Lunar Surface Journal, one of the sites, lists every photo with its file number. So I undertook to make an actual count of every photo taken by astronauts DURING EXTRA-VEHICULAR ACTIVITY (EVA), the time spent on the surface out of the LEM.

Here is my actual count of EVA photos of the six missions:

Apollo 11........... 121
Apollo 12........... 504
Apollo 14........... 374
Apollo 15..........1021
Apollo 16..........1765
Apollo 17..........1986

So 12 astronauts while on the Moon's surface took a TOTAL of 5771 exposures.

That seemed excessively large to me, considering that their TIME on the lunar surface was limited, and the astronauts had MANY OTHER TASKS OTHER THAN PHOTOGRAPHY. So I returned to the Lunar Surface Journal to find how much TIME was available to do all the scientific tasks AS WELL AS PHOTOGRAPHY. Unlike the number of photos, this information is readily available:

Apollo 11........1 EVA .....2 hours, 31 minutes......(151 minutes)
Apollo 12........2 EVAs.....7 hours, 50 minutes......(470 minutes)
Apollo 14........2 EVAs.....9 hours, 25 minutes......(565 minutes)
Apollo 15........3 EVAs...18 hours, 30 minutes....(1110 minutes)

**Apollo 16........3 EVAs...20 hours, 14 minutes....(1214 minutes)
Apollo 17........3 EVAs...22 hours, 04 minutes....(1324 minutes)**

**Total minutes on the Moon amounted to 4834 minutes.
Total number of photographs taken was 5771 photos.**

Hmmmmm. That amounts to 1.19 photos taken EVERY MINUTE of time on the Moon, REGARDLESS OF OTHER ACTIVITIES. (That requires the taking of ONE PHOTO EVERY 50 SECONDS!) Let's look at those other activities to see how much time should be deducted from available photo time:

Apollo 11..........Inspect LEM for damage, deploy flag, unpack and deploy radio and television equipment, operate the TV camera (360 degree pan), establish contact with Earth (including ceremonial talk with President Nixon), unpack and deploy numerous experiment packages, find/document/collect 47.7 pounds of lunar rock samples, walk to various locations, conclude experiments, return to LEM.

Apollo 12..........Inspect LEM for damage, deploy flag, unpack and deploy radio and television equipment (spend time trying to fix faulty TV camera), establish contact with Earth, unpack and deploy numerous experiment packages, walk to various locations, inspect the unmanned Surveyor 3 which had landed on the Moon in April 1967 and retrieve Surveyor parts. Deploy ALSEP package. Find/document/collect 75.7 pounds of rocks, conclude experiments, return to LEM.

Apollo 14..........Inspect LEM for damage, deploy flag, unpack and deploy radio and television equipment and establish contact with Earth, unpack and assemble hand cart to transport rocks, unpack and deploy numerous experiment packages, walk to various locations. Find/document/collect 94.4 pounds of rocks, conclude experiments, return to LEM.

Apollo 15..........Inspect LEM for damage, deploy flag, unpack and deploy radio and television equipment and establish contact with Earth, unpack/assemble/equip and test the LRV electric-powered 4-wheel drive car and drive it 17 miles, unpack and deploy numerous experiment packages (double the scientific payload of first three missions). Find/document/collect 169 pounds of rocks, conclude experiments, return to LEM. (The LRV travels only 8 mph*.)

Apollo 16..........Inspect LEM for damage, deploy flag, unpack and deploy radio and television equipment and establish contact with Earth,

unpack/assemble/equip and test the LRV electric-powered 4-wheel drive car and drive it 16 miles, unpack and deploy numerous experiment packages (double the scientific payload of first three missions, including new ultraviolet camera, operate the UV camera). Find/document/collect 208.3 pounds of rocks, conclude experiments, return to LEM. (The LRV travels only 8 mph*.)

Apollo 17..........Inspect LEM for damage, deploy flag, unpack and deploy radio and television equipment and establish contact with Earth, unpack/assemble/equip and test the LRV electric-powered 4-wheel drive car and drive it 30.5 miles, unpack and deploy numerous experiment packages. Find/document/collect 243.1 pounds of rocks, conclude experiments, return to LEM. (The LRV travels only 8 mph*.)

Let's arbitrarily calculate a MINIMUM time for these tasks and subtract from available photo time:

Apollo 11....subtract 2 hours (120 minutes), leaving 031 minutes for taking photos
Apollo 12....subtract 4 hours (240 minutes), leaving 230 minutes for taking photos
Apollo 14....subtract 3 hours (180 minutes), leaving 385 minutes for taking photos
Apollo 15....subtract 6 hours (360 minutes), leaving 750 minutes for taking photos
Apollo 16....subtract 6 hours (360 minutes), leaving 854 minutes for taking photos
Apollo 17....subtract 8 hours (480 minutes), leaving 844 minutes for taking photos

So do the math:

Apollo 11.......121 photos in 031 minutes............3.90 photos per minute
Apollo 12........504 photos in 230 minutes............2.19 photos per minute
Apollo 14.......374 photos in 385 minutes............0.97 photos per minute
Apollo 15.....1021 photos in 750 minutes............1.36 photos per minute
Apollo 16.....1765 photos in 854 minutes ....2.06 photos per minute
Apollo 17.....1986 photos in 844 minutes ....2.35 photos per minute

Or, to put it more simply:

Apollo 11........one photo every 15 seconds
Apollo 12........one photo every 27 seconds
Apollo 14........one photo every 62 seconds

Apollo 15........one photo every 44 seconds
Apollo 16........one photo every 29 seconds
Apollo 17........one photo every 26 seconds

So you decide. Given all the facts, was it possible to take that many photos in so short a time?

Any professional photographer will tell you it cannot be done. Virtually every photo was a different scene or in a different place, requiring travel. As much as 30 miles travel was required to reach some of the photo sites. Extra care had to be taken shooting some stereo pairs and panoramas. Each picture was taken without a viewfinder, using manual camera settings, with no automatic metering, while wearing a bulky spacesuit and stiff clumsy gloves.

The agency wants the world to believe that 5771 photographs were taken in 4834 minutes! IF NOTHING BUT PHOTOGRAPHY HAD BEEN DONE, such a feat is clearly impossible...made even more so by all the documented activities of the astronauts. Imagine...1.19 photos every minute that men were on the Moon -- that's one picture every 50 SECONDS!

The secret NASA tried to hide has been discovered: The quantity of photos purporting to record the Apollo lunar EVAs could not have been taken on the Moon in such an impossible time frame. So why do these photos exist? How did these photos get made? Did ANY men go to the Moon? Or was it truly the greatest hoax ever?

Editor's Notes: *According to Andrew Chaikin, author of A Man on the Moon the LRV averaged only 5 to 7 miles per hour, which would reduce even further the time available for photography.
http://www.aulis.com/skeleton.html

THERE IS SOMETHING WRONG HERE.

And there is also something wrong here on this Apollo 14 pic. Footprints under the LM and descent nozzle.

AS14-66-9277HR Nasa photograph.

This is photograph AS17-149-22859, a picture taken from **directly underneath** Apollo 17, under the rocket nozzle, on it's lift-off from the "moon". Not a good place to be one would think.

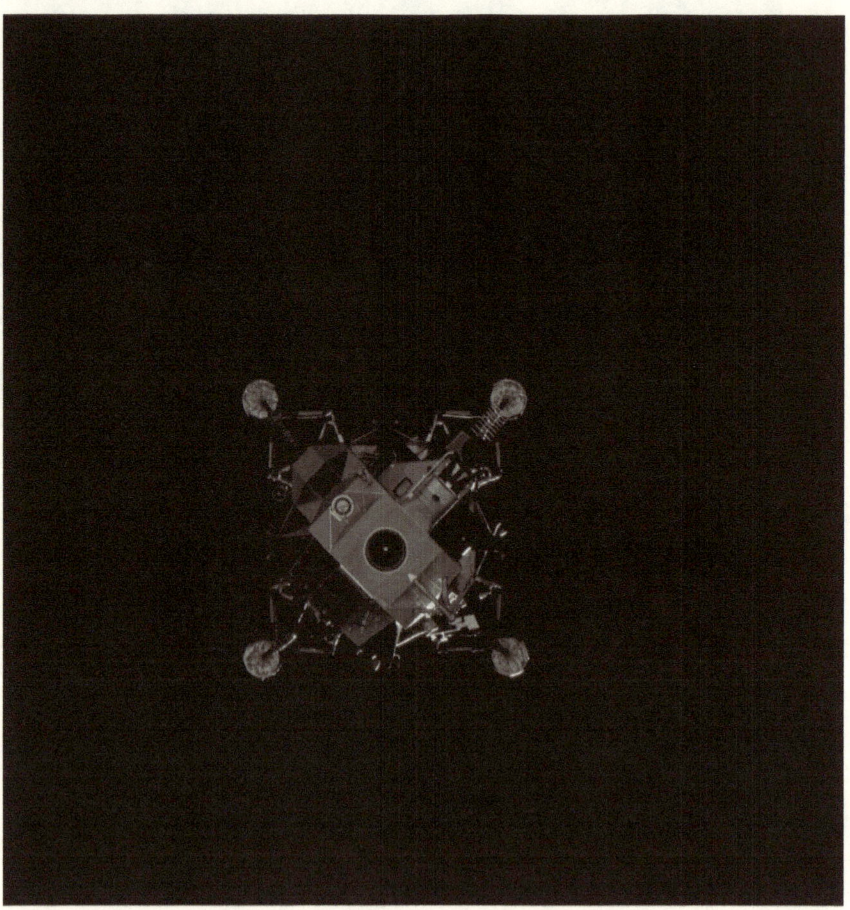

The following picture released by NASA purports to show Apollo 17 in flight. Although the above pic shows landing feet, the one below has "lost" them. Please tell me this pic does not remind you of a kiddies "Cubby House" and not a serious space vehicle.

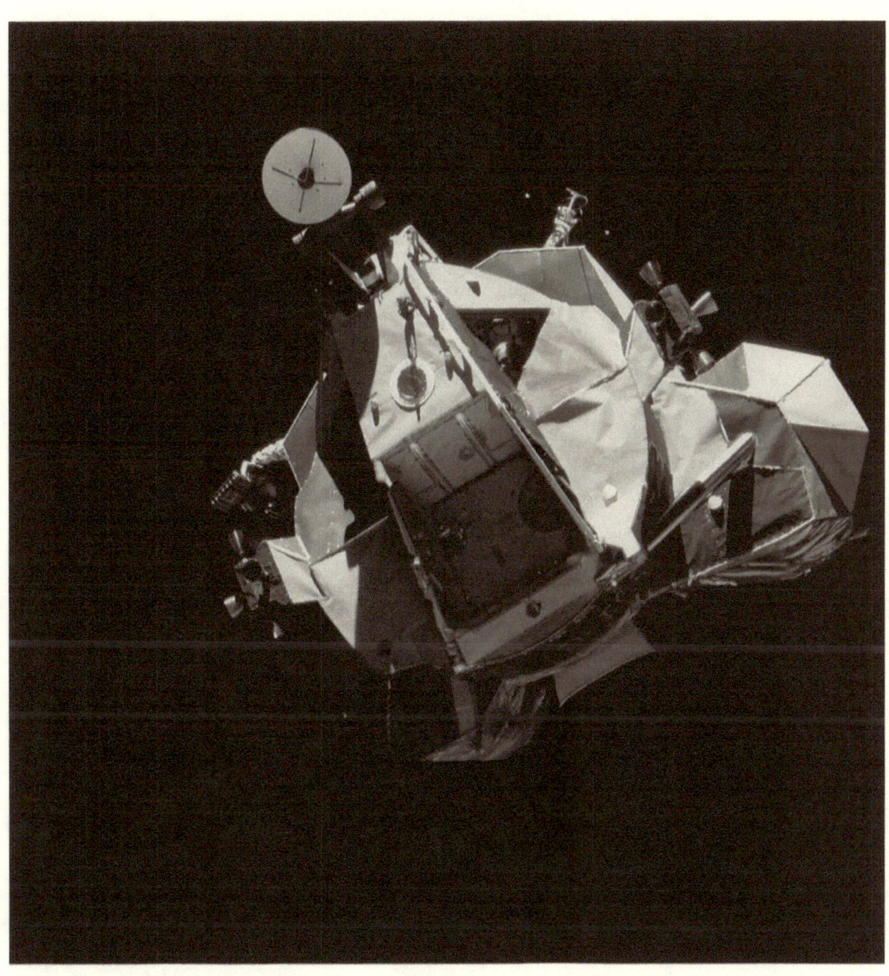

The next photo is number AS11-40-5864HR, an Apollo 11 landed photo. Note is has no crater from the landing rocket engine, and landing feet are gently landed on the surfact, no dust disturbance visible. The same strange effect is seen in all Apollo landed photos.

By 1989 doubtless enough “dust” had hit the fan, along with lack of landing craters and “floating feet.

A solution was presented in the 1989 movie, “Moontrap”. Public re-education could begin, and more people would see and remember the movie that saw or remember the “real” photos. The following pics are from that movie.

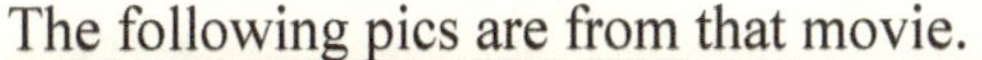

I think we will leave it on this note. There is a vast amount of information on the Internet and documentaries available that firmly establish there is a case to answer. Yet NASA steadfastly refuses to answer any criticism of their official releases, story, and claims.

Very telling is the manner in which all would be “hoax” debunkers handle things. Having no real answers to the problems evidenced from NASA's own source materials, they are left with only the usual ploy and strategy used by such self-promoting experts and authorities.

The speak down to us, wave their credentials, and call us names, such as untrained, uneducated, lacking understanding, not qualified, and of doing shoddy research producing flimsy evidence. Ho Hum.

Unfortunately mere denial of evidence is not an answer.

Get and watch the documentary, available on Internet (Pirate Bay for example) "A Funny Thing Happened on the Way to the Moon".

Apollo 17 svsr, the vehicle.

After looking closely at some Apollo 17 pictures of the craft said to be landed on the moon, and then looking at photos of the vehicle taken with it, like millions of other people I wondered just where and how they got that vehicle to the moon. A reasonable question surely. Pics below show the seeming problem.

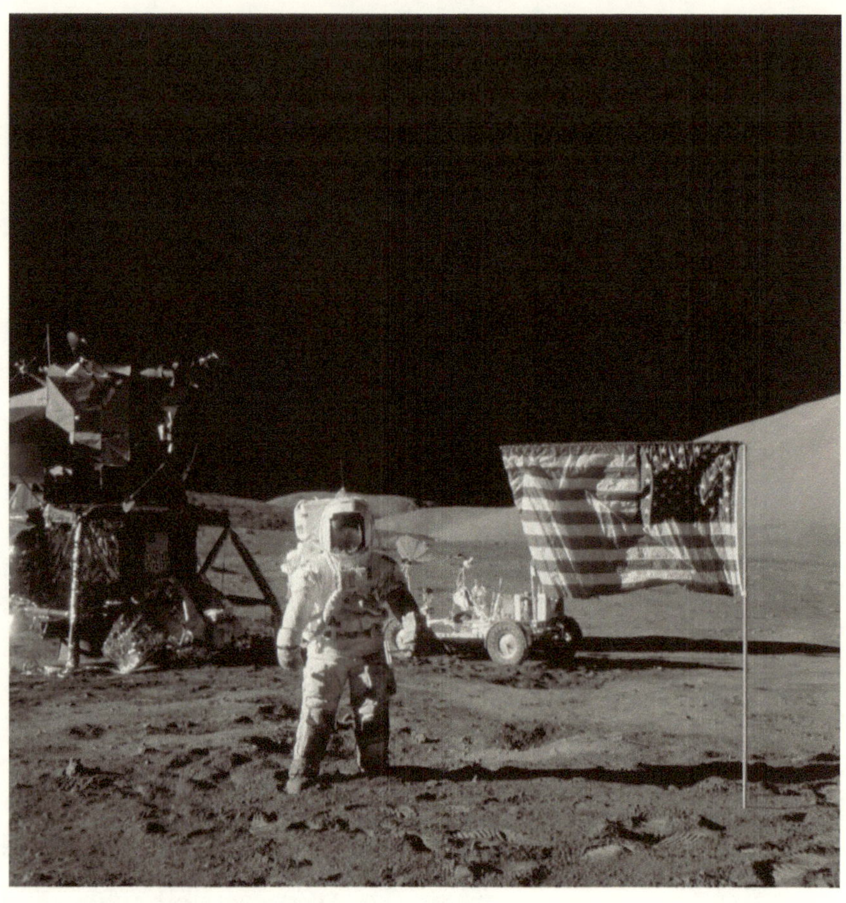

Frankly there just doesn't seem to be enough room to store such a large vehicle under such a low landing module, especially with that HUGE rocket nozzle dominating the underbelly of the craft. Again marvelous that such a large nozzle left no crater or disturbance after landing the combined Apollo 17 vehicle. So how is all this explained?

Well it is not quite a case of it's all done with smoke and mirrors, if one carefully assesses the following RARE picture, we see a carefully folded up svsr attaches to the side of the landing module. Neat, and I guess that accounts for the cost of $38 million 1970s value for each vehicle made. Nice for some.

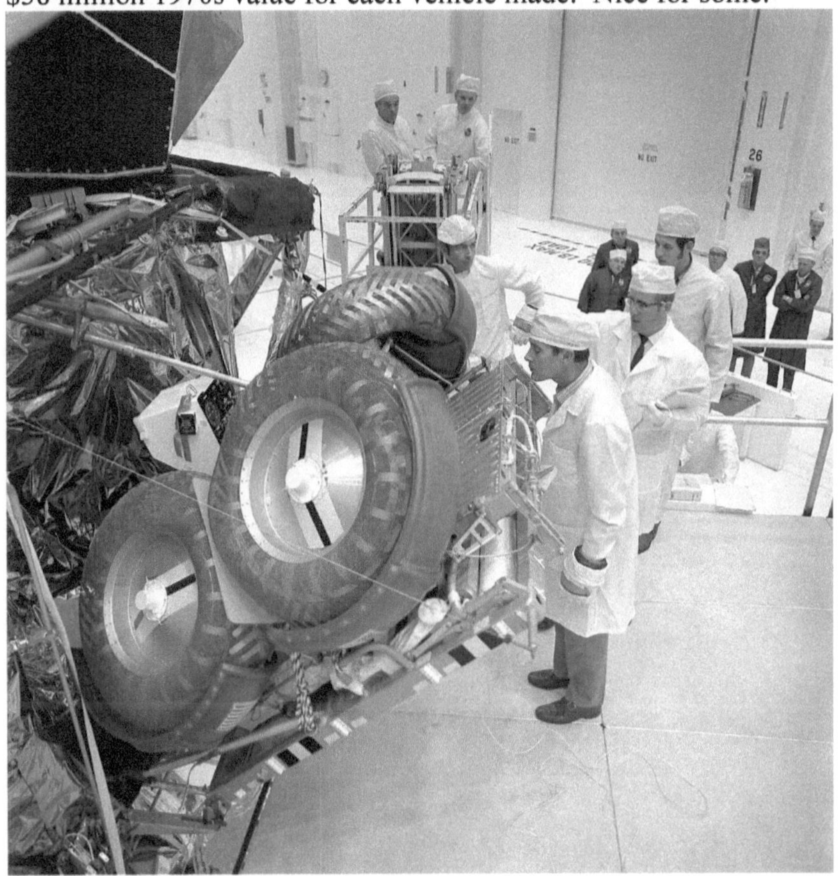

So although this reveals *where* the vehicle was stored, attached to the side of the "cubby house" of the landing module, it fails to reveal exactly *how* the astronauts managed to release and assemble such a complex vehicle clad in those restrictive space suits and gloves. I would love to see that, and a unique opportunity was forever lost to the world of what would have been a billion dollar blockbuster of a movie, had a simple video been made of the astronauts releasing and assembling the vehicle of the moon, then driving off. I understand that Apollo 17 crew took over 5000 photos while on the moon, yet not a single one of such an incredible undertaking as this would have proven. That is if it was in fact done. It leaves a nasty tasting question in the mouth.

401-THEY'RE COMING TO TAKE ME AWAY

"Why is this thus? What is the reason for this thusness?" (Artemus Ward.. Charles Farrar Brown)

Stanley Kubrick's film "2001 A Space Odessy" is memorable for many things, among them is that opening sequence involving the pre-human "ape" creatures. It was intended for the viewers to learn or glean some information from, and there are indeed many lessons within that sequence.

One interesting view is another way of defining the "golden rule". Perhaps more important than the usual "those that got the gold makes the rules" would be "those that have a monopoly on the weapons make all the rules" and that includes deciding not only who gets the gold, but how it is shared and where it's kept.

As well illustrated in those scenes of the movie we learn that if one having weapons, even if only a bone club (still used in the world today) then one can overpower, dispossess, or slaughter who he will providing they are not equally or better armed. And pardon the pun, let's make no bones about it, it is surely a basic element in the makeup of man to want everything for himself alone.

The success to be had in attacking an unarmed group (nation or whatever) or disarming them before an attack is obviously well known since the "beginning".

The United States of America was founded because its resident population having access to arms could thus boldly declare independence from the oppressive British rule. The British not liking that idea sent in its armies and a war of independence ensued. Each side called up their allies and things got very nasty. The former colonists in America won the war, ousted the British and vowed never again to be subject to non local, non resident foreign rule. Recognizing that victory was had because the citizens could take up arms, a constitution was framed declaring a citizen's right to carry arms and the citizens were to be responsible for the government. The citizens were not responsible to the government. Huge big differences there. In any true democracy the people are responsible for the government and the government is responsible to the people. It is when a situation arises that a people fear their government, and that government knowing that, fears its people and is scared of them that scary things happen and are done. By the above definitions there are very few democracies in the world today.

In this world today we find most nations comprised of totally unarmed civilians not only deprived of arms, but also forbidden ownership under various penalties. Such people do NOT control their government but are totally controlled by them and more so by the "departments" that administer the "policy" underlying governments. Departmental bureaucrats are not elected and most often survive changes of government.

Now tying up the matters of large groups or nations of unarmed civilians and mans basic nature, let's look at just recent history to learn some more lessons. Just before we look at case histories, let us ask ourselves just why our governments are so hell bent on disarming its citizens and making ownership of arms so difficult. Globally governments virtually without exception want a monopoly on weapons, and if limited and restricted weapons are permitted insist on knowing who has them and where they are. Had those citizens in the miniseries "V" been as well armed as the American colonists those reptiles would never had achieved what they did. Ask, are we being set up? For what? Let's look at history.

THEY CAME AND THEY TOOK THEM AWAY.

In these examples not only did "they" come and take them away, but there was little if anything anyone was able to do about it. Citizens had been stripped of armaments. These are in no specific order except perhaps chronological order.

USA: We have dealt with how the USA came into being largely because its citizens were allowed to carry arms, but this right did not apply equally to everyone. By the late 1800s there were some 4 million slaves in the USA and about one quarter million free blacks. Disarmament laws were created across the land but were designed to allow only 1 race to freely possess weapons. "No slave may use firearms even in self defence" Louisiana 1806. "Free Negroes may not carry firearms." Florida 1831, Mississippi 1852, Alabama 1866, Louisiana post civil war. "White citizen patrols shall enter into all Negro homes … and lawfully seize … arms, weapons, and ammunition." Florida 1825. "It shall not be lawful for any freed man, Malatto, or free person of color … to own firearms … or other deadly weapons…." Alabama 1866. "Dealers must record the race of all buyers of pistols and ammunition." Mississippi 1906. Current gun laws still require race to be stated and recorded.

Between 1880 and 1965 (See it's not all "stuff" that happened over 100 years ago.) mobs of citizens forcibly took and lynched some 3,500 defenceless unarmed black people. Consider one case where one intended lynch victim was saved because of the presence of just one handgun. A local sheriff was holding a man in his cell and left the jail for a while leaving his 14 year old daughter "in charge". A mob gathered intent on taking the man by force and murdering him by common called lynching. The girl took up a handgun and confronted the mob, vowing to shoot anyone who stepped forward to make good the threats. Not one coward in the mob stood forward. The sheriff returned and the mob disbanded. What a lifesaving difference just one gun can make.

In 1864 those who had the only guns, US troops, gunned down 150 Indians at Sand Creek Colorado, almost all women, children and old men.

December 1890 the US troops caught up with a group of Souix Indian Ghost Dancers at Wounded Knee and forced them to surrender their weapons. Subsequently <u>every</u> Indian was gunned down, some having fled up to 3 miles away.

Couldn't or wouldn't happen now or today you think? During World War 2 the US rounded up its Japanese citizens, many born in the USA. Some 110,000 were taken away and placed in remote "camps" under armed guards. A similar thing happened to US citizens of Italian and German descent. These were men, women, and children, guilty of nothing, disarmed, detained, and placed under armed guard. There for their own safety you think? That's the apologist line.

<u>TURKEY:</u> The events in Turkey involving the Armenian population seems to be lost and forgotten about, drowned out by the horrors of World War 1. It is a tale of willfully premeditated genocide. By 1915 a new government was in power taking over from the old Ottoman Empire. They wanted to rid the country of "wrong and lawless ideas", and to protect "national security". Oh the crimes against humanity committed in the pursuit of such ill defined and often used phrases.

In this case the "wrong and lawless ideas" were held to be those of the Armenians, a Christian minority in a mostly Moslem country. It was decreed that "Since the collective society is endangered … they must all be killed, men, women, and children without discrimination." And to this end a well-established and proven method was used in the genocide. (I will always refuse to grace mass murder and genocide with the words "ethnic cleansing". It is cold-blooded methodical and systematic slaughter on a massive scale. Sadly it is done so often.)

The targeted group were separated from society, this after they had been forced to surrender all weapons after a brutal forced search of their homes. It is said some obtained weapons illegally just to be able to surrender them to the searching troops and hopefully spare their households from the associated brutality.

On 26 June 1915 an order was made to deport them and they were ordered to report 5 days later bearing only what they could carry. As all weapons had been taken the remaining civilians were forced to assemble into groups of from 200 to 4000 and sent off on forced marches under armed troops. I say "remaining civilians" because almost all able bodied men had been forced into the army. They were then completely disposed of out of sight and knowledge of the common civilian within the confines of the army. The groups were then forced to march through the desert were 90% are said to have died. They were simply cut down, shot, butchered, women, children, and the aged and infirm, all defenceless. The small percentage of survivors who finished the march was then also butchered.

Some 1,500,000 Armenian souls perished in that organized massacre of unarmed and defenceless people by their own government.

SOVIET UNION: After the Communists came into power in 1917 following the armed revolution, new and strict laws were put in place to control the ownership and use of firearms. It became almost impossible for ordinary Soviet civilians who were non-Party members to own or access firearms.

Strict licensing laws recorded who had guns and where they were. The country was now defenceless and under the control of a government out of control.

Between 1929 and 1934 the Ukraine was singled out for a purge, and grain production quotas that were impossible to meet were enforced by armed soldiers. After 10 years of forced disarmament and weapons confiscation the civilians were powerless and when food, distribution of which was also controlled by the government, was withheld they were faced with certain starvation. They could not move away from the

Ukraine as travel, yet again rigidly controlled and restricted by the government, was not an option. In effect they were imprisoned within one huge country sized concentration camp under armed guard by soldiers. Their isolation was complete hidden from the world behind an Iron Curtain, and the ensuing starvation, executions, and cannibalism that began took a toll of an estimated 10,000,000. 10 million people killed by their own government.

Worse massacres were to come, but this time at the hands of a former ally, one who would have full knowledge of the unarmed state of the civilians before they treacherously invaded. The Nazi hordes swept almost unopposed across the Soviet territory and mercilessly massacred everyone before them. Nazi intent was to totally eliminate all Russians and take and use an empty country. I believe over 20,000,000 Russian civilians died.

GERMANY: Early in 1933 Hitler was elected into power. Firearm registration laws had been put into effect some 5 years earlier and thus his regime had access to the records of who had the guns and where those people were. It always seems that this is vital information to have to begin to implement the various plans to "rule the world."

The first stage for total control was already in place and stage two was very soon implemented. A mass seizure of guns and weapons from any remotely considered as political opponents, or for those matter undesirables, was started. As a result of raids carried out and obedient civilians surrender of guns, the population was almost totally unarmed and defenceless by 1938.

In 1938 new and very specific gun laws were effected. No Jews were allowed to have ownership or access to guns. It was decreed that no Jew be in possession of weapons, guns, clubs, knives, sharp edged weapons, and those found with such were to be sent immediately to already established concentration camps. Remaining weapons were surrendered. In November of 1938 the so-called "night of broken glass" followed by massive persecution was unleashed upon the defenceless Jews. A "holocaust" followed and millions of European Jews were slaughtered in specially designed death camps and factories. One rare example of resistance arose when some in the Warsaw ghetto, a holding "pen" where people were forced to wait for their turn before transportation to the death factories, got hold of some weapons and said "no more" to the Nazis. Without population disarmament and specifically Jewish disarmament things could have been so very different and so many innocents may not have been mercilessly slaughtered.

Now whereas everywhere else in Europe the populations were able and capable of rising up with arms to resist the Nazi oppression, this was not so in Germany. The government had the monopoly of weapons and the people entirely at its mercy.

Between 1933 and 1945 some 3 million Germans considered political opponents were sent to concentration camps. How could this be? How could this happen? Gun control enabled it.

Over 11,000,000 died in concentration camps. In most cases these were deaths at the hands of their own government.

Over 21,000,000 civilians died in Europe at Nazi hands, a government out of control. Things could have been different.

CHINA: In 1935 a Nationalist government was in power and made the possession of guns "without a good cause" or for military use punishable with imprisonment. The population was of course then disarmed.

1937 to 1939 witnessed the Sino Japanese war and the civilian population was forced into the army. It is estimated 4,000,000 died at both the hands of their own army and that of the invading Japanese. Those refusing to fight for whatever reason were summarily killed. In 1937 the city of Nanking was defended by 300,000 Chinese troops, but they threw down their weapons in the face of an invading army of 225,000 Japanese troops. The unarmed civilians of the city now lay helplessly before the invaders. Hundreds of thousands of unarmed civilians died, having no defence as weapons were forbidden. Their own government and its army were solely to blame.

Between 1942 and 1944 an estimated 4,000,000 died in famine and starvation caused by government troops confiscating crops. An unarmed population is powerless to resist and survive.

In 1949 it was the turn of the Communists to take over. Anyone who supplied their enemy, (opponents) either domestic or foreign, arms or ammunition faced immediate death. Ownership of armaments was strictly forbidden. The communists wanted a smooth ride into power unhindered by an armed citizenry such as hindered the British in its war with American colonists.

Somewhere between 35 and 100 million Chinese perished between 1935 and 1970. It could have been different. (Chairman) Mao stated "political power grows out of the barrel of a gun.

UGANDA: In 1971 Idi Amin led a successful armed revolution against the Obote government and seized power. World leaders thought that was good, as Obote was left wing and "bad". Idi Amin re-defined "bad". In such a country gun control laws were already in place of course. Bad governments take care of that first step as just that, the "first step". Obote's existing laws prohibited the carrying, ownership, or selling of firearms to civilians. Government troops exclusively thank you.

Immediately on gaining control Amin began doing things his way and for him. He began with the slaughter of all or any troops whose loyalty could be doubted. 16,000 were murdered for that cause. He then decreed that all Asians be ejected from Uganda. That causing no immediate problems for him he then selected all English to join them out of Uganda. As the population was unarmed and totally at his mercy, he began his persecution of Christians, and rival tribe's people.

It was illegal for more than 3 people to gather if one had a weapon. The persecutions and killings began.

Somewhere in all this time, a group of non-Ugandans were detained and held at an airport in Entebbe. Their fate looked very bleak indeed, but help came at the hands of a raiding party of Israeli commandos flown in just for the occasion. They were all rescued and flown to safety. Cowards back down or take no action in the face of capable opposition, even if it is a 14 year old girl with just one gun.

After 8 years of slaughter with the worlds governments and powers just watching it happen "over there" in "that place got no oil and don't matter" Uganda, over 300,000 had been murdered by their own government and troops. Finally Amin himself outwore his welcome and had to flee. He went to Saudi Arabia with untold wealth, and lived in luxury, unhindered by thoughts of justice, being held accountable or any conscience.

CAMBODIA: Definitely not the place to be between 1975 and 1979, or even now possibly. This had been a French colony and in 1975 after 5 years of war, communists took control with its Khmer Rouge under Pol Pot. He was a man with a dream. His dream was everyone else's nightmare for it aimed at definite population reduction from 7 million people to around just 1 million people.

He would eliminate all religion and its leaders, all political opponents, all city dwellers, all non-Cambodian ethnic groups, all Western culture, all traces and forms of capitalism, all students and intellectuals (That's anyone with better than a 7th grade level.), professionals of all types, all people who spoke French or English, even people who wore eye glasses. That has to be a bad start.

He imposed worse than tight gun and weapon control from the start of course, and any permits allowed stated that even the lending of a weapon within a family was forbidden. No guns were to be owned, no self-defence allowed, and everyone forced to disarm, except the Khmir rouge of course. Once the disarmament was in place the troubles began.

Total and mass evacuations of cities was ordered and enforced with the entire population forced onto collective farms. Like the Ukraine decades before the entire country became one huge concentration camp. Farming was inadequate and unable to support the entire population of the country. Result, same as Ukraine and China before this, famine and starvation. The administration was brutal. There was no mail, no telephones, no media, no books, no medical care of any sort, family units broken up and disallowed. Common kindness was not only discouraged, but also punishable by immediate death if witnessed. The manner of death was unspeakably and unimaginably brutal and cruel. Special death camps were set up within the greater concentration camp that the country had become.

This out of control insanity endured for 4 years unhindered by the western world or its governments. Over 2,000,000 died in the land at the hands of its own government before a neighboring country took the initiative to end it all. Vietnam invaded "its over there and not important, got no oil" Cambodia and crushed the Khmir within just 1 month. There is an important lesson to be learned here if you can just figure it out.

BANGLADESH (PAKISTAN): In 1971 after President Yahya Khan stated of the "troublemakers", namely Hindus, students, and intellectuals, "kill 3 million of them and the rest will eat out of our hands" 1,500,000 civilians died. Those deaths must be attributed to their own government.

GUATEMALA: Target group: the country's own Mayan Indians. 1981 firearm laws and restrictions provide "only government officials may carry firearms." Now that is very definitely disarming the civilians. Further laws stated that "firearms, even sharp pointed tools or farming implements are forbidden outside of town."

The population was forbidden arms, disarmed by its own government and in the 1980s some 200,000 Mayans died at the hands of their own government.

RWANDA: Herein lies another harrowing tale of government incited mass murder of its own civilians. This incident is more terrible than most if that is possible in as much as government openly broadcast the intent and both government agencies and the public did the mass killings openly together.

In this former Belgian colony the departing government gave all control to the minority Tutsi tribe. 9% of the population were Tutsi, and some 90% were of the Hutu tribe and by 1994 the Hutu had control of power. In that year the Hutu government declared all Tutsi to be rebels and called for the extermination of all Tutsi. In April 1994 it began, and civilians were urged to murder all Tutsi people with army troops participating.

The Ministry of Defence forbids the Tutsi tribe's people to own or carry weapons, and that same department organized the campaign of extermination and total mass murder. In one incident a large group of 5,500 people held out against murderous mobs for a week, holed up in a Christian church complex. It is reported that their defensive weapons included only one gun that had been seized from a soldier. This would indicate that had the Tutsi not been so completely disarmed, such massacres as ensued might not have happened. After one-week grenade using soldiers broke the siege and the entire group were slaughtered without mercy even to babies.

In one town its mayor urged civilians to fully disarm and surrender all weapons. When they failed to do this for obvious reasons he then ordered police to shoot them all. 20,000 Tutsi residents were mercilessly slaughtered in an horrific three-day killing frenzy. Roads were blocked and sealed off, ID cards were in use and checked, and Tutsi people were singled out for murder on the spot. I don't thing the average reader would be able to imagine the horror and terror of such events.

800,000 men women and children were singled out and brutally butchered over a period of just 100 days by fellow civilians and their own government and its agencies.

THE LESSONS TO LEARN

There are many lessons to be learned from the above incidents; some are obvious and some not so obvious. The first and most basic lesson is that no one is really safe from their own government, agencies, or troops when things go "bad". Who would think that merely wearing spectacles, or being a health professional would automatically guarantee your death? Who would imagine that your own country's soldiers of a so-called Ministry of Defence would be ordered to kill their own civilians, yet alone actually carry out that order? Most civilians have no idea that military personnel are carefully trained not to think but to become "automated" killing machines. I was somewhat stunned after I joined one military force that on the first day of training our course NCO yelled at us that he didn't give an "F" why we thought we had joined the service, but that we were now government property, his personal property and that he would teach us how to kill. Rule one therefore, no one is out of possible harm's way when things go bad.

The next rule is probably that warning signs are there for all to see as evidence of a government with the ability to go bad. These signs to not guarantee things will turn bad, but the potential is there and "machinery" is being put into place.

In all of the above cases the civilians are either totally disarmed or forbidden ownership or the right to carry weapons. Additionally any armaments that may be allowed must be registered so that government and its agencies know who has the weapons and where they are. Further and in most cases they will also know the ethnic groups and the numbers of each group in possession of armaments. The possible danger to the people is escalated when those people become required to physically surrender up those weapons to government. There will generally be prosecutions and penalties for breaching firearm laws. And red lights are flashing, sirens screaming, when specific ethnic groups are singled out and discriminated against for ownership of armaments. That would indicate they are marked for some form of special and probably not good attention from government. Never forget that a public without armaments is a vulnerable population, unable to defend itself, and that government or defence forces cannot be totally relied upon for their safety. Nanking found itself deserted by army, and its population was massacred. Ukraine, Tutsi and many others found to their horror that their own army became the instrument of their death.

Then there is the entire issue of citizens being required to use and carry ID cards. I remember decades ago associating the requirement for populations to carry "papers", identity papers that revealed all the information governments demand that enable effective population control including movement, with a total loss of freedom experienced in Nazi occupied Europe or communist Russia. It was simply just not a democratic way of life. We all now compliantly carry ID regardless that our various governments vow that the cards we carry are not intended for or not really ID cards. They are. Effectively our governments know who we all are, where we are, our ages, state of health, our ethnic groups, most likely our religious persuasions, our financial status, and our work classification and certainly if we possess armaments. That's a lot of information that in "bad" government or other hands could spell TROUBLE. But the governments or powers that be know assuredly that as long as the vast majority keep drinking the local tap water, keep watching the TV and media and endlessly consuming products while eating modern foods that we are as controllable as any herd of cattle.

WHAT IT MEANS

When people accept being disarmed they become surprisingly easy to control and to kill. They have surrendered not only any weapons they may have had, but they also surrender their freedom and independence. They become totally dependent on their government and its military. History shows this is not really a safe or good practice. When innocent populations become so defenseless others can slaughter them in mass with the most ordinary of weapons, guns, knives, pitchforks, clubs, or even simple box cutters.

I marveled at the report that a mere handful of alleged highjackers took planes and caused so much havoc and death in the USA, and that there was not a single armed person who could have speedily resolved the issue and brought them down. Likewise I marvel at more recent reports of some rogue going on a shooting and killing spree, generally in schools, in the USA. A single gunman is wandering around and killing at random, often over a fairly lengthy period of time. Gun control alone is responsible for all those sort of deaths as there was not a single person with a gun to resolve the issue by simply bringing the killer down.

Vital lesson: gun control and disarmament of the civilians is not about eliminating crime, or getting guns out of the hands of criminals or the simply crazy. The criminals and the insane will always have and get guns. Recently I saw a documentary about police conducting raids on marihuana growers in New Zealand. Mostly these were simple residential affairs and not large-scale producers in a mega-buck illegal operation, yet in a large number of cases firearms were found, and those invariably unregistered and undeclared. The criminally minded have the guns and will not give them up. It's the trustworthy and honest citizens who lose the firearms and the ability to rightfully defend themselves and their family, or for that matter the community if necessary. I ask what is the point of allowing a situation where a crazed gunman is randomly killing people and a cowering population must either wait their turn or until police with a gun can be located and then arrive on the scene. You think this is emotive claptrap? Tell that to the survivors of any of the above historical and other similar situations.

The way to preserve freedom is to never give any person or persons more power than the people have. If that situation is not adhered to then there is a loss of freedom. It is said that the main reason for the USA forces to be overseas is that other governments have gone "bad" and started killing their people. Those people fear their government and cannot control it, they have no weapons nor access to them. Intervention is needed.

People should not have to live in fear of their government, and there is danger in downplaying the importance of self-defence, or in giving over of all defence, even at a local community level, to a central governmental authority.

None of the above situations or circumstances are now ever likely to change, so don't think that they will. The purpose of this writing is to open your eyes to what is really going on out there in our world, and some of the events are absolute horrifying but could have been prevented.

This is how humans play the game of power and how dictatorships come into power. As a final word on this matter let's ask one question.

If we humans treat our own species and race with such ruthless callousness, how and what could we expect from those not of our species, the "not-us" that probably exist?

402- SOMETHING IN THE WATER

"The crash of the whole solar and stellar systems could only kill you once." -Thomas Carlyle

The following article (in italics) is copied from off the Internet and is freely available. I have reproduced it in full as I doubt that many readers would bother or have the ability or time to search out information for themselves. It of vital concern and relative to the issues being presented as it is indicative of something far more nefarious than mere lunacy. "Something" is afoot and a vast amount of resources, time and energy is being expended to implement just this one universal component action that seems designed solely to debilitate, dumb down, and enslave the entire human species. This one component when added to other similar seemingly unrelated components spell out in very large letters that all the suspicions behind numerous so called conspiracy theories (really conspiracy facts) are almost certainly well founded. The entire human race and species is being wilfully targeted and marked for destruction, or decimation at the least.

Yes, it is time to become very afraid, and to be very alert as to what is really going on, then try to make some preparations as one may see as adequate. Those preparations may at this late stage of the "game" to merely be able to face death with dignity.

Read on, the article is not modified or corrected in any way, except for the making of some text **bold as a highlight.** Almost needless to say this article is not unique in its revealing the effects of sodium fluoride, but it was selected because of its brief and comprehensive nature. The truth really is out there.

FLUORIDE and STUPIDITY

SICKNESS CONTROL 101: FLUORIDE, THE LUNATIC DRUG

"TELL A LIE LOUD ENOUGH AND LONG ENOUGH AND PEOPLE WILL BELIEVE IT." (Adolph Hitler)
"EARTH IS AN INSANE ASYLUM, TO WHICH THE OTHER PLANETS DEPORT THEIR LUNATICS." --Voltaire (Memnon the Philosopher).

Controversial fluoride is one of the basic ingredients in both PROZAC (FLUoxetene Hydrochloride) and Sarin nerve gas (Isopropyl-Methyl-Phosphoryl FLUoride).

Sodium fluoride, a hazardous-waste by-product from the manufacture of aluminum, is a common ingredient in rat and cockroach poisons, anesthetics, hypnotics, psychiatric drugs, and military nerve gas. It's historically been quite expensive to properly dispose of, until some aluminum industries with an overabundance of the stuff sold the public on the terrifically insane but highly profitable idea of buying it at a 20,000% markup, injecting it into our water supplies, and then DRINKING it. Yes, a 20,000% markup: Fluoride-- intended only for human consumption by people under 14 years of age--is injected into our drinking water supply at approx. 1 part-per-million (ppm), but since we only drink 1/2 of one percent of the total water supply, the rest literally goes down the drain as a free hazardous-waste disposal for the chemical industry, where we PAY them so that we can flush their expensive hazardous waste down our toilets. How many salesmen dream of such a deal? (Follow the money.)

Independent scientific evidence repeatedly showing up over the past 50 years reveals that fluoride allegedly shortens our life span, promotes cancer and various mental disturbances, accelerates osteoporosis and broken hips in old folks, and ***makes us stupid, docile, and subservient****, all in one package. There are reports of aluminum in the brain possibly being a causative factor in Alzheimer's Disease, and evidence points towards fluoride's strong affinity for aluminum and also its ability to "trick" the blood-brain barrier by looking like the hydrogen ion, and thus allowing chemical access to brain tissue.*

Scientists who have attempted to blow the whistle on this mega-bucks PR ploy have consistently been given a very unscientific Black-PR treatment, and thus their valid points disputing the current vested interests ***never arrive in the press****. Follow the money to find the control. In 1952 the slick PR campaign which ramrodded the concept of fluoridation through via our Public Health departments and various dental organizations was likened to a highly-emotional "beer-salesman's convention" instead of the objective scientific experiment which it should properly have been. It's continued in that vein right up to present time. To illustrate the emotional vs. the scientific nature of this issue, just look at the response given by people (perhaps yourself included?) when the subject of fluoridation comes up.* ***Ask yourself, "Is this response EMOTIONAL BLUSTER, or is it UNBIASED AND OPENLY- INTERESTED OBJECTIVITY?"***

There is a tremendous amount of emotional, highly unscientific know-it-all attached to fluoridation. Many truly independent (unattached to any vested-interest) scientists who've spent a large portion of their lives studying and working with this subject have been subjected to a surprising amount of uncalled-for and unfair character assassination from strong vested- interest groups who profit from the public's ignorance as well as from their illnesses. (Follow the money.)

Do you have diabetes or kidney disease? There are reportedly more than 11 million Americans with diabetes. Since many diabetics drink more liquids than other people, then according to the Physicians Desk Reference these 11 million Americans probably shouldn't drink fluoridated water, because in doing so, they'll receive an excessive dose of fluoride.

Kidney disease, by definition, lowers the efficiency of the kidneys, which is your main route of fluoride elimination. --So those people with kidney disease also shouldn't drink fluoridated water. Cases are on record (Annapolis, Maryland, 1979) where kidney patients on dialysis machines died, due to a fluoride overdose in the city water supply. *Let's begin at the beginning:*

The first occurrence of fluoridated drinking water on Earth was found in Germany's ***Nazi prison camps****. The Gestapo had little concern about fluoride's supposed effect on children's teeth; their alleged reason for mass-medicating water with sodium fluoride was to sterilize humans and force the people in their concentration camps into* ***calm submission****.* ***(Ref. book: "The Crime and Punishment of I.G. Farben" by Joseph Borkin.)***

The following letter was received by the Lee Foundation for Nutritional Research, Milwaukee Wisconsin, on 2 October 1954, from Mr. Charles Perkins, a chemist:

"I have your letter of September 29 asking for further documentation regarding a statement made in my book, The Truth About Water Fluoridation, to the effect that the idea of water fluoridation was brought to England from Russia by the Russian Communist Kreminoff. "In the 1930's, Hitler and the German Nazi's envisioned a world to be dominated and controlled by a Nazi philosophy of pan-Germanism. The German chemists worked out a very ingenious and far-reaching plan of mass-control which was submitted to and adopted by the German General Staff. This plan was to control the population in any given area through mass medication of drinking water supplies. By this method they could

control the population in whole areas, reduce population by water medication that would produce sterility in women, and so on. In this scheme of mass-control, sodium fluoride occupied a prominent place. ...

"Repeated doses of infinitesimal amounts of fluoride will in time ***reduce an individual's power to resist domination****, by slowly poisoning and narcotizing a certain area of the brain, thus making him* ***submissive*** *to the will of those who wish to govern him. [A convenient light lobotomy]*

"The real reason behind water fluoridation is not to benefit children's teeth. If this were the real reason there are many ways in which it could be done that are much easier, cheaper, and far more effective. The real purpose behind water fluoridation is to ***reduce the resistance of the masses*** *to domination and control and* ***loss of liberty****.*

"When the Nazis under Hitler decided to go into Poland, both the German General Staff and the Russian General Staff exchanged scientific and military ideas, plans, and personnel, and the scheme of mass control through water medication was seized upon by the Russian Communists because it fitted ideally into their plan to communize the world. ...

"I was told of this entire scheme by a German chemist who was an official of the great IG Farben chemical industries and was also prominent in the Nazi movement at the time. I say this with all the earnestness and sincerity of a scientist who has spent nearly 20 years' research into the chemistry, biochemistry, physiology and pathology of fluorine--any person who drinks artificially fluorinated water for a period of one year or more will never again be the same person mentally or physically." CHARLES E. PERKINS, Chemist, 2 October 1954.

Quoting Einstein's nephew, Dr. E.H. Bronner (a chemist who had also been a prisoner of war during WWII) in a letter printed in The Catholic Mirror, Springfield, MA, January 1952:

"It appears that the citizens of Massachusetts are among the 'next' on the agenda of the water poisoners.

"There is a sinister network of subversive agents, Godless 'intellectual' parasites, working in our country today whose ramifications grow more extensive, more successful and more alarming each new year and whose true objective is to demoralize, paralyze and destroy our great Republic--from within if they can, according to their plan--

for their own possession. *"The tragic success they have already attained in their long siege to destroy the moral fiber of American life is now one of their most potent footholds towards their own ultimate* ***victory over us****.*

"Fluoridation of our community water systems can well become their most subtle weapon for our sure physical and mental deterioration. ...

"As a research chemist of established standing, I built within the past 22 years, 3 American chemical plants and licensed 6 of my 53 patents. Based on my years of practical experience in the health-food and chemical field, let me warn: fluoridation of drinking water is criminal insanity, sure national suicide. Don't do it.

"Even in small quantities, sodium fluoride is a deadly poison to which ***no effective antidote has been found****. Every exterminator knows that it is the most efficient rat-killer. ... Sodium fluoride is entirely different from organic calcium-fluoro-phosphate needed by our bodies and provided by nature, in God's great providence and love, to build and strengthen our bones and our teeth. This organic calcium-fluoro-phosphate, derived from proper foods, is an edible organic salt, insoluble in water and assimilable by the human body, whereas the non-organic sodium fluoride used in fluoridating water is instant poison to the body and fully water soluble. The body refuses to assimilate it. "Careful, bonafide laboratory experimentation by conscientious, patriotic research chemists, and actual medical experience, have both revealed that instead of preserving or promoting 'dental health,' fluoridated drinking water destroys teeth, before adulthood and after, by the destructive mottling and other pathological conditions it actually causes in them, and also creates many other very grave pathological conditions in the internal organisms of bodies consuming it. How can it be called a "health" plan? What's behind it?*

"That any so-called "doctors" would persuade a civilized nation to add voluntarily a deadly poison to its drinking water systems is unbelievable. It is the height of criminal insanity. "No wonder Hitler and Stalin fully believed and agreed from 1939 to 1941 that, quoting from both Lenin's Last Will and Hitler's Mein Kampf:

"America we shall demoralize, divide, and destroy from within."

"Are our Civil Defense organizations and agencies awake to the perils of water poisoning by fluoridation? Its use has been recorded in other

countries. Sodium fluoride water solutions are the cheapest and most effective rat killers known to chemists: colorless, odorless, tasteless; no antidote, no remedy, no hope: Instant and complete extermination of rats. ...

"Fluoridation of water systems can be slow national suicide, or quick national liquidation. It is criminal insanity--treason!" Dr. E.H. Bronner, Mfg. Research Chemist, Los Angeles.

Earliest available Russian fluoride evidence:

"I, Oliver Kenneth Goff, was a member of the Communist Party and the Young Communist League, from May 2, 1936, to October 9, 1939. During this period of time, I operated under the alias of John Keats with number 18-B-2. My testimony before the Government is in Volume 9 of the Un-American Activities Report for 1939.

"While a member of the Communist Party, I attended Communist training schools in New York and Wisconsin ... and we were trained in the revolutionary overthrow of the U.S. Government. "... We discussed quite thoroughly the fluoridation of water supplies and how we were using it in Russia as a ***tranquilizer in the prison camps****. The leaders of our school felt that if it could be induced into the American water supply, it* ***would bring about a spirit of lethargy in the nation, where it could keep the general public docile*** *during a steady encroachment of Communism. We also discussed the fact that keeping a store of deadly fluoride near the water reservoir would be advantageous during the time of the revolution, as it would give us opportunity to dump this poison into the water supply and either kill off the populace or threaten them with liquidation, so that they would surrender to obtain fresh water.*

Related Research:
The Crime and Punishment of I. G. Farben
by Joseph Borkin (out of print book search)

Health Effects of Ingested Fluoride
by Bernard Meyer Wagner

Fluoridation : the Great Dilemma
by George L. Waldbott

Fluoride the Aging Factor :
How to Recognize and Avoid the Devastating Effects of Fluoride

by John, Dr. Yiamouyiannis

Scientific Knowledge in Controversy :
The Social Dynamics of the Fluoridation Debate
by Brian Martin

Medical Mafia by Guylaine Lanctot

Racketeering in Medicine by James P. Carter

The Cure for All Diseases by Dr. Hulda Regehr Clark

Censured for Curing Cancer :
The American Experience of Dr. Max Gerson
by S.J. Haught

In a letter abstracted from Fluoridation and Lawlessness, published by the Committee for Mental Health and National Security (with obvious implications) from the aforementioned Charles Perkins, U.S. appointed post-war head of I.G. Farben, to the Lee Foundation for Nutritional Research, Milwaukee, Wisconsin, October 2, 1954, we read the following:

"We are told by the fanatical ideologists who are advocating the fluoridation of the water supplies in this country that their purpose is to reduce the incidence of tooth decay in children, and it is the plausibility of this excuse, plus the gullibility of the public and the cupidity of public officials that is responsible for the present spread of artificial water fluoridation in this country. However - and I want to make this very definite and positive - the real reason behind water fluoridation is not to benefit children's teeth. If this were the real reason, there are many ways in which it could be done which are much easier, cheaper and far more effective. The real purpose behind water fluoridation is to reduce the resistance of the masses to domination, control and loss of liberty."

■■

1944. When a severe pollution incident occurred downwind of the E.I. DuPont de Nemours Company chemical factory in Deepwater, New Jersey. The factory was then producing millions of pounds of fluoride for the Manhattan Project whose scientists were racing to produce the world's first atomic bomb. The farms downwind in Gloucester and Salem counties were famous for their high-quality produce. Their peaches went directly to the Waldorf Astoria Hotel in New York City; their tomatoes were bought up by Campbell's Soup. But in the summer of 1944 the farmers began reporting that their crops were blighted: "Something is burning up the peach crops around here." They said that poultry died after an all-night thunderstorm, and that farm workers who ate produce they'd picked would sometimes vomit all night and into the next day. "I remember our horses looked sick and were too stiff to work," Mildred Giordano, a teenager at the time, told these reporters. Some cows were so crippled that they could not stand up; they could only graze by crawling on their bellies. The account was confirmed in taped interviews with Philip Sadtler (shortly before he died), of Sadtler Laboratories of Philadelphia, one of the nation's oldest chemical consulting firms. Sadtler had personally conducted the initial investigation of the damage. The farmers were stonewalled in their search for information about fluoride's effects on their health, and **their complaints have long since been forgotten**. But they unknowingly left their imprint on history: their complaints of injury to their health reverberated through the corridors of power in Washington and triggered intensive, secret, bomb program research on the health effects of fluoride.

1945. May. Newburgh's water was fluoridated, and over the next 10 years its residents were studied by the New York State Health Department.

1945-1955. Much of the original proof that fluoride is safe for humans in low doses was generated by A-bomb program scientists who had been secretly ordered to provide "evidence useful in litigation" against defense contractors for fluoride injury to citizens. The first lawsuits against the American A-bomb program were not over radiation, but over fluoride damage, the documents show. Human studies were required. Bomb program researchers played a leading role in the design and implementation of the most extensive US study of the health effects of fluoridating public drinking water, conducted in Newburgh, New York, from 1945 to 1955. Then, in a classified operation code-named "Program F", they secretly gathered and analyzed blood and tissue samples from Newburgh citizens with the cooperation of New York State Health Department personnel. The original, secret version (obtained by these reporters) of a study published by Program F scientists in the August 1948 Journal of the American Dental Association1 shows that evidence of adverse health effects from fluoride was censored by the US Atomic Energy Commission (AEC)-considered the most powerful of Cold War agencies-for reasons of "national security". The bomb program's fluoride safety studies were conducted at the University of Rochester-site of one of the most notorious human radiation experiments of the Cold War, in which unsuspecting hospital patients were injected with toxic doses of radioactive plutonium. The fluoride studies were conducted with the same ethical mindset, in which "national security" was paramount.

1995. Dr Phyllis Mullenix, former head of toxicology at Forsyth Dental Center in Boston and now a critic of fluoridation. Animal studies which Mullenix and co-workers conducted at Forsyth in the early 1990s indicated that fluoride was a powerful central nervous system (CNS) toxin and might adversely affect human brain functioning even at low doses. (New epidemiological evidence from China adds support, **showing a correlation between low-dose fluoride exposure and diminished IQ in children.**) Mullenix's results were published in 1995 in a reputable peer-reviewed scientific journal.

1995. The University of Rochester's classified fluoride studies, code-named "Program F", were started during the war and continued up until the early 1950s. They were conducted at its Atomic Energy Project (AEP), a top-secret facility funded by the AEC and housed at Strong Memorial Hospital. It was there that one of the most notorious human radiation experiments of the Cold War took place, in which unsuspecting hospital patients were injected with toxic doses of radioactive plutonium. Revelation of this experiment-in a Pulitzer Prize winning account by Eileen Welsome-led to a 1995 US presidential investigation and a multimillion-dollar cash settlement for victims.

There can no longer be any doubt that fluoride is an extremely dangerous toxic chemical that has almost immediate and long-term deleterious effects on humans. It can no longer be considered that it is ignorance alone that allows or causes the continuance of its use as an additive to water supplies, toothpastes etc. and enforced upon human populations. It also stinks of something far more nefarious than mere greed and the associated corruption. Nor can it be considered "cost effective". I read that somewhere near only half of one percent of such reticulated town water supplies is drunk by the populations, the rest going down the toilets, drains, car washes, etc. I wonder if our pets and gold fish enjoy sharing our madness?

403-This War on Terrorism is Bogus

"Why of course the people don't want war. Why should some poor slob on a farm want to risk his life in a war when the best he can get out of it is to come back to his farm in one piece? Naturally the common people don't want war: neither in Russia, nor in England, nor for that matter in Germany. That is understood. But, after all, it is the leaders of the country who determine the policy, and it is always a simple matter to drag the people along, whether it is a democracy, a fascist dictatorship, a parliament, or a communist dictatorship. Voice or no voice, the people can always be brought to the bidding of the leaders. That is easy. All you have to do is tell them they are being attacked and denounce the peacemakers for lack of patriotism and exposing the country to danger. It works the same in any country." -- Hermann Goering

New -- 11 September 2003

The following is a British Member of Parliament's take on U. S. President George Bush and his administration's attempt to manipulate the events of 9-11-2001 in an effort to create justification for his incursions into Afghanistan, and Iraq (and other dominoes to come). It is presented as a pretext for waging his Oil Wars and/or Bush Wars.

For the sake of credibility, it should be noted that the author, Michael Meacher MP, was British Environment Minister from May 1997 to June 2003, Much of the information discussed herein is completely available in the public domain, but Mr. Meacher does an excellent job of putting every-thing into context, and in a relatively confined space.

(7/4/05) For a more American perspective, Senator Ernest F. Hollings has his own views about "**Why We're in Iraq**." It's worth reading as well.

This War on Terrorism is Bogus

The 9/11 attacks gave the US an ideal pretext to use force to secure its global domination

Michael Meacher
Saturday September 6, 2003
http://www.guardian.co.uk *The Guardian*

Massive attention has now been given - and rightly so - to the reasons why Britain went to war against Iraq. But far too little attention has focused on why the US went to war, and that throws light on British motives too. The conventional explanation is that after the Twin Towers were hit, retaliation against al-Qaida bases in Afghanistan was a natural first step in launching a global war against terrorism. Because Saddam Hussein was alleged by the US and UK governments to retain weapons of mass destruction, the war could be extended to Iraq as well. However this theory does not fit all the facts. The truth may be a great deal murkier.

We now know that a blueprint for the creation of a global Pax Americana was drawn up for Dick Cheney (now vice-president), Donald Rumsfeld (defence secretary), Paul Wolfowitz (Rumsfeld's deputy), Jeb Bush (George Bush's younger brother) and Lewis Libby (Cheney's chief of staff). The document, entitled Rebuilding America's Defences, was written in September 2000 by the neoconservative think tank, Project for the New American Century (PNAC).

The plan shows Bush's cabinet intended to take military control of the Gulf region whether or not Saddam Hussein was in power. It says "while the unresolved conflict with Iraq provides the immediate justification, the need for a substantial American force presence in the Gulf transcends the issue of the regime of Saddam Hussein."

The PNAC blueprint supports an earlier document attributed to Wolfowitz and Libby which said the US must "discourage advanced industrial nations from challenging our leadership or even aspiring to a larger regional or global role". It refers to key allies such as the UK as "the most effective and efficient means of exercising American global leadership". It describes peace-keeping missions as "demanding American political leadership rather than that of the UN". It says "even should Saddam pass from the scene", US bases in Saudi Arabia and Kuwait will remain permanently... as "Iran may well prove as large a threat to US interests as Iraq has". **It spotlights China for "regime change", saying "it is time to increase the presence of American forces in SE Asia".**

The document also calls for the creation of "US space forces" to dominate space, and the total control of cyberspace to prevent "enemies" using the internet against the US. It also hints that the US may consider developing biological weapons "that can target specific genotypes [and] may transform biological warfare from the realm of terror to a politically useful tool".

Finally - written a year before 9/11/2001 - it pinpoints North Korea, Syria and Iran as dangerous regimes, and says their existence justifies the creation of a "worldwide command and control system". **This is a blueprint for US world domination.** But before it is dismissed as an agenda for rightwing fantasists, it is clear it provides a much better explanation of what actually happened before, during and after 9/11 than the global war on terrorism thesis. This can be seen in several ways.

First, it is clear the US authorities did little or nothing to pre-empt the events of 9/11. It is known that at least 11 countries provided advance warning to the US of the 9/11 attacks. Two senior Mossad experts were sent to Washington in August 2001 to alert the CIA and FBI to a cell of 200 terrorists said to be preparing a big operation (*Daily Telegraph*, September 16 2001). The list they provided included the names of four of the 9/11 hijackers, none of whom was arrested. (*This could of course be a totally bogus or fictitious report, and inculcated to actually implicate "non-existent" terrorists, but cover an "inside job". – Bob.)*

It had been known as early as 1996 there were plans to hit Washington targets with aeroplanes. Then in 1999 a US national intelligence council report noted that "al-Qaida suicide bombers could crash-land an aircraft packed with high explosives into the Pentagon, the headquarters of the CIA, or the White House".

Fifteen of the 9/11 hijackers obtained their visas in Saudi Arabia. Michael Springman, the former head of the American visa bureau in Jeddah, has stated that since 1987 the CIA had been illicitly issuing visas to unqualified applicants from the Middle East and bringing them to the US for training in terrorism for the Afghan war in collaboration with Bin Laden (BBC, November 6 2001). It seems this operation continued after the Afghan war for other purposes. It is also reported that five of the hijackers received training at secure US military installations in the 1990s (*Newsweek*, September 15 2001).

Instructive leads prior to 9/11/01 were not followed up. French Moroccan flight student Zacarias Moussaoui (now thought to be the 20th hijacker) was arrested in August 2001 after an instructor reported he showed a suspicious interest in learning how to steer large airliners. When US agents learned from French intelligence he had radical Islamist ties, they sought a warrant to search his computer, which contained clues to the September 11 mission (*Times*, November 3 2001). But they were turned down by the FBI. One agent wrote, a month before 9/11, that Moussaoui might be planning to crash into the Twin Towers (*Newsweek*, May 20 2002).

All of this makes it all the more astonishing - on the war on terrorism perspective - that there was such slow reaction on September 11 itself. The first hijacking was suspected at not later than 8.20am, and the last hijacked aircraft crashed in Pennsylvania at 10.06am. Not a single fighter plane was scrambled to investigate from the US Andrews Air Force base, just 10 miles from Washington DC, until after the third plane had hit the Pentagon at 9.38 am. Why not? There were standard FAA intercept procedures for hijacked aircraft before 9/11. Between September 2000 and June 2001 the US military launched fighter aircraft on 67 occasions to chase suspicious aircraft (*Associated Press*, August 13 2002). It is a US legal requirement that once an aircraft has moved significantly off its flight plan, fighter planes are sent up to investigate. *(All of these allegations and data indicate there most certainly was something really "wrong" with the entire incident and the fact that there are contradictory statements and claims, warnings etc. make the whole thing stink worse than dead fish on a summer beach. There is calculated confusion that is either the greatest muck up in history, or intentional. It must convince anyone that nothing that is claimed can be believed, other than that mayhem and destruction did happen, just like at Pearl Harbour. But who to blame and for what motive is our mission to discover. –Bob.)*

Was this inaction simply the result of key people disregarding, or being ignorant of, the evidence? Or could US air security operations have been deliberately stood down on September 11? If so, why, and on whose authority? The former US federal crimes prosecutor, John Loftus, has said: "The information provided by European intelligence services prior to 9/11 was so extensive that it is no longer possible for either the CIA or FBI to assert a defense of incompetence."

Nor is the US response after 9/11 any better. No serious attempt has ever been made to catch Bin Laden. In late September and early October 2001, leaders of Pakistan's two Islamist parties negotiated Bin Laden's extradition to Pakistan to stand trial for 9/11. However, a US official said, significantly, that "casting our objectives too narrowly" risked "a premature collapse of the international effort if by some lucky chance Mr Bin Laden was captured". *(to me it seems they really did not want him, but wanted what came and happened afterwards. –Bob.)* The US chairman of the joint chiefs of staff, General Myers, went so far as to say that **"the goal has never been to get Bin Laden"** (*Associated Press*, April 5 2002). The whistleblowing FBI agent Robert Wright told ABC News (December 19 2002) that **FBI headquarters wanted no arrests. And in November 2001 the US Air Force complained it had had al-Qaida and Taliban leaders in its sights as many as 10 times over the previous six weeks, but had been unable to attack because they did not receive permission quickly enough (*Time Magazine*, May 13 2002).** None of this assembled evidence, all of which comes from sources already in the public domain, is compatible with the idea of a real, determined war on terrorism.

The catalogue of evidence does, however, fall into place when set against the PNAC blueprint. From this it seems that the so-called "war on terrorism" is being used largely as bogus cover for achieving wider US strategic geopolitical objectives. Indeed Tony Blair himself hinted at this when he said to the Commons liaison committee: "To be truthful about it, there was no way we could have got the public consent to have suddenly launched a campaign on Afghanistan but for what happened on September 11" (*Times*, July 17 2002). Similarly **Rumsfeld was so determined to obtain a rationale for an attack on Iraq that on 10 separate occasions he asked the CIA to find evidence linking Iraq to 9/11; the CIA repeatedly came back empty-handed (*Time Magazine*, May 13 2002).**

In fact, 9/11 offered an extremely convenient pretext to put the PNAC plan into action. The evidence again is quite clear that plans for military action against Afghanistan and Iraq were in hand well before 9/11. A report prepared for the US government from the Baker Institute of Public Policy stated in April 2001 that "the US remains a prisoner of its energy dilemma. Iraq remains a destabilizing influence to... the flow of oil to international markets from the Middle East". Submitted to Vice-President Cheney's energy task group, the report recommended that because this was an unacceptable risk to the US, "military intervention" was necessary (*Sunday Herald*, October 6 2002).

Similar evidence exists in regard to Afghanistan. The BBC reported (September 18 2001) that Niaz Niak, a former Pakistan foreign secretary, was told by senior American officials at a meeting in Berlin in mid-July 2001 that "military action against Afghanistan would go ahead by the middle of October". Until July 2001 the US government saw the Taliban regime as a source of stability in Central Asia that would enable the construction of hydrocarbon pipelines from the oil and gas fields in Turkmenistan, Uzbekistan, Kazakhstan, through Afghanistan and Pakistan, to the Indian Ocean. **But, confronted with the Taliban's refusal to accept US conditions, the US representatives told them "either you accept our offer of a carpet of gold, or we bury you under a carpet of bombs" (*Inter Press Service*, November 15 2001).**

Given this background, it is not surprising that some have seen the US failure to avert the 9/11/2001 attacks as creating an invaluable pretext for attacking Afghanistan in a war that had clearly already been well planned in advance. There is a possible precedent for this. The US national archives reveal that President Roosevelt used exactly this approach in relation to Pearl Harbor on December 7 1941. Some advance warning of the attacks was received, but the information never reached the US fleet. The ensuing national outrage [of the attack, not the lack of warning] persuaded a reluctant US public to join the second world war. Similarly the PNAC blueprint of September 2000 states that the process of transforming the US into "tomorrow's dominant force" is likely to be a long one in the absence of "some catastrophic and catalyzing event - like a new Pearl Harbor". The 9/11 attacks allowed the US to press the "go" button for a strategy in accordance with the

PNAC agenda which it would otherwise have been politically impossible to implement.

The overriding motivation for this political smokescreen is that the US and the UK are beginning to run out of secure hydrocarbon energy supplies. By 2010 the Muslim world will control as much as 60% of the world's oil production and, even more importantly, 95% of remaining global oil export capacity. As demand is increasing, so supply is decreasing, continually since the 1960s.

This is leading to increasing dependence on foreign oil supplies for both the US and the UK. The US, which in 1990 produced domestically 57% of its total energy demand, is predicted to produce only 39% of its needs by 2010. A DTI minister has admitted that the UK could be facing "severe" gas shortages by 2005. The UK government has confirmed that 70% of our electricity will come from gas by 2020, and 90% of that will be imported. In that context it should be noted that Iraq has 110 trillion cubic feet of gas reserves in addition to its oil.

A report from the commission on America's national interests in July 2000 noted the most promising new source of world supplies was the Caspian region, and this would relieve US dependence on Saudi Arabia. To diversify supply routes from the Caspian, one pipeline would run westward via Azerbaijan and Georgia to the Turkish port of Ceyhan. Another would extend eastwards through Afghanistan and Pakistan and terminate near the Indian border. This would rescue Enron's beleaguered power plant at Dabhol on India's west coast, in which Enron had sunk $3bn investment and whose economic survival was dependent on access to cheap gas.

Nor has the UK been disinterested in this scramble for the remaining world supplies of hydrocarbons, and this may partly explain British participation in US military actions. Lord Browne, chief executive of BP, warned Washington not to carve up Iraq for its own oil companies in the aftermath of war (*Guardian*, October 30 2002). And when a British foreign minister met Gadaffi in his desert tent in August 2002, it was said that "the UK does not want to lose out to other European nations already jostling for advantage when it comes to potentially lucrative oil contracts" with Libya (*BBC Online*, August 10 2002).

The conclusion of all this analysis must surely be that the “global war on terrorism” has the hallmarks of a political myth propagated to pave the way for a wholly different agenda - the US goal of world hegemony, built around securing by force command over the oil supplies required to drive the whole project. Is collusion in this myth and junior participation in this project really a proper aspiration for British foreign policy? If there was ever need to justify a more objective British stance, driven by our own independent goals, this whole depressing saga surely provides all the evidence needed for a radical change of course.

Michael Meacher MP was environment minister from May 1997 to June 2003

mailto:meacherm@parliament.uk>meacherm@parliament.uk

404-EDUCATION

"I am not arguing with you – I am telling you." –J.A. McNeill Whistler

The following article presents a recent history of education in a most succinct, clear, and readable manner and some real issues and problems in dealing with it. It is of course straight off the Internet and is unchanged and unedited from how it appears there. Please note the source and authorship at the foot of the article. That site and source is 100% recommended for everyone, and is an excellent launching pad for additional research not only on the issues of this section, but of this entire book.

The Public School Nightmare:

Why fix a system designed to destroy individual thought?

by John Taylor Gatto [Two time New York State "Teacher of the Year"]

I want you to consider the frightening possibility that we are spending far too much money on schooling, not too little. I want you to consider that we have too many people employed in interfering with the way children grow up -- and that all this money and all these people, all the time we take out of children's lives and away from their homes and families and neighborhoods and private explorations -- gets in the way of education.

That seems radical, I know. Surely in modern technological society it is the quantity of schooling and the amount of money you spend on it that buys value.

And yet last year in St. Louis, I heard a vice-president of IBM tell an audience of people assembled to redesign the process of teacher certification that in his opinion this country became computer-literate by self-teaching, not through any action of schools. He said 45 million people were comfortable with computers who had learned through dozens of non-systematic strategies, none of them very formal; if schools had pre-empted the right to teach computer use we would be in a horrible mess right now instead of leading the world in this literacy.

Now think about **Sweden**, a beautiful, healthy, prosperous and up-to-date country with a spectacular reputation for quality in everything it produces. It makes sense to think their schools must have something to do with that.

Then what do you make of the fact that you can't go to school in Sweden until you are 7 years old? The reason the unsentimental Swedes have wiped out what would be first and seconds grades here is that they don't want to pay the large social bill that quickly comes due when boys and girls are ripped away from their best teachers at home too early. It just isn't worth the price, say the Swedes, to provide jobs for teachers and therapists if the result is sick, incomplete kids who can't be put back together again very easily.

The entire Swedish school sequence isn't 12 years, either -- it's nine. Less schooling, not more. The direct savings of such a step in the US would be $75-100 billion, a lot of unforeclosed home mortgages, a lot of time freed up with which to seek an education.

Who was it that decided to force your attention onto **Japan** instead of Sweden? Japan with its long school year and state compulsion, instead of Sweden with its short school year, short school sequence, and free choice where your kid is schooled? Who decided you should know about Japan and not **Hong Kong**, an Asian neighbour with a short school year that outperforms Japan across the board in math and science? Whose interests are served by hiding that from you?

One of the principal reasons we got into the mess we're in is that we allowed schooling to become a very profitable monopoly, guaranteed its customers by the police power of the state. Systematic schooling attracts increased investment only when it does poorly, and since there are no penalties at all for such performance, the temptation not to do well is overwhelming. That's because school staffs, both line and management, are involved in a **guild system**. And in that ancient form of association no single member is allowed to outperform any other member, none are allowed to advertise or to introduce new technology or improvise without the advance consent of the guild. Violation of these precepts is severely sanctioned--as Marva Collins, Jaime Escalante and a large number of once-brilliant teachers found out.

The guild reality cannot be broken without returning primary decision-making to parents, letting them buy what they want to buy in schooling, and encouraging the entrepreneurial reality that existed until 1852. That is why I urge any business to think twice before entering a cooperative relationship with the schools we currently have. Cooperating with these places will only make them worse.

The structure of American schooling, 20th century style, began in 1806 when **Napoleon's** amateur soldiers beat the professional soldiers of Prussia at the battle of Jena. When your business is selling soldiers, losing a battle like that is serious. Almost immediately afterwards a German philosopher named **Fichte** delivered his famous "Address to the German Nation" which became one of the most influential documents in modern history.

In effect he told the Prussian people that the party was over, that the nation would have to shape up through a new Utopian institution of forced schooling in which everyone would **learn to take orders.**

So the world got compulsion schooling at the end of a state bayonet for the first time in human history; modern forced schooling started in Prussia in 1819 with a clear vision of what centralized schools could deliver:

1. Obedient soldiers to the army;

2. Obedient workers to the mines;

3. Well subordinated civil servants to government;

4. Well subordinated clerks to industry

5. Citizens who thought alike about major issues.

Schools should create an artificial national consensus on matters that had been worked out in advance by leading German families and the head of institutions. Schools should create unity among all the German states, eventually unifying them into Greater Prussia.

Prussian industry boomed from the beginning. She was successful in warfare and her reputation in international affairs was very high. Twenty-six years after this form of schooling began, the King of Prussia was invited to North America to determine the boundary between the United States and Canada. Thirty-three years after that fateful invention of the central school institution, as the behest of Horace Mann and many other leading citizens, we borrowed the style of Prussian schooling as our own.

You need to know this because over the first 50 years, our school's Prussian design -- which was to create a form of state socialism -- gradually forced out our traditional American design, which in most minds was to prepare the individual to be **self-reliant**.

In Prussia the purpose of the Volksshule [work school], which educated 92 percent of the children, was not intellectual development at all, but socialisation in obedience and subordination. Thinking was left to the Real Schulen, [Real School] in which 8 percent of the kids participated. But for the great mass, intellectual development was regarded with managerial horror, as something that caused armies to lose battles.

Prussia concocted a method based on complex fragmentation to ensure that its school products would fit the grand social design. Some of this method involved dividing whole ideas into school subjects, each further divisible, some of it involved short periods punctuated by a horn so that self-motivation in study would be muted by ceaseless interruptions.

There were many more techniques of training, but all were built around the premise that isolation from first-hand information, and fragmentation of the abstract information presented by teachers, would result in obedient and subordinate graduates, properly respectful of arbitrary orders.

"Lesser" men would be unable to interfere with policy makers because, while they could still complain, **they could not manage sustained or comprehensive thought. Well-schooled children cannot think critically, cannot argue effectively.**

One of the most interesting by-products of Prussian schooling turned out to be the two most devastating wars of modern history.

Erich Maria Ramarque, in his classic "All Quiet on the Western Front" tells us that the **First World War** was caused by the tricks of schoolmasters, and the famous Protestant theologian Dietrich Bonhoeffer said that the **Second World War** was the inevitable product of good schooling.

It's important to underline that Bonhoeffer meant that literally, not metaphorically -- schooling after the Prussian fashion **removes the ability of the mind to think for itself**. It teaches people to wait for a teacher to tell them what to do and if what they have done is good or bad. Prussian teaching paralyses the moral will as well as the intellect. It's true that sometimes well-schooled students sound smart, because they **memorise** many opinions of great thinkers, but they actually are badly damaged because their own ability to think is left rudimentary and undeveloped.

We got from the United States to Prussia and back because a small number of very passionate ideological leaders visited Prussia in the first half of the 19th century, and fell in love with the order, obedience and efficiency of its system and relentlessly proselytised for a translation of Prussian vision onto these shores.

If Prussia's ultimate goal was the unification of Germany, our major goal, so these men thought, was the unification of hordes of immigrant Catholics into a national consensus based on a northern European cultural model. To do that children would have to be removed from their parents and from inappropriate cultural influence.

In this fashion, compulsion schooling, a bad idea that had been around at least since Plato's Republic, a bad idea that New England had tried to enforce in 1650 without any success, was finally rammed through the Massachusetts legislature in 1852.

It was, of course, the famous "Know-Nothing" legislature that passed this law, a legislature that was the leading edge of a famous secret society which flourished at that time known as "The Order of the Star Spangled Banner," whose password was the simple sentence, "I know nothing" -- hence the popular label attached to the secret society's political arm, "The American Party."

Over the next 50 years state after state followed suit, ending schools of choice and ceding the field to a new government monopoly. There was one powerful exception to this -- the children who could afford to be privately educated. [Although it may be relevant that not ALL private schools are geared to a "real" education, but are simply more of the same as the public schools, but are promoted as being for the elite.]

It's important to note that the underlying premise of Prussian schooling is that the government is the true parent of children -- the State is sovereign over the family. At the most extreme pole of this notion is the idea that biological parents are really the enemies of their own children, not to be trusted.

How did a Prussian system of **dumbing children down** take hold in American schools?

Thousands and thousands of young men from prominent American families journeyed to Prussia and other parts of Germany during the 19th century and brought home the Ph. D. degree to a nation in which such a credential was unknown. These men pre-empted the top positions in the academic world, in corporate research, and in government, to the point where opportunity was almost closed to those who had not studied in Germany, or who were not the direct disciples of a German PhD, as John Dewey was the disciple of G. Stanley Hall at Johns Hopkins. Virtually every single one of the founders of American schooling had made the pilgrimage to Germany, and many of these men wrote widely circulated reports praising the Teutonic methods. Horace Mann's famous 7th Report of 1844, still available in large libraries, was perhaps the most important of these.

By 1889, a little more than 100 years ago, the crop was ready for harvest. It that year the US Commissioner of Education, William Torrey Harris, assured a railroad magnate, Collis Huntington, that American schools were "scientifically designed" to prevent "over-education" from happening. The average American would be content with his humble role in life, said the commissioner, because he would not be tempted to think about any other role.

My guess is that Harris meant he would not be able to think about any other role.

In 1896 the famous John Dewey, then at the University of Chicago, said that independent, self-reliant people were a counter-productive anachronism in the collective society of the future. In modern society, said Dewey, people would be defined by their associations --not by their own individual accomplishments. In such a world **people who read too well or too early are dangerous because they become privately empowered, they know too much, and know how to find out what they don't know by themselves, without consulting experts. [emphasis added]**

Dewey said the great mistake of traditional pedagogy was to make reading and writing constitute the bulk of early schoolwork. He

advocated the phonics method of teaching reading be abandoned and replaced by the whole word method, not because the latter was more efficient (he admitted that it was less efficient), but because independent thinkers were produced by hard books, thinkers who cannot be socialised very easily.

By socialisation Dewey meant a program of social objectives administered by the best social thinkers in government. This was a giant step on the road to state socialism, the form pioneered in Prussia, and it is a vision radically disconnected with the American past, its historic hopes and dreams.

Dewey's former professor and close friend, G. Stanley Hall, said this at about the same time, "Reading should no longer be a fetish. **Little attention should be paid to reading."**

Hall was one of the three men most responsible for building a gigantic administrative infrastructure over the classroom. How enormous that structure really became can only be understood by comparisons: New York State, for instance, employs more school administrators than all of the European Economic Community nations combined.

Once you think that the control of conduct is what schools are about, the word "reform" takes on a very particular meaning. It means making adjustments to the machine so that young subjects will not twist and turn so, while their minds and bodies are being scientifically controlled. Helping kids to use their minds better is beside the point.

Bertrand Russell once said that American schooling was among the most radical experiments in human history, that **America was deliberately denying its children the tools of critical thinking.**

When you want to teach children to think, you begin by treating them seriously when they are little, giving them responsibilities, talking to them candidly, providing privacy and solitude for them, and making them readers and thinkers of significant thoughts from the beginning. That's if you want to teach them to think. There is no evidence that this has been a State purpose since the start of compulsion schooling.

When Frederich Froebel, the inventor of kindergarten in 19th century Germany, fashioned his idea he did not have a "garden for children" in mind, but a metaphor of teachers as gardeners and children as the vegetables.

Kindergarten was created to be a way to break the influence of mothers on their children. I note with interest the growth of daycare in the US and the repeated urgings to extend school downward to include 4-year-olds. The movement toward state socialism is not some historical curiosity, but a powerful dynamic force in the world around us.

The state socialism movement is fighting for its life against those forces which would, through vouchers or tax credits, deprive it of financial lifeblood, and it has countered this thrust with a demand for even more control over children's lives, and even more money to pay for the extended school day and year that this control requires. A movement as visibly destructive to individuality, family and community as government-system schooling has been, might be expected to collapse in the face of its dismal record, coupled with an increasingly aggressive shake down of the taxpayer, but this has not happened.

The explanation is largely found in the transformation of schooling from a simple service to families and towns to an enormous, centralized corporate enterprise. While this development has had a markedly adverse effect on people and on our democratic traditions, it has made schooling the single largest employer in the United States, and the largest grantor of contracts next to the Defense Department.

Both of these low-visibility phenomena provide monopoly schooling with powerful political friends, publicists, advocates and other useful allies. This is a large part of the explanation **why no amount of failure ever changes things in schools**, or changes them for very long. **School people are in a position to outlast any storm and to keep short-attention-span public scrutiny thoroughly confused.**

An overview of the short history of this institution reveals a pattern marked by intervals of public outrage, followed by enlargement of the monopoly in every case. After nearly 30 years spent inside a number of public schools, some considered good, some bad, I feel certain that management cannot clean its own house. It relentlessly marginalizes all significant change.

There are no incentives for the "owners" of the structure to reform it, nor can there be without outside competition. What is needed for several decades is the kind of wildly-swinging free market we had at the beginning of our national history.

It cannot be overemphasised that no body of theory exists to accurately define the way children learn, or which learning is of most worth. By pretending the existence of such we have cut ourselves off from the information and innovation that only a real market can provide. Fortunately our national situation has been so favourable, so dominant through most of our history, that the margin of error afforded has been vast.

But the future is not so clear. Violence, narcotic addictions, divorce, alcoholism, loneliness... all these are but tangible measures of a poverty in education. Surely schools, as the institutions monopolising the daytimes of childhood, can be called to account for this. In a democracy the final judges cannot be experts, but only the people.

Trust the people, give them choices, and the school nightmare will vanish in a generation.

****

This article is not a favourite among public school teachers. Unless, of course, they are truly interested in teaching, and much more importantly, their students learning to be self-reliant, thinking human beings -- as was Mr. Gatto, before his voluntary departure from the system. Teachers of the latter stripe, of course, have very little options when school administrators and bureaucrats dictate in the tradition of evil dictators. It is, of course, the latter who are so excessively compensated, while the front line teachers are just attempting to make ends meet. Perhaps parents should consider joining forces, finding good teachers, and hiring them on a private basis for a multiple home schooling scenario.

The Library of ialexandriah

And elsewhere on the Internet the following is found.

I would go so far as to say that all the “world problems/solutions” derive from a stupid populace, one who has no chance for freedom because they don’t know the truth, or even suspect that they don’t know it. There is, in fact, no effort being made to continually question everything, including one’s old opinions and Paradigms. Even if one solved a world problem -- e.g. outlawed vaccinations -- within a couple of generations, the same problem would rear its head again, and because the children’s education was such a low priority, they would not know the history of why the people did away with it before.

If, for example, one indoctrinates children into believing that "Father Knows Best", that all Vaccines are good for you, that the "authorities" really know what they're doing, and will always do things with your best interests in mind... Then there's no hope for freedom. Just keep in mind that freedom includes the right to fail, to get burned by the stove.

It is no wonder every State arrogates to itself the power to oversee and influence the education of its citizens. Compulsory education is the premiere means by which change for the better (for the citizens) can be nipped in the bud, and thereby imbue to the benefit of the State. It is those individuals in charge of the state, and who wish to maintain absolute dominance, that the educational system is inevitably designed to protect. Compulsory education is mind-control par excellence. In Western so-called "democracies" it is a subtle part of the system that few intellectuals -- let alone the masses -- can or are willing to see (the latter being, effectively, wilful ignorance), and therefore no remedy is thought by the so-called intellectual elite to be needed.

****

Judith: Janus, would you consider that what needs to be compulsory is not a State-supervised education system, but a State-supervised minimum education standard? By this, I mean, that the State is strictly LIMITED to administering tests to ensure that all "young minds" are acquainted with certain fundamentals -- such as reading, writing, mathematics, basic understandings of human rights and the constitution, and so forth. The idea is to ensure that education is not withheld by parents or groups to some minimum basis, but at the same time, to place the actual educational process itself -- along with all the other materials deemed appropriate by the student's parents and community -- in the hands (control) of the parents and community. ....

Judith: I find it curious that communists, religious fundamentalists, and other very-control-oriented groups have invariably adopted the strategy that the best idea (i.e. the highest priority) is to educate children in such a way that within a generation (say 15 years), they will have a group of fanatic, true believers that will do anything but exercise critical thinking, discriminate between conflicting ideas, and so forth. They will thus avoid touching the stove, not because of its inadvisability, but because of a traumatic order.

405-Education 2

"Education makes a people easy to lead, but difficult to drive; easy to govern but impossible to enslave." Lord Brougham, 1778-1868.

Updated 9 September 2004

To educate is "to draw forth...". Unfortunately, public and most private educational systems -- along with the various other forms of communicating the culture of a society to its members -- i.e. the mainstream Media, corporate and public relation firms, religious teaching organizations, and indoctrination institutions of every kind -- are less about "drawing forth" and a great deal more about "imposing upon". As Albert Einstein is reputed to have observed, "It is a miracle that curiosity survives formal education." Accordingly, these "educators" of every stripe (in the convict tradition) have become a blight ("blit/ *n.* & *v.* 1 any obscure force which is harmful or destructive...") upon civilization, and threaten the liberties and inalienable rights of... well, just about everyone.

G. I. Gurdjieff (himself, an object of study) has said, "There do exist inquiring minds, which long for the truth of the heart, seek it, strive to solve the problems set by life, try to penetrate to the essence of things and phenomena and to penetrate into themselves. If a man reasons and thinks soundly, no matter which path he follows in solving these problems, he must inevitably arrive back at himself, and begin with the solution of the problem of what he is himself and what his place is in the world around him." Or has been said elsewhere: "Know thyself." Educate yourself. (It's a lead pipe cinch no one can do it for you!) Never cease to be amazed at the phenomena which is you. Or more succinctly: O2BNAWE.

From a broader perspective, all civilizations can be said to have five primary aspects:

1) The basic technologies which allow the work of maintaining the civilization to proceed -- which at a fundamental level is often based and dependent upon the energy sources (be they domesticated animals, human slavery, fossil fuel, nuclear

industrial based, and/or which use advanced, Connective Physics, energy technologies),

2) The art or culture of the civilization -- the creative thinking and continuation of previous creative thinking (i.e. traditions) of individuals and groups,

3) The philosophy or spiritual basis of the civilization -- which is often based upon the energy source (human slavery, for example, having a dramatic spiritual effect),

4) The social contract between the sovereign entities of the civilization -- including the relationships between individuals and the rules which govern them, and

5) How a people passes on its technology, culture, philosophy and social contract to its children and dependents (as in the case of Indigo Children).

All five aspects are interconnected and interdependent, but given a particular energy source and its related science/technology (along with the art, culture, spiritual basis, and philosophy of the civilization), the key to the civilization's sustainability and quality of life is its social contract. The social contract is, in turn, dependent upon the constitutional structure of the civilization, the degree of liberty and freedom allowed, and the guarantees provided by the collective binding agreement of the society.

Fundamentally the extent to which the social contract can be *maintained* is dependent upon education. If the sovereign members of a civilization are unaware of their liberties (or the withholding of previously held liberties), then they are no longer players in the greater scheme of things. An uninformed citizenry is a contradiction in terms.

The same might be said of an uninformed sentient being... particularly in a universe of such immense wonders -- again, O2BNAW. After, all "Any education is the process of learning how little you know." [1]

Alternatively, it can also be the process whereby you learn how little others know! There is for example much to be said about knowing thyself. Admittedly, Thoreau took this to something of an extreme in bailing out of society and heading for Walden, where he wrote: "Most men, even in this comparatively free country, through mere ignorance and mistake, are so occupied with the facetious cares and superfluously coarse labors of life that its finer fruits cannot be plucked by them."[2] This is the essence of why the so-called Work Ethic has become an example of extremism taken too far in the other direction. Some time has to be taken out in order to be able to see anything from a reasonable perspective.

There is, on the other hand, the fact that we learn primarily from others -- even if most of the time it's "don't do what I just did because..." As Peter Temes has pointed out, "Thoreau's problem is that he can't see beyond himself." "He's missing regard for other people. He's missing love. That's the price you pay for being all alone in the woods." From a more positive perspective, "To find transcendence from the ordinary [which includes the everyday chores of life] is the highest spiritual place to be." [2]

An essential aspect of learning is thus being around others. This does not, however, imply that large families are better. As *Time Magazine* has reported [3], the median net worth of individuals born between 1957 and 1964 was $62,000 for an only child, $49,000 for someone growing up with one sibling, and $6,000 for someone growing up with six or more siblings. While it is very probable that anyone having seven or more children is not passing on high intelligence genes -- or simply doesn't care one way of the other -- it is also notable that the child living with two adults is likely to do better than the child living with two adults and another child. The average intelligence/experience level of the latter group is inevitably lower than the former.

Education is nevertheless often facilitated by exposure to the diversity of others -- particularly the diversity of wealth, status, and family background, albeit in a room of one teacher and thirty students, the average intellectual level is somewhat less than ideal. This is of course the genius -- or lack thereof -- of an educational system. It is also the kind of diversity which causes cliques, gangs, and groupings of modern high schools -- and the inevitable problems and positive opportunities resulting from such mixings.

Which of course brings us to the arena of public education -- in all of its myriad forms. It is important, however, to remember than education and public education are not only NOT synonymous, but quite likely at serious odds with one another.

At the same time, to complain about the Media (press, television, radio, mailings, and portions of the Inter Net), or wonder where we've gone wrong with public education, is basically missing the point. **The Media and public education are accomplishing precisely what they were intended to accomplish. It's just that 99% of the population is unaware of the actual agenda under which both the Media and public education are laboring -- said agenda being to produce functional cogs for the Corporate State machine.** In that sense, therefore, public education -- in all of its many and diverse forms -- is doing a bang up job (pardon the pun). The good news about the public education *system* is that:

> "Nearly 64% of children ages 3 to 5 were in pre-school or kindergarten in 2001, well up from the 37% enrolled in 1970. [4] [Put those little buggers to work! No sense in having them lollygagging around.]
>
> "Over 84% of Americans over age 25 finished high school by 2002. [4] [Or presumably were finished with high school!]
>
> "More than 25% of Americans have a bachelor's degree." [4]
>
> "In 1971 just 9% of medical degrees, 7% of law degrees and 4% of M.B.A.s were awarded to women; 30 years later, the respective figures were 43%, 47%, and 41%." [5]
>
> "Washington, D.C., Maryland, and Colorado lead the nation in the number of college-educated residents, all with more than 35%." [4]

On the other hand...

> The average tuition at a four-year public university has risen from $2,535 per year in 1993 to $4,694 in 2003. [6] The average cost of attending college as a full-time

undergraduate for one year, including room and board, was $14,710 in 1999-2000. [4]

"One in four college freshman at four-year universities do not return for their sophomore year. Community colleges fare even worse with half not returning for the second year." [7] Stanford University referred to the situation as: "Betraying the college dream."

"Nationally, 15% of children ages 5 to 19 are overweight, triple the rate of 20 years ago. Research suggests that fat adolescents have a 70% to 80% chance of becoming fat adults." "Many obesity experts argue that the lunchroom and gym are the spots where schools should focus their energies." [8] The former is called vending machines in elementary and other schools -- a lucrative source of finances to the school, but at the cost of the health and well being of the students.

While "parents naturally want their kids to get a good education, trouble is, with so many failing schools they have to be selective about where they live." Parents are under enormous pressures to keep their kids competitive in the marketplace, and to this end spend huge sums to ensure the child has all the right opportunities. The result is that having a child becomes "'the single best predictor' of financial ruin." [9]

It gets worse. There's the looming spectre of ever more testing, for example.

Gerald W. Bracey [10] has noted that "the SAT 9 now comes with instructions on what to do if a child vomits on the answer sheet."

It's hard to top that one, but consider the fact that new Regents Examinations introduced in 1999 in Fairport, New York have had the following consequences:

"Passing rates have declined;

"The performance gap between large urban centers and other public school districts have widened;

> "Students have been moved into GED programs to hide the dropout rate;
>
> "Staff development now addresses test scoring and alighnment of curricula to tests, not teacher improvements;
>
> "Teachers in the tested grades are fleeing the profession or asking for transfers to other grades;
>
> "The New York City dropout rate has increased 2% each year; and
>
> "The dropout rate for English-language learners increased 12% in 2001." [10]

Testing has become the new fad, and it bodes ill for everyone. John Taylor Gatto -- a New York City Teacher of the Year and New York State Teacher of the Year (and author of two books, *Dumbing Us Down* and *The Exhausted School*) -- has said, "The first thing I discovered was that the types of learning that are measured by standardized tests are not real learning at all. To do well on a standardized reading test, for example, does not mean that you read well. There are approximately 150 categories of information that a complex passage of reading delivers, and the standardized test covers approximately six of those categories over and over again." [11]

Testing organizations also require statistics as a means to claim legitimacy of their tests. In this regard, the "Bell Curve" has been sacrosanct in defining a specific and very limiting aspect of people. The assumption is that the public as a whole has precise percentages of dumb people, smart people, and the various levels in between. The problem is that many excellent teachers have been demonstrating that such a curve is inherently flawed.

Schoolteacher Jaime Escalante, for example, "taught the sons and daughters of Mexican immigrants advanced mathematics so well that they won all of the major prizes in the State of California." His achievement was not welcome, however. As Taylor as said, "teaching disadvantaged kids advanced mathematics is OK as long as you are not successful at it. The minute you *are* successful, you have demonstrated that the bell curve is a crock, and the system itself could not survive the elimination of the bell curve."

Taylor goes on to ask: "What if we operated on the premise that mathematics in its most advanced state is, in fact, fairly easy to learn, so that people with no experience at all with heavy thinking can pretty much master it in about a year? What if we generalized that to difficult reading, to writing, to public argumentation? What kind of society would emerge? Certainly one unlike any that has ever existed in human history, and the people who have a material stake and a status to defend are not interested in knowing such a society." [11]

Ah yes, the status quo. The "Let's not rock the boat, lest the elite find themselves to be in an egalitarian society." But fear not, gravy-train-riders of society. President George Bush and is neo-conservative, Dominionism sidekicks are hot on the case to prevent such equalities in education, and are proceeding on at least two fronts.

In the first line of attack, Bush "has a clear penchant for what might be called conservative social engineering." Bush wants to encourage family formation with tax credits -- even a marriage bonus -- keep children with their families at all costs, and distributing "money to voluntary groups that promote fatherhood and marriage education." [12] [What? No promotion of motherhood? Is motherhood sort of essential for the continuation of the race?]

President Bush also wants "to encourage responsible parenting by strengthening character education in the nation's schools." [12] This enthusiasm for social engineering shows up most notably in the No Child Left Behind (NCLB) law -- landmark legislation which has come under mounting opposition from the National Education Association (NEA). Of course, as suggested by Education Secretary Roderick R. Paige, the NEA is a "terrorist organization". There may be more than a few students of the system who would agree!

Bracey [10] has noted that, "Some superintendents and principals have begun to feel a Big Brother aspect to NCLB." It's not the testing per se, but provisions which allow military recruiters to have information on students or for the Boy Scouts to have access to using school property despite the organization's anti-homosexual stance. Bracey goes on to say that the "NCLB is a trap, a Trojan's Horse..." The intent, he believes, is to crash the public school system and hand it over to privatization, school vouchers, and tuition tax credits. Ultimately, big business moves in to achieve enormous profits and provide minimal actual "bringing forth", i.e. education.

Vouchers has been pushed by such luminaries as Milton Friedman, who argued that such vouchers would introduce market competition into the system. The assumption is that "competition has made progress possible in every area of economic life." The flaw is of course that those with financial interests would just be cooking the books in the same manner as Enron, WorldCom, etceteras, etceteras, etceteras... ad infinitum.

There is also the religion angle. Vouchers could be used to support narrow-minded religions. "A state judge found the Florida state constitution 'clear and unambiguous' in prohibiting public money from being spent on church schools or any sectarian institution and declared Florida's voucher program unconstitutional. Florida Gov. Jeb Bush immediately appealed the decision." [10] One wonders how Jeb would feel about the Saudi Arabian's government defacto support for the extremist Islam cults education and brainwashing schools.

In some respects the vouchers for religious schools may be something of a red herring. The real issue is about "The 500-Pound Gorilla" lurking the hallways outside of class. As Alfie Kohn [13] has noted, "Corporations are not shy about trying to make over the schools in their own image." And the evidence of their efforts is becoming more than a little bit clear -- which brings us to the Bush Administration's second covert push.

The College Entrance Examination Board which owns the SAT is changing the SAT to include among other things, an essay. Suddenly any pretense toward total objectivity silently slips away. But the Board has gone a bit further in order to undertake "an unprecedented effort to push local school districts to alter their curriculums." "In short, the dreaded SAT could actually help produce a national curriculum, a sweeping education reform enacted without the passage of a single law." [14] In fact, the very nature and purpose of the SAT is being changed. "The goal of influencing school curriculums has become the overriding preoccupation of the new test's development."

Why make such curriculum changes? **The most obvious answer is to make the graduate more useful -- without extensive and expensive training -- to corporations. But this is not particularly new. This has always been the goal of public education.**

Prior to the advent of "public education", the literacy rate for Americans was considerably higher than after the introduction of what amounted to a Prussian system of education. An important reference to this aspect of the U. S. educational system is contained in John Taylor Gatto's exceptional article on The Public School Nightmare. Gatto, it might be noted is a two-time winner of New York State's "Teacher of the Year". His article is *absolutely essential reading for anyone who thinks education is always beneficial.*

The importation of the Prussian educational System was not without a reason.

THE "SYSTEM"

Briefly, the Prussians divide all student into three groups: The children of society's elite, comprising 0.5% of the society, who are actually educated; a second, more open category, comprising 5.5% of the remaining children, who were sent to schools where they were partially taught to think, and the remaining 94% who were sent to work schools to learn "harmony, obedience, freedom from stressful thinking and how to follow orders." The Prussian system does not specify it as such, but Americans have evolved a fourth category.

The idea is now to take the great mass of people within the jurisdiction of a governmental authority, and split them into four groups:

THE "ELITE"

1) The first group are the elite (as in the Prussian model) -- almost always the children of the previous generation of the elite (i.e. governmental and societal leaders), who do not go to public schools, but instead are sent to *selected* private schools where they are taught to be the leaders of society. On rare occasion, someone enormously talented (primarily in gathering power) is granted admission into the elite, and thereafter (a few generations), their children may become full-fledged elitists themselves.

(This leaves the *public* education system to deal with the rest of humanity. As such it is designed to identify and separate the three other groups, i.e.:)

THE "CREATIVES"

2) A second group, known as the Creatives, are those individuals who provide the brain power to advance technology and the quality of life. These are "accidents of birth" where a particular, random gene pattern produces a genius. Said genius can be manifested as talents in everything from art and music to science and engineering -- no one in the elite cares as long as the "genius" creates a better technology and an improved quality of life for the elite. These children are sent to what the Prussians called "real" schools, but where the so-called education is geared to "thinking in a box" (i.e. specialization), and avoids universality or generality (from whence derives the word, "university").

THE "MASSES" AND THEIR PURPOSE.

3) A third group, known affectionately as the Masses, are those individuals who are expected to obey authority, never question the controllers, and seldom think (and even then almost never take action based on their few errant thoughts). Instead, they believe in the bastardized "Work Ethic" as something marvelously good. These are the targets of all the advertising and marketing -- including government

and media propaganda. These are the people who in order to demonstrate their patriotism, go to wars that benefit only the banksters and corporations. These are the believers in authority who dutifully pay their taxes to an immoral government. They are the clear majority, but recent trends suggest the Creatives may be gaining in percentage population of the mass of humanity.

THE "UNWANTED" FREE THINKERS

4) A fourth group, an American invention known as the Rabble Rousers, are those individuals who early on question authority, think independently, and are uncooperative in becoming mere cogs in the machine. The principle goal of the American educational system is prevent these types from gaining the ability to learn how to affect the system, and if necessary, to simply weed them out all together -- preferably by getting them completely out of the school system. These are the same individuals who **market analysts consider "fringe members" of society. They -- as well as the Creatives -- are never pitched to in advertisements, in that they either don't have the money to buy anything (particularly the "Fringees"), or even if they have the money, they are not as subject to manipulation by the advertisers as others, and thus by extension are not easily marketed to. They are ignored, or as politicians, advertisers, and other liars refer to it: "marginalized". Anyone reading this website is in serious danger of becoming a Rabble Rouser!**

Meanwhile, the Media's job is to cater to the second and third group, keeping the masses in line, and hopefully, keeping the Creatives from getting too creative in terms of new and interesting ways of becoming independent of their government and the elite's leadership.

A vitally important part of the educational system is to break the link between reading and the young child, because a child who reads too well becomes knowledgeable and independent from the system of instruction and is capable of finding out anything. In order to have an efficient policy-making class and a sub-class beneath it, you've got to remove the power of most people to make anything out of available information. <u>This can be done by discouraging reading, or making distractions as entertaining as possible. This explains in part</u>

the enormous success of television and radio. Don't read; don't think; watch television (or in a pinch, listen to radio). Above all: Don't think. That's called in politics: "Staying on message."

Another technique is the use of "positive language" (i.e. a "no-contradiction principle"). This is one that can be and is encouraged by the (Prussian-style) authorities to be used in the home. According to this concept, when one parent tells a child something, the other parent (or "significant other") must concur. Unfortunately, in this manner, the parents are (unknowingly?) avoiding a situation where two diverse viewpoints could be rationally presented, and the child then obliged to use (and/or develop) an ability to judge between alternatives, and thus access avenues of critical thinking, self-reliance, autonomy, and ultimately, flexibility. The "no-contradiction principle" motivates children to become used to following authority figures, rather than questioning received wisdom and thus thinking for themselves. Such a system is anti-Discrimination -- in the negative sense.

The idea children need "positive language" in order to learn the truths that are necessary for survival is not the issue. Any such *truths* must be taught by acts, not by words! The idea is to insist that a child not blindly follow the words of an authority figure (the parent), but to model themselves on how the parent acts, even if their actions are in contradiction with their words. A mind is a terrible thing to waste in a dysfunctional educational system.

But it gets worse.

1963 saw the culmination of research by a combination of German, chemical medicine and Wundt psychology on American upbringing. This group of scientific researchers centered at John Hopkins University, and supported by the General Education Board, reached the point where they concluded that they could use amphetamines like Dextrines and Ritalin to "treat" children that were considered "difficult" or hyperactive -- see, for example, "The Myth of the Hyperactive Child, and other Means of Child Control" by Divoky and Schrag.

The subsequent national terror occasioned by the wide-spread use of Ritalin, et al and Mandatory Vaccines constitute a tragedy far in excess of 9-11-2001. All, of course, in the name of control.

But more than educational methods is at stake. There is also the matter of curriculum -- what is taught and the depth to which a student is allowed to plume. On the one hand, the inability or unwillingness of education to address any and all issues is a strong deterrent to the education of anyone, including a young Calvin.

In the Calvin and Hobbes [<http://www.ucomics.com/calvinandhobbes/>] cartoon strip, Calvin, one of the more incorrigible of resistant learners, is asked by his father at one point as to why Calvin doesn't do better in school. His father points out that he doesn't believe Calvin is stupid, if only because Calvin knows every dinosaur that ever lived, knows their habits and locales, even how to pronounce their names accurately. And thus Calvin can't be dumb. So why doesn't the young man do better in school? Calvin's answer is classic: "They don't teach dinosaurs." This, in a nutshell, is why there are a lot of Calvins in our school systems. The system does not teach anything in which they are interested. Worse yet, it often (but not always -- in rare occasions) discourages any one from any probing question for which the system does not have a ready or willing answer.

In one classic case, an honor student wrote a paper on whether or not new discoveries in history would be able to change our understanding of history -- whether in fact, such revelations would even be tolerated. The purpose of the paper was to simply ask the question, but in the process of using a notorious example (i.e. the reconstruction of the ancient history of Egypt as suggested by Immanuel Velikovsky), the student encountered such wrath from the irate history teacher, that the student's thesis was partially proven correct, that history cannot be changed by new evidence. The system does not tolerate inconvenient questions.

Nor does it tolerate inconvenient topics for discussion and essays. The new SAT tests, for example, will include reading passages from various novels -- but will specifically avoid such title as often assigned in good English classes (*Animal Farm*, *Catch 22*, and the like). Alternatively, as the new SAT selections become obvious, then the whole direction of English classes will have to cater to those novels in order to ensure there students have an equal chance to score high marks. Which is, of course, the fatal flaw in the essay portion: If one takes a test and gets a topic they really love, they are doing to do much better on their score than if they are discovering some subject for the first time.

But it gets worse. The new SAT does far more damage than creating scores which will be religiously followed by college admissions and which are fundamentally biased. **The SAT will also change from a test of general-reasoning abilities to a measure of what the student learned in school.** "For decades, the purpose of the test has been to try to measure students' general-reasoning abilities, not their specific knowledge of algebra or the extent to which they have written practice essays." "The more you challenge yourself intellectually, the more you condition your brain; your academic achievements are less impressive if you don't have the conditioning to build upon them. As the SAT becomes more an assessment of one's achievements, it will less sensitively guage these underlying skills". [14]

Part of the problem is that the new SAT takes the "Old Paradigm of Education" and takes it to an extreme, whereas what is essential for anyone to consider themselves educated, a "New Paradigm of Learning" is essential. The Aquarian Conspiracy [15] defines these two opposing, fundamental assumptions:

Assumptions of Old Paradigm	Assumptions of New Paradigm
Emphasis on *content*, acquiring a body of 'right' information, once and for all.	Emphasis on learning how to learn, how to ask good questions, pay attention to the right things, be open to and evaluate new concepts, have access to information. What is now 'known' may change. Importance of *context*.
Learning as a *product*, a destination.	Learning as a *process*, a journey.
Hierarchical and authoritarian structure. Rewards conformity, discourages dissent.	Egalitarian. Candor and dissent permitted. Students and teachers see each other as people, not roles. Encourages autonomy.
Relatively rigid structure, prescribed curriculum, emphasis on "appropriate" ages for certain activities, age segregation, compartmentalized.	Relatively flexible structure, belief there are many ways to teach a subject. Integration of age groupings, individuals not limited by age to certain subject matter.

Priority on performance. Priority on self-image as the generator of performance.

Emphasis on external world. Inner experience considered inappropriate in school setting. Inner experience seen as context for learning, use of imagery, storytelling, dream journals, and exploration of feelings encouraged.

Guessing and divergent thinking discouraged

Guessing and divergent thinking encouraged as part of the creative process

Emphasis on analytical, linear, left-brain thinking.

Strives for whole-brain education. Augments left-brain rationality with holistic, nonlinear, and intuitive strategies. Fusion emphasized.

Labeling of students (remedial, gifted, minimally brain dysfunctional, etc.) contributes to self-fulfilling prophecy.

Labeling used only in minor prescriptive role and not as fixed evaluation that dogs the individual's educational career.

Concern with norms. Concern with individual performance in terms of potential, testing outer limits, transcending perceived limitations.

Primary reliance on theoretical, abstract "book knowledge".

Theoretical and abstract knowledge heavily complemented by experiment/experience, both in and out of classroom. Field trips, apprenticeships, demonstrations, visitors.

Classroom designed for efficiency, convenience.

Concern for the environment of learning; needs for privacy and interaction, quiet and exuberant activities.

Bureaucratically determined, resistant to community input.

Encourages community input, even community control.

Education seen as a social necessity for a certain period of time, to inculcate minimum skills and specific role training.

Education seen as lifelong process, one only tangentially related to schools.

Increasing reliance on technology (audio visual equipment, computers, tapes, texts), dehumanization.

Appropriate technology, human relationships between teacher and learners of primary importance.

Teacher imparts knowledge; one-way street.

Teacher is learner, too, learning from students.

One must ultimately ask which of the paradigms will make create the most wisdom and understanding of the individual. It should be clear that the corporate elite would prefer the old paradigm in that it leads to more cooperative and rule-abiding slaves. The key ingredient of such thinking is that one follows the rules, even when one is not sure about the whole rule.

For example, when a first grade teacher collected well known proverbs, gave each child in her class the first half of a proverb (the "rule") and then asked them to come up with the remainder of the proverb, these first graders... "6" year-olds... came up with the following:

Better to be safe than.............................punch a 5th grader.

Strike while the ....bug is close.

It's always darkest before........................Daylight Saving Time.

Never underestimate the power of............termites.

You can lead a horse to water but...........how?

Don't bite the hand that...........................looks dirty.

No news is...impossible.

A miss is as good as a.........................Mr.

You can't teach an old dog new.............math.

If you lie down with dogs, you'll..............stink in the morning.

Love all, trust.......................................me.

The pen is mightier than the..................pigs.

An idle mind is....................................the best way to relax.

Where there's smoke there's................pollution.

Happy is the bride who...........................gets all the presents.

A penny saved is................................not much.

Two's company, three's........................the Musketeers.

Don't put off till tomorrow what..............you put on to go to bed.

There are none so blind as...................Stevie Wonder.

Children should be seen and not...........spanked or grounded.

If at first you don't succeed...................get new batteries.

When the blind leadeth the blind...........get out of the way.

Better late than..................................pregnant!!!!

It is worth noting that in the Old Paradigm these answers and indeed the whole exercise would be considered a horrid waste of time and something to be discouraged, while the New Paradigm would welcome such creativity and innovation. As John Taylor Gatto has pointed out, education is more than just lectures; it is also apprenticeships, community service, field curriculum, independent study, parent partnerships, work study, critical-thinking exercises, and solitude. [11]

It is an unfortunate, perhaps, but very true reality that knowledge is power. Acquiring that knowledge is what education is about -- with the added caveat of course that the ability to use said knowledge is even more important. Without the knowledge in the first place, however, logical manipulations of limited data has minimal value. Public education is thus the prime example of diverting attention from the more interesting forms of knowledge (and shunning wisdom altogether!). It does train its students, however, in being able to read ball scores quite well.

Ultimately, the best thing that can be said for public education is that... hmmmmm... I'll get back to you on that one.

Just don't tell your teacher what you're doing! (*No kidding!*)

References:

{1] Richard Corliss, "Hook, Line and Thinker," *Time Magazine*, May 26, 2003.

[2] Peter Temes, "Thoreau in the Bronx", *Utne Reader*, May-June 2003.

[3] Notebook, *Time Magazine*, September 2002.

[4] What's New Netscape, April 6, 2004.

[5] Claudia Wallis, "The Case for Staying Home", *Time Magazine*, March 22, 2004.

[6] Notebook, *Time Magazine*, November 3, 2003.

[7] What's New Netscape, April 5, 2004.

[8] Education, *Time Magazine*, September 15, 2003.

[9] "Maryanne Murray Buechner, "Parent Trap", *Time Magazine*, October 2003.

[10] Gerald W. Bracey, "The 12th Bracey Report on the Condition of Public Education," *Phi Delta Kappan*, October 2002.

[11] Ellen Becker, "School's Out: An Interview With John Taylor Gatto," *The Sun*.

[12] "George Bush, big-government conservative", *The Economist*, April 21, 2001.

[13] Alfie Kohn, "The 500-Pound Gorilla", *Phi Delta Kapan*, October 2002.

[14] John Cloud, "Inside the New SAT," *Time Magazine*, October 27, 2003.

[15] Marilyn Ferguson, *The Aquarian Conspiracy; Personal and Social Transformation in the 1980s,* Jeremy P. Tarcher/Perigee, 1980.

406-MEET OUR TRUE LEADERS

"When one has been threatened with a great injustice, one accepts a smaller as a favour." –Jane Welsh Carlyle

No our true leaders, owners if you will, are not the various governments around the globe, of whatever professed colour, democratic, fascist, communist, etc. Simple thinking and reasoning should expose this self-evident truth. If you haven't already thought about the above claim consider this: Governments of all nations and countries come and go, even the great communist Soviet Union has been disbanded. Their various leaders change even more frequently, whereas those who have almost total control over all governments remain forever, unchanged with the exceptions of becoming wealthier, more powerful, and greedier.

When we consider the position and state of the world's corporate bodies we should not forget to include in our thinking that most, if not all, of the bureaucratic governmental or semi-government institutions are basically the same. They are non-elected ruling bodies with effective direct control over a country's population. They control health, housing, defence, economy, in fact just about everything you can name that has some or any effects upon mankind.

Now let me ask, if you wanted to rule the world, you would want the powerful corporates on your side wouldn't you? Or is there something really basically wrong with that question? Should it read, If you were the corporate bodies **already** ruling the world you would want ownership of governments, and all government departments and bureaucrats totally under your control, wouldn't you? Yes start to squirm uncomfortably right now, because it is probably too late to change anything now. We can only hope that for some brief time in perhaps a post-apocalyptic future, some of humanity may live in freedom for a brief time before yet another new empire is set up.

The article below records some facts about such bodies and is freely available on the Internet. Again, this is but one sample of a great wealth and abundance of material available.

A History of Corporate Rule
and Popular Protest

A new populist movement has emerged to challenge corporate power and call for a more equitable economic order that protects traditional cultures and ecosystems and promotes sustainability.

by Richard Heinberg © 2002
Editor/Publisher
MuseLetter
1604 Jennings Avenue
Santa Rosa, CA 95401, USA
Email: heinberg@museleter.com
Website: http://www.museletter.com/

The corporation was invented early in the colonial era as a grant of privilege extended by the Crown to a group of investors, usually to finance a trade expedition. The corporation limited the liability of investors to the amount of their investment--a right not held by ordinary citizens. Corporate charters set out the specific rights and obligations of the individual corporation, including the amount to be paid to the Crown in return for the privilege granted.

Thus were born the East India Company, which led the British colonisation of India, and Hudson's Bay Company, which accomplished the same purpose in Canada. Almost from the beginning, Britain deployed state military power to further corporate interests--a practice that has continued to the present. Also from the outset, corporations began pressuring government to expand corporate rights and to limit corporate responsibilities.

The corporation was a legal invention - a socio-economic mechanism for concentrating and deploying human and economic power. The purpose of the corporation was and is to generate profits for its investors. As an entity, it has no other purpose; it acknowledges no higher value.

Many people understood early on that since corporations do not serve society as a whole, but only their investors, there is therefore always a danger that the interests of corporations and those of the general populace will come into conflict. Indeed, the United States was born of a revolution not just against the British monarchy but against the power of corporations. Many of the American colonies had been chartered as corporations (the Virginia Company, the Carolina Company, the Maryland Company, etc.) and were granted monopoly power over lands and industries considered crucial to the interests of the Crown.

Much of the literature of the revolutionaries was filled with denunciations of the "long train of abuses" of the Crown and its instruments of dominance, the corporations. As the yoke of the Crown corporations was being thrown off, Thomas Jefferson railed against "the general prey of the rich on the poor". Later, he warned the new nation against the creation of "immortal persons" in the form of corporations. The American revolutionaries resolved that the authority to charter corporations should lie not with governors, judges or generals, but only with elected legislatures.

At first, such charters as were granted were for a fixed time, and legislatures spelled out the rules each business should follow. Profit-making corporations were chartered to build turnpikes, canals and bridges, to operate banks and to engage in industrial manufacture. Some citizens argued against even these few, limited charters, on the grounds that no business should be granted special privileges and that owners should not be allowed to hide behind legal shields. Thus the requests for many charters were denied, and existing charters were often revoked. Banks were kept on a short leash, and (in most states) investors were held liable for the debts and harms caused by their corporations.

All of this began to change in the mid-19th century. According to Richard Grossman and Frank Adams in Taking Care of Business: "Corporations were abusing their charters to become conglomerates and trusts. They were converting the nation's treasures into private fortunes, creating factory systems and company towns. Political power began flowing to absentee owners intent upon dominating people and nature."1

Grossman and Adams note that: "In factory towns, corporations set wages, hours, production processes and machine speeds. They kept blacklists of labour organisers and workers who spoke up for their rights. Corporate officials forced employees to accept humiliating conditions, while the corporations agreed to nothing."

The authors quote Julianna, a Lowell, Massachusetts, factory worker, who wrote: "Incarcerated within the walls of a factory, while as yet mere children, drilled there from five till seven o'clock, year after year what, we would ask, are we to expect, the same system of labour prevailing, will be the mental and intellectual character of future generations a race fit only for corporation tools and time-serving slaves?... Shall we not hear the response from every hill and vale: 'Equal rights, or death to the corporations'?"

Industrialists and bankers hired private armies to keep workers in line, bought newspapers and (quoting Grossman and Adams again): "painted politicians as villains and businessmen as heroes. Bribing state legislators, they then announced legislators were corrupt, that they used too much of the public's resources and time to scrutinise every charter application and corporate operation. Corporate advocates campaigned to replace existing chartering laws with general incorporation laws that set up simple administrative procedures, claiming this would be more efficient. What they really wanted was the end of legislative authority over charters."

During the Civil War, government spending brought corporations unprecedented wealth. "Corporate managers developed the techniques and the ability to organise production on an ever grander scale," according to Grossman and Adams. "Many corporations used their wealth to take advantage of war and Reconstruction years to get the tariff, banking, railroad, labour, and public lands legislation they wanted."

In 1886, the US Supreme Court declared that corporations were henceforth to be considered "persons" under the law, with all of the constitutional rights that designation implies.

The Fourteenth Amendment to the Constitution, passed to give former slaves equal rights, has been invoked approximately ten times more frequently on behalf of corporations than on behalf of African Americans. Likewise the First Amendment, guaranteeing free speech, has been invoked to guarantee corporations the "right" to influence the political process through campaign contributions, which the courts have equated with "speech".

If corporations are "persons", they are persons with qualities and powers that no flesh-and-blood human could ever possess--immortality, the ability to be in many places at once, and (increasingly) the ability to avoid liability. They are also "persons" with no sense of moral responsibility, since their only legal mandate is to produce profits for their investors.

Throughout the late 19th and early 20th centuries, corporations reshaped every aspect of life in America and much of the rest of the world. The factory system turned self-sufficient small farmers into wage earners and transformed the family from an interdependent economic production unit to a consumption-oriented collection of individuals with separate jobs. Advertising turned productive citizens into "consumers". Business leaders campaigned to create public schools to train children in factory-system obedience to schedules and in the performance of isolated, meaningless tasks. Meanwhile, corporations came to own and dominate sources of information and entertainment, and to control politicians and judges.

During two periods, corporations faced a challenge: the 1890s (a depression period when Populists demanded regulation of railroad rates, heavy taxation of land held only for speculation, and an increase in the money supply), and the 1930s (when a profound crisis of capitalism led hundreds of thousands of workers and armies of the unemployed to demand government regulation of the economy and to win a 40-hour week, a minimum-wage law, the right to organise, and the outlawing of child labour). But in both cases, corporate capitalism emerged intact.

In the words of historian Howard Zinn: "The rich still controlled the nation's wealth, as well as its laws, courts, police, newspapers, churches, colleges. Enough help had been given to enough people to make Roosevelt a hero to millions, but the same system that had brought depression and crisis remained."2

World War II, like previous wars, brought huge profits to corporations via government contracts. But following this war, military spending was institutionalised, ostensibly to fight the "Cold War". Despite occasional regulatory setbacks, corporations seized ever more power, and increasingly transcended national boundaries, loyalties and sovereignties altogether.

GLOBAL PILLAGE

In the 1970s, capitalism faced yet another challenge as post-war growth subsided and profits fell. The US was losing its dominant position in world markets; the production of oil from its domestic wells was peaking and beginning to fall, thus making America increasingly dependent upon oil imports from Arab countries; the Vietnam War had weakened the American economy; and Third World countries were demanding a "North & South dialogue" leading towards greater self-reliance for poorer countries. President Nixon responded by doing away with fixed currency exchange rates and devaluing the dollar, largely erasing US war debts to other countries. Later, newly elected President Reagan, at the 1981 Cancún, Mexico, meeting of 22 heads of state, refused to discuss new financial arrangements with the Third World, thus effectively endorsing their further exploitation by corporations.

Meanwhile, the corporations themselves also responded with a new strategy. Increased capital mobility (made possible by floating exchange rates and new transportation, communication and production technologies) allowed US corporations to move production offshore to "export processing zones" in poorer countries. Corporations also undertook a restructuring process, moving toward "networked production"--in which big firms, while retaining and consolidating power, hired smaller firms to take over aspects of supply, manufacture, accounting and transport. (Economist Bennett Harrison defined networked production as "concentration of control combined with decentralisation of production".) This restructuring process is also known as "downsizing", because it results in the shedding of higher-paid employees by large corporations and the hiring of low-wage contingent workers by smaller subcontractors.

Jeremy Brecher and Tim Costello write in Global Village or Global Pillage that: "As the economic crisis deepened, there gradually evolved a 'supra-national policy arena' which included new organisations like the Group of Seven (G7) industrial nations and NAFTA and new roles for established international organisations like EU, IMF, World Bank, and GATT. The policies adopted by these international institutions allowed corporations to lower their costs in several ways. They reduced consumer, environmental, health, labour, and other standards. They reduced business taxes. They facilitated the move to lower wage areas and threat of such movement. And they encouraged the expansion of markets and the 'economies of scale' provided by larger-scale production."3

All of this has led to a globalised economy in which (again quoting Brecher and Costello): "All over the world, people are being pitted against each other to see who will offer global corporations the lowest labour, social, and environmental costs. Their jobs are being moved to places with inferior wages, lower business taxes, and more freedom to pollute. Their employers are using the threat of 'foreign competition' to hold down wages, salaries, taxes, and environmental protections and to replace high-quality jobs with temporary, part-time, insecure, and low-quality jobs. Their government officials are justifying cuts in education, health, and other services as necessary to reduce business taxes in order to keep or attract jobs."

Corporations, no longer bound by national laws, prowl the world looking for the best deals on labour and raw materials. ***Of the world's top 120 economies, nearly half are corporations, not countries****. Thus the power of citizens in any nation to control corporations through whatever democratic processes are available to them is receding quickly.*

In November 1999, tens of thousands of students, union members and indigenous peoples gathered in Seattle to protest a meeting of the World Trade Organisation (WTO). This mass demonstration seemed to signal the birth of a new global populist uprising against corporate globalization. In the three years since then, more mass demonstrations - some larger, many smaller – have occurred in Genoa, Melbourne, Milan, Montreal, Philadelphia, Washington and other cities.

In January 2001, George W. Bush and Dick Cheney took office, following a deeply flawed US election. With strong ties to the oil industry and to the huge energy-trading corporation Enron, the new administration quickly proposed a national energy policy that focused on opening federally protected lands for oil exploration and on further subsidising the oil industry.

Enron, George W. Bush's largest campaign contributor, was the seventh largest corporation in the US and the 16th largest in the world. Despite its reported massive profits, it had paid no taxes in four out of the previous five years. The company had thousands of offshore partnerships, through which it had hidden over a billion dollars in debt. When this hidden debt was disclosed in October 2001, the company imploded. Its share price collapsed and its credit rating was slashed. Its executives resigned in disgrace, taking with them multimillion-dollar bonuses, while employees and stockholders shouldered the immense financial loss. Enron's bankruptcy was the largest in corporate history up to that time, but its creative accounting practices appear to be far from unique, with dozens of other corporations poised for a similar collapse.

Following the outrageous and tragic attacks of September 11, Bush launched a "War on Terror", raising the listed number of potential target countries from three to nearly 50, most having exportable energy resources. With Iraq (holder of the world's second-largest proven petroleum reserves) high on the list of enemy regimes to be violently overthrown, the Bush administration's Terror War appeared to be geared toward making the world safe for the expanded reach of US oil corporations. Meanwhile, new laws and executive orders curtailed constitutional rights and erected screens of secrecy around government actions and decision-making processes.

It remains to be seen how the American populace will react to these new developments. Here again, a little history may help us understand the options available.

HURDLES IN THE PATH

The Populism of the 1890s failed for two main reasons: divisiveness within, and co-optation from without. While many Populist leaders saw the need for unity among people of different racial and ethnic backgrounds in attacking corporate power, racism was strong among many whites. Most of the Alliance leaders were white farm owners who failed in many instances to support the organising efforts of poor rural blacks, and poor whites as well, thus dividing the movement.

"On top of the serious failures to unite blacks and whites, city workers and country farmers," writes Howard Zinn, "there was the lure of electoral politics. Once allied with the Democratic party in supporting William Jennings Bryan for President in 1896 the pressure for electoral victory led Populism to make deals with the major parties in city after city. If the Democrats won, it would be absorbed. If the Democrats lost, it would disintegrate. Electoral politics brought into the top leadership the political brokers instead of the agrarian radicals... In the election of 1896, with the Populist movement enticed into the Democratic party, Bryan, the Democratic candidate, was defeated by William McKinley, for whom the corporations and the press mobilised, in the first massive use of money in an election campaign."4

Today, a new populist movement could easily fall prey to the same internal divisions and tactical errors that destroyed its counterpart a century ago. In the recent American presidential election, populists faced the choice of supporting their own candidate (Ralph Nader) and thereby contributing to the election of the far-right, pro-corporate Republican candidate (Bush), or supporting the centrist Gore and seeing their movement co-opted by pro-corporate Democrats.

Meanwhile, though African Americans, Asian Americans, Hispanic Americans, European Americans and Native Americans have all been victimised by corporations, class divisions and historical resentments often prevent them from organising to further their common interests. In recent elections, ultra-right candidate Pat Buchanan appealed simultaneously to "populist" anti-corporate and anti-government sentiments among the working class, as well as to xenophobic white racism. Buchanan's critique of corporate power was shallow, but it was often the only such critique permitted in the corporate controlled media. One cannot help but wonder: were the corporations looking for a lightning rod to rechannel the anger building against them?

While Buchanan had no chance of winning the presidency, his candidacy did raise the spectre of another kind of solution to the emerging crisis of popular resentment against the system - a solution that again has roots in the history of the past century.

A FALSE REVOLUTION

In the early 1900s, workers in Italy and Germany built strong unions and won substantial concessions in wages and work conditions; still, after World War I they suffered under a disastrous post-war economy, which fanned unrest. During the early 1920s, heavy industry and big finance were in a state of near-total collapse. Bankers and agribusiness associations offered financial support to Mussolini -who had been a socialist before the war - to seize state power, which he effectively did in 1922 following his march on Rome. Within two years, the Fascist Party (from the Latin fasces, meaning a bundle of rods and an axe, symbolising Roman state power) had shut down all opposition newspapers, crushed the socialist, liberal, Catholic, democratic and republican parties (which had together commanded about 80 per cent of the vote), abolished unions, outlawed strikes and privatised farm co-operatives.

In Germany, Hitler led the Nazi Party to power, then cut wages and subsidised industries.

In both countries, corporate profits ballooned. Understandably, given their friendliness to big business, Fascism and Nazism were popular among some prominent American industrialists (such as Henry Ford) and opinion shapers (like William Randolph Hearst).

Fascism and Nazism relied on centrally controlled propaganda campaigns that cleverly co-opted the language of the Left (the Nazis called themselves the National Socialist German Workers Party while persecuting socialists and curtailing workers' rights). Both movements also made calculated use of emotionally charged symbolism: scapegoating minorities, appealing to mythic images of a glorious national past, building a leader cult, glorifying war and conquest, and preaching that the only proper role of women is as wives and mothers.

As political theorist Michael Parenti points out, historians often overlook Fascism's economic agenda - the partnership between Big Capital and Big Government - in their analysis of its authoritarian social program. Indeed, according to Bertram Gross in his startlingly prescient Friendly Fascism (1980), it is possible to achieve fascist goals within an ostensibly democratic society. 5 Corporations themselves, after all, are internally authoritarian (courts have ruled that citizens give up their constitutional rights to free speech, freedom of assembly, etc., when they are at work on corporate-owned property); and as corporations increasingly dominate politics, media and economy, they can mould an entire society to serve the interests of a powerful elite without ever resorting to stormtroopers and concentration camps. No deliberate conspiracy is necessary, either: each corporation merely acts to further its own economic interests. If the populace shows signs of restlessness, politicians can be hired to appeal to racial resentments and memories of national glory, dividing popular opposition and inspiring loyalty.

In the current situation, "friendly fascism" works somewhat as follows. Corporations drive down wages and pay a dwindling share of taxes (through mechanisms outlined above), gradually impoverishing the middle class and creating unrest. As corporate taxes are cut, politicians (whose election was funded by corporate donors) argue that it is necessary to reduce government services in order to balance the budget. Meanwhile, the same politicians argue for an increase in the repressive functions of government (more prisons, harsher laws, more executions, more military spending). Politicians channel the middle class's rising resentment away from corporations and toward the government (which, after all, is now less helpful and more repressive than it used to be) and against social groups easy to scapegoat (criminals, minorities, teenagers, women, gays, immigrants).

Meanwhile, debate in the media is kept superficial ***(elections are treated as sporting contests),*** *and right-wing commentators are subsidised while left-of-centre ones are marginalised.* ***People who feel cheated by the system turn to the Right for solace, and vote for politicians who further subsidise corporations, cut government services, expand the repressive power of the state and offer irrelevant scapegoats for social problems with economic roots. The process feeds on itself.***

Within this scenario, George W. Bush (and similar ultra-right figures in other countries) are not anomalies but, rather, predictable products of a strategy adopted by economic elites - harbingers of a less-than-friendly futures - the more "moderate" tactics for the maintenance and consolidation of power founder under the weight of corporate greed and resource exhaustion.

CAUSE FOR HOPE?

These circumstances are, in their details, unprecedented; but in broad outline we are seeing the re-enactment of a story that goes back at least to the beginning of civilisation. Those with power are always looking for ways to protect and extend it, and to make their power seem legitimate, necessary or invisible so that popular protest seems unnecessary or futile. If protest comes, the powerful always try to deflect anger away from themselves. The leaders of the new populist movement appear to have a good grasp of both the current circumstances and the historical ground from which these circumstances emerge. They seem to have realised that, in order to succeed, the new populism will have to:
¥ avoid being co-opted by existing political parties;
¥ heal race, class and gender divisions and actively resist any campaign to scapegoat disempowered social groups;
¥ avoid being identified with an ideological category - "communist", "socialist" or "anarchist" - against which most of the public is already well inoculated by corporate propaganda;
¥ direct public discussion toward the most vulnerable link in the corporate chain of power: the legal basis of the corporation;
¥ internationalise the movement so that corporations cannot undermine it merely by shifting their base of operations from one country to another.

As Lawrence Goodwyn noted in his definitive work, The Populist Moment, the original Populists were "attempting to construct, within the framework of American capitalism, some variety of co-operative commonwealth". This was "the last substantial effort at structural alteration of hierarchical economic forms in modern America".6

In announcing the formation of the Alliance for Democracy, in an article in the August 14, 1996 issue of The Nation, activist Ronnie Dugger compiled a list of policy suggestions which comprise some of the core demands of the new populist movement. These include: a prohibition of contributions or any other political activity by corporations; single-payer national health insurance with automatic universal coverage; a doubling of the minimum wage, indexed to inflation; a generic low-interest-rate national policy, entailing the abolition of the Federal Reserve System; statutory reversal of the court-made law that corporations are "persons"; establishment of a national public oil company; limitations on ownership of newspapers, magazines, radio and TV stations to one of any kind per person or owning entity; and the halving of military spending. The new populists are, in Ronnie Dugger's words, "ready to resume the cool eyeing of the corporations with a collective will to take back the powers they have seized from us".7

The new populism draws some of its inspiration from the work of the Program on Corporations, Law and Democracy (POCLAD), a populist "think-tank" that explores the legal basis of corporate power. POCLAD believes that it is possible to control - and, if necessary, dismantle corporations by amending or revoking their charters.8

Since the largest corporations are now transnational in scope, the new populism must confront their abuses globally. The International Forum on Globalization (IFG) was founded for this purpose in 1994, as an alliance of 60 activists, scholars, economists and writers (including Jerry Mander, Vandana Shiva, Richard Grossman, Ralph Nader, Helena Norberg-Hodge, Jeremy Rifkin and Kirkpatrick Sale), to stimulate new thinking and joint action along these lines.

In a position statement drafted in 1995, the International Forum on Globalization said that it: "views international trade and investment agreements, including the GATT, the WTO, Maastricht and NAFTA, combined with the structural adjustment policies of the International Monetary Fund and the World Bank, to be direct stimulants to the processes that weaken democracy, create a world order in the control of transnational corporations and devastate the natural world. The IFG will study, publish and actively advocate in opposition to the current rush toward economic globalization, and will seek to reverse its direction. Simultaneously, we will advocate on behalf of a far more diversified, locally controlled, community-based economics. We believe that the creation of a more equitable economic order-based on principles of diversity, democracy, community and ecological sustainability will require new international agreements that place the needs of people, local economies and the natural world ahead of the interests of corporations."9

Leaders of the new populism appear to realise that anti-corporatism is not a complete solution to the world's problems; that the necessary initial focus on corporate power must eventually be supplemented by a more general critique of centralising and unsustainable technologies, money-based economics and current nation-state governmental structures, by efforts to protect traditional cultures and ecosystems, ***and by a renewal of culture and spirituality.***

It would be foolish to underestimate the immense challenges to the new populism from the current US administration and from the jingoistic, bellicose post September 11 public sentiment fostered by the corporate media. Nevertheless, POCLAD, the Alliance for Democracy and the IFG (along with dozens of human rights, environmental and anti-war organisations around the world) provide important rallying points for citizens' self-defence against tyranny in its most modern, invisible, effective and even seductive forms.

Endnotes:

1. Grossman, Richard and Frank Adams, Taking Care of Business: Citizenship and the Charter of Incorporation, pamphlet, 1993, available at http://www.poclad.org/resources.html.

2. Zinn, Howard, A People's History of the United States: 1492 to Present, Harper Perennial, 2001.

3. Brecher, Jeremy and Tim Costello, Global Village or Global Pillage: Economic Reconstruction from the Bottom Up, South End Press, 1998.

4. Zinn, op. cit.

5. Gross, Bertram, Friendly Fascism: The New Face of Power in America, South End Press, 1998.

6. Goodwyn, Lawrence, The Populist Moment: A Short History of the Agrarian Revolt in America, Oxford University Press, 1978.

7. The Alliance for Democracy website, http://www.theallancefordemocracy.org/.

8. POCLAD website, http://www.poclad.org/.

9. IFG pamphlet, 1995; revised position statement at IFG website, http://www.ifg.org/.

About the Author:

Richard Heinberg is a journalist, educator, editor, lecturer and musician. He has lectured widely and appeared on national radio and TV in five countries. He is a core faculty member of New College of California, where he teaches courses on Culture, Ecology and Sustainable Community.

He is the author of: "Memories and Visions of Paradise"; "Celebrate the Solstice"; "A New Covenant with Nature"; and "Cloning the Buddha: the

Moral Impact of Biotechnology". His next book, "The Party's Over: Oil, War and the Fate of Industrial Societies", is to be published by New Society in March 2003. His essays have been featured in The Futurist, Intuition, Brain/Mind Bulletin, Magical Blend, New Dawn and elsewhere.

Richard is also author/editor/publisher of MuseLetter, a highly regarded monthly, subscription-only, alternative newsletter which is now in its tenth year of publication. MuseLetter's purpose is "to offer a continuing critique of corporate-capitalist industrial civilisation and a re-visioning of humanity's prospects for the next millennium". His article, "A History of Corporate Rule and Popular Protest", was originally published in MuseLetter in 1996 as "The New Populism", and was revised in August 2002. Visit the MuseLetter website at http://www.museletter.com/.

407-Corporate Rule

"Though this be madness, yet there is method in't" Shakespeare

Updated -- 21 September 2003

*Corporations tend to get a bad rap. The good news is that they don't necessarily receive a lot of bad press. This is because they tend to control the media, and in some cases, are the **Media**. Clearly the Media companies are corporations, and/or are owned by corporations. Duh. Meanwhile, those individuals who work for corporations, even the **Transnational Corporations**, may think that it's not as bad as it appears from the outside. But this may only be a case of a gross inability to see the forest for the trees. (Or simply the wholesale, wanton and irresponsible destruction of the same trees, which when piled up alongside the river bank, inhibit any alternative viewpoint.)*

But even those who see corporations as potentially the good guys, sometimes see the dark side, and suspect that there may be some need for reform. One case in point is Robert Monks, a graduate of Harvard University and Harvard Law School, who operates an investment fund with more than $250 million under management, and who has written books about wealth and power, including, The Emperor's Nightingale: Restoring the Integrity of the Corporation in the Age of Shareholder Activism (Addison Wesley, 1998).

Monks has been concerned about the adverse effects of corporate power on society. But by and large, he is still a corporate reformer, i.e. not a corporate abolitionist. He does believe that it is the owners of the corporation who should determine its destiny. In that mode, he—as the controller of a lot of stock ownership—is not above flexing his investment group's stock muscles from time to time.

*Monks did join forces with George Soros, in a campaign against Waste Management Inc., a company with a long history of corporate wrongdoing. Monks identified Waste Management as a company with low stock value, sloppy management and misleading accounting. He then organized shareholders on the **Inter Net**, and forced the resignation of two CEOs. Eventually, the Monks group convinced the third CEO that the level of rot within* *the company was so great that the only way out was a merger with a cleaner, smaller company. Eventually, Waste Management merged with the smaller USA Waste.*

In his book, Monks describes what he calls the ***four corporate dangers: unlimited life, unlimited size, unlimited power and unlimited license. (Which, of course, is coincident with limited liability. And therein lies the problem!)*** *Only shareholder activism, Monk argues, can bring about the four "solutions", i.e.: long-term life, appropriate size, balanced power, and accountability to long-term owners. He believes other efforts at controlling corporate power—such as corporate chartering, investigative reporting, regulation, independent boards—are bound to fail. From his viewpoint, only an activated shareholder movement can bring corporations into line. And as a major stockholder, this provides him with a considerable base of power, i.e. there might be a slight conflict of interest there.*

Even if Monks is right, even if well intentioned, the odds are that his efforts will be in vain.

When the U. S. Supreme Court in 1886 (Santa Clara County v. Southern Pacific Railroad) held that a private corporation was a natural person under the U. S. Constitution [the 14th Amendment to the ***Constitution for the United States of America****], the way was opened for corporations to dominate the public and private life of society, define the economic, cultural and political agenda for humans and all other living things, and perhaps most importantly, allow corporations to assume* ***The First Amendment*** *rights, wherein they can say anything, and yet... still have limited liability.*

An absolutely fundamental difference between living, sentient beings and corporations is that human beings have unlimited liability, and corporations have limited liability. Human beings can forfeit their freedom, their property, even their lives for everything from libel and slander to mistaken judgment to accidents to intentional acts. Corporations can not even be driven out of business for the most questionable acts*—except perhaps in rare cases of public outrage. The latter, however, is wholly unacceptable in that an uninformed or misinformed public opinion could have been molded by—you guessed it—another corporation or group of corporations, also known as the media.*

For those who've seen the movie, Erin Brockervich, the question to ask is: "Did anyone at the corporate utility (employees, directors, consultants), once it had agreed to pay thru the nose for destroying the health and lives of hundreds of people, ever do a single day of jail time? Did the corporation then recoop its liability losses by simply raising the rates for its services? Now, THAT's limited liability, the "corporate veil" at work!

Enter the United States Government, with the where-with-all to curb corporate tyrannies. Unfortunately, this hasn't worked out too well for anyone but the corporations and their highly paid (bribed?) political operatives. Current campaign financing laws, combined with the "revolving door" between corporations and government, are the means by which corporations control absolutely the politics of nations. The degree to which corporations are out of control is best illustrated by the recent massive theft at the upper levels by ***CEOs****—aided and abetted by* ***Independent Accounting Firms****. As the CEO of the New York Stock Exchange departs after eight years, he takes a mere $168 million home with him! Gee, I hope he can find another job.*

Corporate Politics *is in all respects the prime mover in the destruction of Constitutional principles and the idealized American way of life. Whether it is Oprah providing aid and soft money (i.e. free, extensive publicity) to the Republican and Democratic presidential candidates—while ignoring completely the qualified and legitimate third party candidates—or the international banksters being responsible for the* ***US Bankruptcy****, and its after-math, corporations rule the world from the unethical base of the "bottom line".*

About the only saving grace in Corporate Rule is that different corporations have different agendas in many situations, and thus the competition of the greedy keeps some aspects at bay. At the same time, however, organizations which support non-profitable institutions, such as public schools and environmental causes, have a very reduced ability to compete with the 900 pound gorillas.

[Suggestions whereby the non-profit organizations should have access to the same power that Corporations possess, show a lack of foresight. Making still more 900 pound gorillas is not the way to solve the problem of 900 pound gorillas.]

Prisons, for example, once the purview of government (and non-profit), are increasingly being taken over by (for profit) corporations. The corporations are, in most cases, paid on the basis of the number of convicts or prisoners in their "care". The result is that these prison corporations, in order to bolster the bottom line, i.e. have more prisoners, begin to bribe judges and courts so they will levy heavier sentences, and even send individuals to prison who don't deserve it, or who are simply not even guilty—all in the name of profit for the shareholders. [But of course, the stockholders see precious little, as the CEOs and top ranking executives skim off the top most of the ill-gotten gains in the form of compensation, Golden Parachutes, fringe benefits, stock options, etceteras.]

*Corporations not only strongly influence domestic policies within the United States, but also affect American foreign policy, by the simple expedient of an unbridled support of **Capitalism** by the American Government (or governments, when State governments are included). This in turn leads to tremendous inequities in the way the Military Industrial Complex (which President Eisenhower warned us about so many years ago) treats the people of the world. What we have is a Corporate Pac Man Conglomerate gobbling up resources from countries all over the world, and in the end so infuriating (often rightly so) the people of those foreign lands, that terrorist organizations the world over have easy access to disgruntled and disillusioned individuals. And not just from third world countries, as the presence of American and European proxy-terrorists have made clear.*

*The American Government's wholehearted support of Corporate **Capitalism**, combined with what amounts to general greed, stupidity, and power-at-any-cost, has led to a **Crisis of American Government**, where **The Art of Cover Ups** has risen to new heights of sophistication, and Americans and non-Americans alike no longer trust anything the US government says, does, or in which it becomes involved.*

The final horror is a situation where the awesome power of the United States, its military and industrial base, and its collective citizenry, is not used to curb Corporate abuse, but instead to aid it in its nefarious agenda to accumulate ever greater power in the hands an elite. Instead, all indication are that governments throughout the world are slowly handing over the reins of power to what can only be called a ***Corporate State****.*

So... What are the victims of Corporate tyrannies to do?

As Carolyn Chute, author of the 1985 classic, The Beans of Egypt, Maine, and a staunch corporate abolitionist phrased it: "...if faceless financiers and their tool, the corporation, aren't completely out of our lobbies, out of all our campaigns, and out of our constitution soon, looks like We, the People are going to come in the middle of the night with a can of gasoline in every hand and one fat match."

This is not necessarily a recipe for a ***Revolution****, or even* ***Anarchy*** *on a massive scale. But inasmuch as it might have a deleterious effect on the bottom line of the Corporations, they might consider taking Ms. Chute's sentiments with a certain amount of concern. Just as labor unions forced their way into the mainstream, despite the government's assistance to the corporations in attempting to destroy them during the first half of the Twentieth Century, there is a point where it's no longer profitable to forge the* ***Corporate State****.*

A curious, highly ironic article appeared in the Wall Street Journal on ***9-11-2001*** *(i.e. the morning edition written and published before the more notorious events of that date). The article was about the many protests against the World Trade Organization, and had a quote from a person of considerable authority who stated confidently that, "DC will not burn!" Admittedly, the Pentagon is across the river from "DC", but I think the point is made that protests cannot be automatically dismissed out of hand.*

408-Transnational Corporations

"The least pain in our little finger gives us more concern and uneasiness than the destruction of millions of our fellow beings" Williiam Hazlitt.

Transnational corporations -- those corporations which operate in more than one country or nation at a time -- have become some of the most powerful economic and political entities in the world today. From Joshua Karliner, in his book, The Corporate Planet: Ecology and Politics in the Age of Globalization [Sierra Club Books, 1997], we can gleam a host of fundamental realizations, including the fact that many of these companies have far more power than the nation-states across whose borders they operate.

For example, the combined revenues of just General Motors and Ford -- the two largest automobile corporations in the world -- exceed the combined Gross Domestic Product (GDP) for all of sub-Saharan Africa. The combined sales of Mitsubishi, Mitsui, ITOCHU, Sumitomo, Marubeni, and Nissho Iwai, Japan's top six Sogo Sosha or trading companies, are nearly equivalent to the combined GDP of all of South America. Overall, fifty-one of the largest one-hundred economies in the world are corporations. The revenues of the top 500 corporations in the U.S. equal about 60 percent of the country's GDP. Transnational corporations hold ninety percent of all technology and product patents worldwide, and are involved in 70 percent of world trade. More than thirty percent of this trade is "intra- firm"; in other words, it occurs between units of the same corporation.

The number of transnational corporations in the world has jumped from 7,000 in 1970 to 40,000 in 1995. While global in reach, these corporations' home bases are concentrated in the Northern industrialized countries, where ninety percent of all transnationals are based. More than half come from just five nations: France, Germany, the Netherlands, Japan and the United States. But despite their growing numbers, power is concentrated at the top. i.e., the 300 largest corporations account for one-quarter of the world's productive assets.

The United Nations has justly described these corporations as "the productive core of the globalizing world economy." Their 250,000 foreign affiliates account for most of the world's industrial capacity, technological knowledge, international financial transactions, and ultimately the power of control. In terms of energy, they mine, refine and distribute most of the world's oil, gasoline, diesel and jet fuel, as well as build most of the world's oil, coal, gas, hydroelectric and nuclear power plants. They extract most of the world's minerals from the ground. They manufacture and sell most of the world's automobiles, airplanes, communications satellites, computers, home electronics, chemicals, medicines and biotechnology products. They harvest much of the world's wood and make most of its paper. They grow many of the world's major agricultural crops, while processing and distributing much of its food.

Given their dominance of politics, economics and technology, it is not surprising to find the big transnationals deeply involved in most of the world's serious environmental crises.

Transnational corporations exert significant influence over the domestic and foreign policies of the Northern industrialized government that host them. Surprise! Indeed, the interests of the most powerful governments in the world are often intimately intertwined with the expanding pursuits of the transnationals that they charter. At the same time, transnational corporations are moving to circumvent national governments. The borders and regulatory agencies of most governments are caving in (or being paid off) to the New World Order of globalization, allowing corporations to assume an ever more stateless quality, leaving them less and less accountable to any government anywhere.

These corporations, together with their host governments, are reorganizing the world economic structures -- and thus the balance of political power -- through a series of intergovernmental trade and investment accords. These treaties serve as the frameworks within which globalization is evolving -- allowing international corporate investment and trade to flourish across the Earth. They include:

- *The Uruguay Round of the General Agreement on Tariffs and Trade (GATT)*

- *The World Trade Organization, which was created to enforce the GATT's rules.*

- *The proposed Multilateral Agreement on Investment. (MAI)*
- *The North American Free Trade Agreement (NAFTA).*
- *The European Union (EU).*

These international trade and investment agreements allow corporations to circumvent the power and authority of national governments and local communities, thus endangering workers' rights, the environment and democratic political processes.

A legitimate question is: Is the World Trade Organization an arm of the United Nations?

In a "Substantive session", 16 July 1996, the United Nations provided a discussion and a "draft decision submitted by the President of the Council on the basis of informal consultations" on the topic of one item on their agenda, i.e. "Non-Governmental Organizations" (i.e. Transnational and National Corporations):

> *"The Economic and Social Council, reaffirming the importance of the contributions of non-governmental organizations to the work of the United Nations, taking into account the contributions made by non-governmental organizations to recent international conferences, [emphasis added]*
>
> *"Decides to recommend that the General Assembly examine, at its fifty-first session, the question of the participation of non-governmental organizations in all areas of the work of the United Nations, in the light of the experience gained through the arrangements for consultation between non-governmental organizations and the Economic and Social Council."*

In many respects, the logic is inescapable. If as few as 300 Transnational Corporations (TNCs) do indeed represent 51% of the largest one-hundred economies, then it is basic economic (and therefore political) logic that the TNCs should be represented at the United Nations!!! Or else, one needs to remove from the General Assembly all the smaller economies (countries). But then this also implies that at some point a TNC or two will become a member of the Economic and Social Councils, and then, perhaps, the General Assembly. And then, of course, the Security Council?

Is this altogether bad news? If we are replacing tyrannical, autocratic, or undemocratic member nations with TNCs (who are theoretically answerable to shareholders, and where anyone in the world can become a shareholder), then it might not sound so bad. However, there are shareholders and there are shareholders. Holding a hundred shares of stock does not count for quite as much as holding ten million shares. And the reality is that if TNCs do become effectively sovereign nations, then it's inevitable that the rich shareholders will use their power to take over the TNCs and become the world's elite governing force. What the United Nations is perhaps not taking into account, is the unequal distribution of control over the destiny of a TNC. The fact that CEOs and high executive officers have been taking any and most all corporate shareholders to the cleaners for decades is just more fuel to the raging inferno of the lack of corporate accountability.

To counter this trend, Dave Hartley has suggested "four wisdoms of de-globalization":

Think for Yourself

Question Authority

Globalize Consciousness

Localize Economies

There is a slight problem, however. The TNCs and domestic major corporations already control the governments of the United States and most of the industrialized nations. This is done by Corporate Politics, and the wholesale purchase of politicians. Already in place is a Corporate State, which wields the economic power of Money. This, by the way, is not a new development. Even in the United States, the dye was cast by the Fourteenth Amendment to the Constitution for the United States of America, whereby private corporations were given the status of persons -- and thus allowed to enjoy the benefits of the Bill of Rights, without the commensurate unlimited liabilities of life. In other words, it's pretty much a done deal.

The problem, of course, is not merely the greed, power-hungry, and irresponsible actions of certain CEOs and the elite of the TNCs -- but also those who support them. The latter clearly includes bribed governments, but they also include the average person who buys the products and uses the services that these corporations provide. The fact that there is seemingly little choice -- due to governmental intervention -- makes it less of a charge of negligence to the average individual. But these same individuals also often take the path of least resistance, and choose the quality of life which includes telephones, Inter net connections, utilities and services at their beck and call, easy transportation, vacations at remote locations, and so forth and so on. An intriguing discussion of this idea is included in One World Order -- a conservation among several interested observers.

E.g., one highlight of the One World Order is the number of allegedly "high-ranking" individuals who have made it clear that there are forces in the world (potentially for the last several hundred years) that control the destiny of the planet. That there's a Hierarchy which far transcends the mundane, worldly powers. These forces may have initiated all manner of questionable acts -- from initiating wars and famines to encouraging plagues and other members of the Four Horsemen. Are they have simply responded to these incidents, but in a way that mundane ethics and morality might have found shocking.

If we assume TNCs are simply a portion of the problem, that there are in fact powers far in excess of the average multi-billion dollar corporate CEOs, and that these higher powers are intent upon a One World Order, then we have to wonder if perhaps, just perhaps, this is a good thing in the grander scheme of things. If one was in an overcrowded lifeboat, for example, and in order to save everyone else, the toughest dude in the boat tossed three of the weaker members overboard and allowed them to drown... what would you do?

Being in charge sometimes requires tough decisions. The key is the deeper intentions of whoever is in charge. If, for example, someone or some group has much more power than President George Bush, would this be all that bad? Would that perhaps be wonderful news? Doesn't it depend upon what the tougher dude's intentions are? And how those stack up to Bush's intentions (which seems to be clearly: "more money through O.I.L.")? Let's face it: Could things be worse than having John Ashcroft as Attorney General?

Conspiracies do not generally have a great deal of credibility with the average individual. But there is just the possibility that the "Mother of Conspiracies" may be one that has an agenda that is not all that bad. Perhaps, in our Creating Reality, we can manifest exactly that sort of situation.

So, how about getting on that project immediately!

The Library of Halexandria

409-Aristocracy

"Custom reconciles us to everything." Edmund Burke

Updated 15 September 2004

Aristocracy is a name associated with Western Europe and the old days. There was, for example, up until the midpoint of the Twentieth Century, an automatic assumption among the powers that be that there were two distinct classes: Aristocracy and Commoners. This attitude is reflected in Pope Leo XIII's **Encyclical of Condemnation**, *which was issued against* **Freemasonry** *for holding, practicing and encouraging such heretical ideas as equality of human beings (male and female) in all aspects of life; including the right of commoners to remove heads of state and other high ranking officials by the rules of law. Can you imagine his chagrin at the American* **Revolution**!

It's important to note, however, that the Founding Fathers of the United States -- that bastion of radical thoughts such as "all men are created equal" -- these people were in the main, aristocrats. Locally grown, perhaps, and without titles, but pretty much the elite. They were also often Freemasons, but that's another issue. For the most part, they were the people with the money, power, social standing, connections, and family bloodlines. Same old, same old. And any pretense of the American Revolution being merely about the struggle of commoners against an outrageous aristocracy is flawed by a lack of not being the whole truth. In fact, the American revolution was about an American aristocracy using commoners for cannon fodder in a common goal to supplant the English aristocracy and replace them with a home-grown version. Some of the American aristocrats may have been doing what they thought best for the commoner, but these were rare individuals.

There has since arisen an equally ludicrous concept that the aristocracies of old no longer have any real power. In reality, aristocracies exist today which are in fact the powers that be -- even when for various and sundry diplomatic and deceptive reasons they may choose to not be addressed as lords, ladies, knights, earls, and esquires. They may be positively demure in exhibiting themselves on the street or calling attention to themselves in any way.

Dirk Wittenborn, in his delightful and entertaining novel, Fierce People [1], has one of his characters say, "The brilliance of the American aristocracy is they've convinced the world they don't exist... It's safer that way. Unlike us [the old European aristocracy], they're invisible targets..." Furthermore, "It's very clever -- you teach them in America anyone can become rich, so that when they hate the rich, they hate themselves. It paralyzes them. All they can do is eat."

Wittenborn's book is as Candace Bushnell has noted, "A riveting page-turner that offers a haunting and fascinating glimpse into the lives of the super-rich." In effect, the characters in the book are the American aristocracy, and also equivalent to the Fierce People -- more cutthroat than anything in any ghetto, slum, or grimy downtown street. Think of Bill Gates with premenstrual syndrome and a government inspired migraine.

The fascinating possibility is that Wittenborn was allegedly raised in a community similar to the one he describes in his fictional story. In effect, the story is realistic to the core, and there is, according to Wittenborn and others, an American aristocracy which does in fact control all significant wealth and power in the country, regularly pass this wealth and power on to its progeny (including those who constitute the demented results of years of in-breeding), and in general do whatever they damn well please -- any and all laws, rules, regulations, and whatever notwithstanding. They are the Lords of the Flies, and rules applicable to the commoners need never apply.

These are the people whose children are raised in very private, exclusive schools, where their ***Education*** *is about assuming their rightful places at the apexes of power. These are the children who get to meet the astronauts up close and personal, who can receive an abortion by the best trained doctors in the world (regardless of the parents' public stand on the issue or relevant State or local laws), who can avoid the draft with the dismissal wave of their hand, who will go to the best colleges and universities, and who ultimately can become most anything they want (barring the desires of their equally aristocratic peers) -- including becoming President of the United States.*

[As interesting aside is that President Bush (aka Shrub) recently had his administration argue in a court of law against any form of quota systems in college admissions. The college, in question, was attempting to allow minorities (just by virtue of being a member of a minority) an extra 20 extra points out of a maximum of 150 -- as a means of giving them a headstart in the admissions lottery. But when Shrub went to Yale, the admissions program

was equivalent to giving him (as a son of President, CIA Director, etceteras) an extra 250 points out of, say, a total of 150. Basically, Shrub got into college on a quota system -- his minority being his stereos in the aristocracy. Irony is a wonderful thing.]

*And speaking of presidents -- and their so-called elections -- the late breaking news (as of August 16, 2004) is that John Kerry will be elected in November 2004. Why? Because he has "more royal connections than his Republican [sic] rival". Straight from London, we have the news that **Royal Researchers** are predicting a Democratic [sic] victory simply "because of the fact that every presidential candidate with the most royal genes and chromosomes has always won the November presidential election", and that based on 42 previous presidents, the coming election "will go to John Kerry." Talk about an American Aristocracy! Of course, the polls since then have not been encouraging (or discouraging, depending upon your viewpoint). But who ever said that polls -- or for that matter, votes -- were the deciding factor in elections! I mean, get serious!*

*In a related touch of madness, **Forbes Magazine** has noted that -- if elected -- John Kerry will only be the third richest president in history, falling behind front runner George Washington and second place finisher, John F. Kennedy. [What might blow your mind is that good-ole-boy, Lyndon Baines Johnson, placed fourth in the presidential-financial sweepstakes! Never let it be said that Congress is not an enriching experience!]*

*So where are the surprises? The reality is that the **Republic** called the United States of America was founded by Aristocrats. The great distinguishing feature, however, was that these founders were a very rare breed of aristocrats -- individuals with true vision and a curious thing called a consciousness. Gordon S. Wood [1] has written:*

> *"In fact, these 18th century figures were extraordinary men, products of a peculiar moment in our history when the forces of aristocracy and democracy were nicely balanced. Although most all of them were men of relatively modest origins, they were unabashed elitists who had a contempt for electioneering and popular politics. [Can you imagine their reaction to the presidential mud slinging bath of 2004?] They rejected*

> *blood and family as sources of status, however, and were eager to establish themselves by principles that could be acquired through learning and **education**. They struggled to internalize the new, Enlightened Man-made standards that had come to define what Jefferson called the 'natural aristocracy' -- politeness, sociability, compassion, virtue, disinterestedness and an aversion to corruption and courtlike behavior."*

*Obviously, American politics has been corrupted to a horrific extent -- but don't assume that they're done yet. With **Dominionism** rampant in the higher echelons, it can only get worse. There seem to be no constraints on the immorality of those in power. Which is a curious thing. As Jefferson once said [1]: "State a moral case to a ploughman and a professor; the ploughman will decide it as well, and often better, because he has not been led astray by artificial rules." It might perhaps be well for American politics to replace Ken Rove with a ploughman as President Bush's chief political advisor.*

*Meanwhile, the American Aristocracy has joined the old aristocracies in being the leaders of any and all generations. It is perhaps to their credit (or to their practicality) that those born of lesser parents can sometimes be reluctantly admitted to the aristocratic club. But such admissions are generally to the advantage of the club's elder guard, who perceive a use or value in so and so enterprising (and likely manageable) fellow's activities, and in no way need be considered to be a gesture of genuine **Philanthropy**.*

*(6/22/06) A decidely more conspiratorial view of "the elite" is Nick Sandburg's "**Blueprint for a Prison Planet**". While this view may appear to be somewhat extreme, it is nevertheless built on some logical and rational foundation. Much of it is quite plausible. If nothing else it makes for a good, horror story. One of its more classic statements is: "America is the ultimate control fantasy - **consensual incarceration** - whole groups of people slowly driven to believe that there exists no way of securely living together other than by the giving up of personal freedom bit by bit." [**emphasis** added]*

Fortune Magazine in 2002 had an interesting article on the aspect of the American and other aristocracies, whose premier members have finally caught on to the wisdom of Wittenborn's character. In effect, what was once considered a perk for CEOs (Chief Executive Officers of major corporations -- those who likely attained their position by family connection and wealth), to flaunt their success with new Trophy World Corporate Headquarters buildings, has now gone slightly out of fashion. Such arrogance has proven to be a bit of an Achilles heel.

How Trophies Lost Their Allure

These days, brand-name buildings are out and anonymity is in.

FORTUNE

Monday, October 28, 2002

By Devin Leonard

CEOs used to have an important rite of passage: When a company reached a certain level of grandeur, the chief executive hired a major architect to design a new headquarters, a trophy building symbolizing the company's financial might and corporate style. The skyline of every major American city is adorned with these wonders. Chicago has the Sears Tower. San Francisco boasts the Transamerica Pyramid. New York City has monuments erected by Woolworth, Citicorp, Seagram, and Chrysler, to name just a few.

*Now corporate America's love affair with the trophy building is waning. Obviously there's a concern that after **9-11-2001**, buildings calling attention to themselves may be terrorist targets. [And yet the City of New York is planning an even bigger target!] Earlier this year DEGW, a consulting and design firm, surveyed real estate executives at large U.S. companies and found 41% were less interested in occupying high-profile buildings after Sept. 11. Nearly a fifth said they would be happier in "anonymous" buildings. Morgan Stanley, for example, passed on plans to move into a posh midtown tower, instead relocating hundreds of employees to Jersey City, where the tony firm is just another tenant in a nondescript building.*

*It would be a mistake, however, to blame Sept. 11 entirely for the decline of trophy buildings. Matthew Cullen, head of real estate for GM and chairman of CoreNet Global, an association of top real estate executives, says companies began forsaking trophy buildings nearly a decade ago when technological advances like the **Inter Net** and cell phones allowed employees to work anywhere, lessening the need for a fancy tower.*

Sandy Apgar, a director at the Boston Consulting Group who advises corporations on real estate, sees the same trend. He counsels clients to follow the lead of AT&T, which cut costs and increased productivity in the early '90s when it left its posh, Philip Johnson-designed Manhattan headquarters, moved its executive offices to the Jersey suburbs, and encouraged employees to telecommute. Sept. 11, says Apgar, has only accelerated things. Though a handful of companies like Ernst & Young and Reuters have recently moved into trophy-worthy digs, "the age of the corporate-icon building is passing," he says. "One of the great challenges facing the real estate industry is how to rethink and reuse the traditional corporate headquarters building for other purposes. I know some people who are thinking, very quietly to be sure, about converting buildings in New York and elsewhere into condos." Some companies--in particular, Random House and AOL Time Warner, parent of FORTUNE's publisher--are relinquishing the top floors of their under-construction Manhattan trophy buildings so that they can be sold as luxury condominiums rather than set aside for executive offices. That tells you how much things have changed.

So is this the end of a great corporate tradition? Probably not. Trophy buildings go up in boom times, not in downturns. Merrill Lynch recently noted that there was currently a "complete absence of demand" for office space and expected paralyzing market conditions to last at least though 2003. However, once the economy picks up, that sentiment may change. For all their downsides, trophy buildings offer CEOs a form of immortality. Few of the cars built by Walter Chrysler in the '20s survive, but the art deco spire he commissioned still looms majestically over Manhattan. You'll never make history for moving your employees into a nondescript suburban office building.

The fact that the upcoming American Aristocracy may be less obvious in their corporate digs is not necessarily cause for celebration. A covert conspiracy is often worse than an overt action -- although increasingly, there is no apparent attempt to conceal anything.

Other than perhaps an aristocratic contempt for commoners. As The Little King once said, when informed that the peasants were revolting, "They always have been."

References

[1] Gordon S. Wood, "Where Are the Jeffersons of Today?" Time Magazine, July 5, 2004.

References:

[1] Dirk Wittenborn, Fierce People, Bloomsbury, New York and London, 2002.

410-INTERESTING INTERPRETATION

"The greater the power, the more dangerous the abuse." Edmund Burke.

(The following article is for interest only, NOT for education. I see David Icke named in the credits, and feel he is known "nutter", now notorious for declaring the English royal family is comprised of Alien Lizard vampire species who never die, but re-incarnate as the next royal, but actually the same person each incarnation. He also sees the Anunnaki as purely reptilian aliens. That probably says enough about the legitimacy of such articles and authors. Bob M.)

The Anunnaki, the Vampire, and the Structure of Dissent

by Marcus LiBrizzi
from Reconstruction Website

The vampire, an archetypal figure who pops up in many myths from around the world, is most familiar to Western audiences in the form Bram Stoker's Dracula and Anne Rice's Lestat - aristocratic bloodsucking immortals of unholy origin. In more paranoid circles, vampires have been re-imagined as a race of alien beings called the Anunnaki, who have traveled from beyond to control and colonize the planet Earth (in fact, they've been in control for quite a while now).

Looking at the conspiracy theories of underground celebrity David Icke, Marcus LiBrizzi offers his own theory about the meaning of these horrific beings for a world caught in the grip of a grand economic reorganization. Linking these myths to the realities of transnational capital and the Network Society, LiBrizzi is able to craft his own compelling narrative about the horrors of the New World Order.

<1> The latest incarnation of the vampire - in the conspiracy theories of David Icke - reveals the critical, revolutionary heart of the vampire legend. Discourse on the vampire appears above all to provide a structure

of dissent, a metaphorical means of representing and soliciting critiques of the social order. The Anunnaki form of the vampire - in its immersion in the constellation of contemporary conspiracy theories, in its reflection on global capitalism, and in its blurring of historical and fictional narratives - has moved this structure of dissent from the cloak of darkness to the light of day.

<2> Considered by some to be the reigning conspiracy theorist in the US, David Icke (who is British) formulates his theories of a worldwide, age-old conspiracy around an extraterrestrial race of beings called the Anunnaki. Self-styled the "most controversial author and speaker in the world," David Icke has been subject to much ridicule but has nonetheless become an industry, publishing eleven books, producing video and audiotapes, embarking on a worldwide lecture circuit, and creating a website that allegedly attracts 10,000 visitors a day (Canadian Par. 13).

A former soccer player from a working-class family, Icke became a household name in the UK as a national sports and news reporter for the BBC and as the spokesperson for the Green Party ("About" Par. 7-8). Starting a full-time writing career in the early 1990s, Icke began with New Age inspired works like Truth Vibrations (1991), which combines accounts of his self transformation with psychically-imparted warnings on the imminent destruction of the earth, from there moving towards conventional conspiracy theories, and finally, beginning with his 1999 book The Biggest Secret, focusing his conspiracy theories around the Anunnaki and their nefarious involvement in human history.

<3> The Anunnaki, whose name is Sumerian, meaning "those who from Heaven to Earth Came" (Icke 5), refer to a reptilian race that originated from the legendary planet known as Nibiru (Planet X), or the place of the crossing, which has a 3,600 year elliptical orbit that takes it between Jupiter and Mars and then out into space (5). For the past 450,000 years, according to Icke, the Anunnaki have been ruling earth in different guises and from different dimensions. Through genetic engineering, the Anunnaki have manipulated the evolution of humans as a slave race.

"The Anunnaki created bloodlines to rule humanity on their behalf," he writes, "and these ... are the families still in control of the world to this day" (9).
The interbreeding of the rich and powerful (primarily, for Icke, the European aristocracy and the Eastern Establishment of the US) is not done for reasons of snobbery but rather, "to hold a genetic structure that

gives them certain abilities, especially the ability to 'shape-shift' and manifest in other forms" (9).
Working with these crossbreeds are full-blooded Anunnaki, some physically present on earth, others influencing individuals and events psychically from what Icke calls "the lower fourth dimension" (25). Forming a "Brotherhood" or secret society network, the Anunnaki have effectively "hijacked the planet" (46).

<4> The recurring motif in the discourse on the Anunnaki is vampirism. In fact, so strong is this component in their depiction that it's safe to say that Icke's work represents one of the most recent developments in the discourse of the vampire.
"While vampire beliefs are varied," writes James Craig Holte, "certain elements of the vampire myth are consistent. The most important are the inability to experience death, the importance of blood, and the sexual connection between vampire and victim" (246).

Other structural similarities between the traditional vampire and the Anunnaki include shape-shifting, hypnotism, and links to secret societies. After establishing the Anunnaki as a manifestation of the vampire, we'll unpack the implications of this figure, using the tools of a Marxist critical practice.

<5> The Anunnaki, like traditional vampires, enjoy eternal or extenuated life spans. Icke claims that, "the fourth dimensional reptilians wear their human bodies like a genetic overcoat and when one body dies the same reptilian 'moves house' to another body and continues the Agenda into another generation" (46).
One type of creature Icke describes is a reptilian "inside" a human physical body;
"it seems that … the Anunnaki need to occupy a very reptilian dominated genetic stream to do this, hence certain bloodlines always end up in the positions of power. Other less pure crossbreed human-reptilians are those bodies which are possessed by a reptilian consciousness from the fourth dimension and these are people who psychics see as essentially human, but 'overshadowed' by a reptilian" (46).

Crossbreeding to infuse reptilian genetics into human bloodlines, the Anunnaki gain the means to defy death, as we conceive it.

<6> In respect to blood drinking, Icke is very clear: The Anunnaki drink blood, which they need in order to exist in this dimension and hold a

human form (288). Embedded in this need lies another parallel between the Anunnaki and the figure of the vampire - the power to shape-shift (from reptilian to human form for the Anunnaki, and usually from vampire form to that of bat or even mist for the traditional vampire). But the Anunnaki also feed off fear, aggression, and other negative emotions. Thus, while blood is needed as a vital life force, the Anunnaki are also addicted to "adrenalchrome," a hormone released in the human body during periods of extreme terror (290, 331). Rather than sucking the blood directly from the necks of their victims, the Anunnaki apparently slash the throats of their victims from left to right and consume the blood out of goblets (303). Icke claims that the origin of the vampire stories are the blood drinking and "energy sucking" rituals of the Anunnaki (26).

"In India," he writes, "it was called soma and in Greece it was ambrosia, some researchers suggest. This was said to be the nectar of the gods and it was - the reptilian gods who are genetic blood drinkers" (288).

<7> In the sexual connection between slayer and victim, the Anunnaki also share another similarity with the traditional vampire. However, depictions of the Anunnaki by Icke contain none of the erotic allure and seductiveness that distinguish many vampire texts. Instead, the sexual bond between the Anunnaki and their victims is characterized by violence - rape, murder, and Satanic ritual.

"Satanism at its core is about the manipulation and theft of another person's energy and consciousness," writes Icke, who states that "sex is so common in Satanic ritual because at the moment of orgasm, the body explodes with energy which the Satanists and the reptiles can capture and absorb" (295).

For Icke, of course, the demons honored or appeased by satanic sex rituals are none other than the reptilian Anunnaki (34). Sex is also a fundamental tool of the Anunnaki mind control program and, more prosaically, it figures prominently as a means of blackmail. The picture that emerges is one involving vast networks of sexual abuse and ritual murder - graphic accounts of satanic practices at the playgrounds for world leaders, such as the Bohemian Grove, a 2,700 acre compound north of San Francisco - mass graves for victims drained of their blood and libidinal energies - and the cultivation of sexual crimes to create an energy field that nourishes these rapacious ETs.

<8> There are other shared traits between the traditional vampire and the Anunnaki, for example, the role of secret societies. One of Icke's chief contributions to the discourse on the vampire lies in his immersion of this figure into a vast web of clandestine organizations, from ancient mystery schools and cults like the Brotherhood of the Snake to the Knights Templar and the Masonic Order, from global entities like the UN, the Trilateral Commission, and the Council on Foreign Relations to drug cartels, satanic churches, and the Black Nobility. A keystone in this architecture of conspiracy is the Order of Draco, which conjures up the most famous of all vampires - Count Dracula - and underscores his demonic, draconian, and reptilian associations.

"According to Laurence Gardner, the name Dracula means 'Son of Dracul' and was inspired by Prince Vlad III of Transylvania-Wallachia, a chancellor of the Court of the Dragon in the 15th century. This prince's father was called Dracul within the Court" (56).

In their network of secret societies, of which the Order of Draco is but a single manifestation, the Anunnaki highlight the conspiratorial dimension of all vampires. Finally, the Anunnaki share with the traditional vampire the capacity to hypnotize: Icke writes that reptilian bloodlines, "have the ability to produce an extremely powerful hypnotic stare, just like a snake hypnotizing its prey and this is the origin of giving someone the 'evil eye'" (42).

<9> Icke's paradigm displays more than the vitality, persistence, and adaptive qualities of the vampire legend. His theories reveal the dissident energies contained already in the vampire legacy.

<10> To begin with, Icke's work represents a major fusion of the vampire cult and the field of conspiracy theories. Richard Hofstadter, in his famous essay "The Paranoid Style in American Politics" (1963) claims that the,
"distinguishing thing about the paranoid style is not that its exponents see conspiracies or plots here and there in history, but that they regard a 'vast' … conspiracy as the motive force in historical events. History is a conspiracy" (29).
Conspiracies, even when they're not construed as vast, over-arching plots, however, have an internal, integrative logic. In other words, there is a momentum in conspiracy theories to pull in all other theories, and **finally to arrive at a state in which everything is connected.**

Part of Icke's popularity lies in his ability to integrate most contemporary American conspiracy theories into one over-arching framework. Situated squarely in the center of this design is the ancient figure of the vampire. Thus, the vampire (or, more specifically, the Anunnaki Vampire) has colonized the field of conspiracy theories - government-sponsored alien cover-ups, the New World Order, suspicious deaths, the secret government, suppressed research, the intrigues of the CIA, and the list goes on indefinitely.

<11> From a Marxist perspective, of course, this development is more than just a formal or aesthetic innovation, for many of the conspiracy theories now circulating in the cultural medium of the US contain, at their core, critical, dissenting, and rebellious points of view (encompassing both extreme right and left) that are articulated in opposition to the social, political, and cultural status quo. While Hofstadter claims that the US has no monopoly on conspiracism, other scholars like Peter Knight hold that conspiracy theories hold an indispensable place in American ideology formation, and that current,
"conspiracy theories can be read in part as panicked responses to the increasing multiculturalism and globalization of the present" (5).

Revolutionary or reactionary, however, these theories are inimical to the governing elite and represent a tradition of oppositional practice. As Knight puts it, "conspiracy theory has become the lingua franca of a countercultural opposition that encompasses a vast spectrum of political thinking from the committed to the casual" (6-7).

<12> An initial difficulty in seeing the vampire as a symbol of the ruling class - capitalist or otherwise - lies in the diverse variations taken on by vampires in different places and times. As Brian Frost puts it, "the vampire is a polymorphic phenomenon with a host of disparate guises to its credit" (1).

Among the various legendary "guises" of the vampire inventoried by Frost are spirit vampires, astral vampires, psychic vampires, animal vampires, and real-life vampires who are, "sadistic criminals … urged on by a physical craving for blood" (15).

Complicating the picture is the fact that Bram Stoker's character of Count Dracula, who for many encapsulates the aristocratic ethos of the vampire,

"lacks precisely what makes a man 'noble': servants. Dracula stoops to driving the carriage, cooking the meals, making the beds, cleaning the castle" (Moretti 90).
Furthermore, in some of the earliest European vampire legends, the undead feed off the living members of their own families (Murgoci 18), which at first glance mitigates the social-class dynamic often conjured up in the image of aristocratic vampires draining the lifeblood of their locals.

<13> There is, nevertheless, a critical and even radical dimension to the figure of the vampire, who, as a parasite, circulates as a political metaphor. The word vampire has from the start been used in oppositional literature as a symbol of an exploiting class, government, industry, or institution. A decade, "after the introduction of the word 'vampire' in an English publication in 1732, (an account of the investigation of Arnold Paul in Serbia) ... a serious utilization of the vampire as a political metaphor occurred in Observations on the Revolution of 1688 (... published in 1741)"which identified foreign investors as "'Vampires of the Publick'" (Melton 538).

Only "a few years later, in 1764, Voltaire, in his Philosophical Dictionary," refers to "vampires" as "'stock-jobbers, brokers, and men of business who sucked the blood of the people in broad daylight'" (538).

<14> But it was Marx who first suggested that the vampire can be interpreted as a metaphor of capitalism and who also implied a method for this interpretation. In volume one of Capital (1867), he writes that, **"capital is dead labour, which, vampire-like, lives only by sucking living labour, and lives the more, the more labour it sucks" (342).**

Extrapolating on this analogy, Franco Moretti provides a reading of Bram Stoker's Dracula, writing, "If the vampire is a metaphor for capital, then Stoker's vampire, who is of 1897, must be the capital of 1897" (92). Accordingly, Moretti sees Count Dracula as the expression or figure of monopoly capitalism, which, to the 19th century bourgeoisie, could not be recognized as an emerging force but only as a relic of the past displaced into the present (93). Whether or not one agrees with Moretti's reading of the Count, it is his method that's of most value. As Rob Latham pus it, "Moretti stresses that, while the vampire is a perfect general image for the basic mechanism of capitalist development, individual vampire texts illuminate specifically the historical phases of capitalism in which they are produced" (129).

<15> Applying Moretti's method, we can perceive the Anunnaki as metaphorical of the unique forms capitalism has taken by the 21st century. Certainly, Anunnaki vampires embody the market for genetic engineering as well as space exploration. These dimensions, in fact, are projected back into the origins of Anunnaki control over earth and its resources: travel from another planet, interdimensional traffic, and a crossbreeding agenda coterminous with the evolution of the human race. Anunnaki vampires also control finance, which was undergoing a tremendous transformation and development during the time when Icke was writing that, **of all the spheres of Anunnaki domination, "the most important ... in terms of control, is banking" (207).**

Electronic banking, credit, and the demediation of stock exchange through on-line trading are some of the key elements in the recent development of the finance industry (Castells 152-53). But we can go deeper than this kind of analysis, and discover in the discourse on the Anunnaki examples of remarkable changes, not in select markets, but rather in the very structure of the economy.

<16> In this, more significant sense, the Anunnaki are linked to present-day capitalism through their association with global control. Icke consistently depicts these alien bloodsuckers as monopolizing world leadership positions in government, finance, religion, and the media. In this sense, Anunnaki vampires represent a demonized expression of the unique form capitalism has taken during the very period in which Icke's theories were formulated, published, and popularized. The late 1990s issued in - for the first time in history - a global economy, defined by Manuel Castells as, "an economy whose core components have the institutional, organizational, and technological capacity to work as a unit in real time, or in chosen time, on a planetary scale" (102).
Thus, "this is a new brand of capitalism, technologically, organizationally, and institutionally distinct" (160-61).

<17> The forces spearheading this change derive in part from key industries, notably information technology - centering on the Internet - finance, and biotechnology (Castells 161). Other contributing factors in the formation of the global economy are government policies that restructured capitalism through laws deregulating and liberalizing economic activity (148). The global economy has, of course, catapulted the scale of capitalism; "for the first time in history the whole planet is capitalist or dependent on its connection to global capitalist networks" (160-61).

However, as Castells points out, the global economy, "is not a planetary economy … because it does not embrace all economic processes in the planet, it does not include all territories, and it does not include all people in its workings, although it does affect directly or indirectly the livelihood of all humankind" (132).

Thus the global economy is significant, not only for it inclusivity, but also for its significant and shifting exclusions, marginalizations, and hidden bypasses fraught through its great grid or network of power relations.

<18> Anunnaki vampires are perfectly suited to, and a perfect representation of, a global economy in the scope of their engagement and their profile in emergent industries, but there are other ways as well. This is because their secret agenda has always already been the creation of a one-world government - a New World Order - bypassing nations and creating a system or web from which there is no escape. The New World Order figures prominently in conspiracy theories and in literature such as Aldous Huxley's Brave New World (1932) and George Orwell's Nineteen Eighty-Four (1949). But during the millenium and start of the 21st century, demonstrations against globalism have been on the rise, responding to rapid developments in transnationalism.

Another aspect of the Anunnaki relevant here is their multicultural image. The Anunnaki have been written retroactively into all mythological systems, making them true transnationals. For example, they people the pages of the Indian Vedas, Babylonian myths, as well as the books of the Bible, and they are at the heart of ancient snake-worshipping cults worldwide. Moreover, they are literally seeded into the human genome through the Anunnaki engineering of the race, interbreeding alien genetics into all peoples, symbolized, for example in Genesis, as the saliva Jehovah mixes with clay to form the first man.

<19> Not surprisingly, Anunnaki narratives have a lot to say in terms of the location, construction, and commodification of the self. Unlike traditional vampires who feed solely off a victim's blood or soul, the Anunnaki thrive off of negative energies such as fear and aggression. These ETs drain individuals of their sense of wellbeing through the manipulation and absorption of libidinal energies and - ultimately - the theft of consciousness and agency. On the one hand, the location of the self that the Anunnaki attack seems closely linked to consumerist notions. For example, New Age self-actualization products as well as the market for energy drinks - even caffeine-enhanced water - not to mention designer drugs - are only a few of the new industries catering profitably to the very malady Icke derives from Anunnaki domination.

And, of course, Icke's works themselves represent a (profitable) venture in a multi-million dollar market for conspiracy theories in American popular culture. On the other hand, discourse on the Anunnaki is not necessarily complicit with the capitalist system that produces such effects. A current line of cultural theory,
"has alleged that the modalities of consumer culture - and the forms of subjectivity they enable - do not necessarily integrate seamlessly into the capitalist society which has mobilized them but may instead be potentially subversive of its purposes" (Latham 132).

The consumption of Icke's works - in fact, the growing market for conspiracism in the US - would seem to be a case in point here, disseminating and perpetuating an oppositional worldview, a "hermeneutics of suspicion," while contributing to the accumulation of capital.

<20> Another revealing dimension of Anunnaki vampires lies in their collective depiction; unlike many accounts of the vampire, Icke's theories do not revolve around distinct Anunnaki individuals but rather focuses on them as a class or group; in this sense the Anunnaki do not convey the same individualistic focus so often encountered in vampire narratives. Even Anunnaki forms of consciousness are best described as a "groupthink" mentality. On this, Icke writes that,
"the reptilians seek … to influence everyone by stimulating the behavioral patterns of the reptile region of the brain:
hierarchical thinking
aggression
conflict
division
lack of compassion
a need for ritual" (46)

Symbolic of contemporary capitalism, this collective depiction of the Anunnaki reflects the rise of networks, and their decentering development, which have instrumentally caused - and are themselves produced by - the new global economy. The network supersedes the individual as the subject of the vampire narrative. Here Castells, speaking on the network society of global economics, is instructive:

"For the first time in history, the basic unit of economic organization is not a subject, be it individual (such as the entrepreneur ...) or collective (such as the capitalist class, the corporation, the state)" (214). Instead, "the unit is the network, made up of a variety of subjects and organizations, relentlessly modified as networks adapt" (214).

<21> In their networked, post-subjective form of the vampire, the Anunnaki are metaphorical of the precise trajectory assumed by contemporary capitalism. Network is the same term Icke uses to describe the reptilian base of operations today, writing, "after thousands of years of evolution, the reptilian network is now a vast and often unfathomable web of interconnecting secret societies, banks, businesses, political parties, security agencies, media owners, and so on" (259).

Discourse on the Anunnaki vampire is in step with broader trends in American conspiracy theories, themselves responses to ideological crises associated with post-modernism and the growth of a network society. Writing on conspiracy theories in the postwar US, Timothy Melley points out that, "the term 'conspiracy' rarely signifies a small, secret plot any more. Instead, it frequently refers to the workings of a large organization, technology, or system, a powerful and obscure entity so dispersed that it is the very antithesis of the traditional conspiracy" (59).

Melley argues that conspiracy theories in the US have historically been an ideological means of validating individualism. And this new, impersonal breed of conspiracism reflects anxiety over the loss of individuality and agency and stands as both "an acknowledgment, and rejection, of postmodern subjectivity" (65).

<22> Perhaps most revealing of all is the dissolution of the boundary between fantasy and reality - the presentation of the vampire as an historical agent rather than a fictional character. Deeply ironic and radical, this slippage of fact and fantasy drives the vampire legacy much closer to its critical core. If the traditional vampire articulates dissent, it also distorts the representation of real relations, which are displaced into the realm of the imaginary. In the form of the Anunnaki, however, vampires have infiltrated the field of conspiracy theories, spilling from the page onto the pavement, as it were. Moving from metaphor to a kind of mimesis of the grotesque, the vampire legacy shape-shifts - its implicit charge evolving into an explosive critique.

Works Cited

"About David Icke, the Man, His Philosophy, and His Work." N.d. Online. Internet. 3 January 2003. Available http.//davidicke.com/icke/index1a.html

Canadian Association for Free Expression. David Icke's Telling the Truth Archives: Conspiracies, CoverUps, Truths, Facts, Oddities, Research: "Dante's Infernal Guide to Human Rights and Wrongs."

Castells, Manuel. The Rise of the Network Society. 2nd ed. Vol. 1. Oxford: Blackwell, 2000.

Frost, Brian J. The Monster with a Thousand Faces: Guises of the Vampire in Myth and Literature. Bowling Green, OH: Bowling Green State U Popular P, 1989.

Hofstadter, Richard. "The Paranoid Style in American Politics." In The Paranoid Style in American Politics and Other Essays. 1963. Cambridge: Harvard UP, 1996.

Holte, James Craig. "The Vampire." Malcolm South, ed. Mythical and Fabulous Creatures: A Source Book and Research Guide. New York: Greenwood, 1987. 243-64.

Icke, David. The Biggest Secret: The Book That Will Change the World. Scottsdale, AZ: Bridge of Love, 1999.

Knight, Peter. "Introduction: A Nation of Conspiracy Theorists." In Conspiracy Nation: The Politics of Paranoia in Postwar America. Ed. Peter Knight. New York: New York UP, 2002. 1-17.

Latham, Rob. "Consuming Youth: The Lost Boys Cruise Mallworld." Blood Read: The Vampire as Metaphor in Contemporary Culture. Joan Gordon and Veronica Hollinger, eds. Philadelphia: U of Pennsylvania P, 1997. 129-47.

Marx, Karl. Capital. Vol. 1. 1867. Harmondworth, UK: Penguin, 1976.

Melley, Timothy. "Agency Panic and the Culture of Conspiracy." In Conspiracy Nation: The Politics of Paranoia in Postwar America. Ed. Peter Knight. New York: New York UP, 2002. 57-81.

Melton, J. Gordon. The Vampire Book: The Encyclopedia of the Undead. Detroit: Visible Ink, 1999.

Moretti, Franco. "The Dialectic of Fear." Signs Taken for Wonders: Essays in the Sociology of Literary Forms. 1983. New York: Verso, 1997. 83-108.

Murgoci, Agnes. "The Vampire in Roumania." Alan Dundes, ed. The Vampire: A Casebook. Madison: U of Wisconsin P, 1998. 12-34.

FINALLY, some articles of a more modern relevance that clearly demonstrate that things are not what they often appear to be, and that frankly we are being told plain lies.

411-THE BITTER TRUTH ABOUT ARTIFICIAL SWEETENERS

412-VACCINES

413-WHERE IS THE GOLD?

414-The Population Control Agenda

415-Me – Here and Now.

411-THE BITTER TRUTH ABOUT ARTIFICIAL SWEETENERS

"A man may fish with the worm that hath eat of a king, and eat of the fish that hath fed of that worm." - Shakespeare, Hamlet

Aspartame sugar substitutes cause worrying symptoms from memory loss to brain tumours. But despite US FDA approval as a 'safe' food additive, aspartame is one of the most dangerous substances ever to be foisted upon an unsuspecting public.

Phone: (617) 497 7843,
E-mail: mgold@holisticmed.com
Web page: http://www.holisticmed.com/aspartame/

Aspartame is the technical name for the brand names, NutraSweet, Equal, Spoonful, and Equal-Measure. Aspartame was discovered by accident in 1965, when James Schlatter, a chemist of G.D. Searle Company was testing an anti-ulcer drug. Aspartame was approved for dry goods in 1981 and for carbonated beverages in 1983. It was originally approved for dry goods on July 26, 1974, but objections filed by neuroscience researcher Dr John W. Olney and Consumer attorney James Turner in August 1974 as well as investigations of G.D. Searle's research practices caused the US Food and Drug Administration (FDA) to put approval of aspartame on hold (December 5, 1974). In 1985, Monsanto purchased G.D. Searle and made Searle Pharmaceuticals and The NutraSweet Company separate subsidiaries.

Aspartame is, by far, the most dangerous substance on the market that is added to foods. **Aspartame accounts for over 75 percent of the adverse reactions to food additives** reported to the US Food and Drug Administration (FDA). Many of these reactions are very serious including seizures and death as recently disclosed in a February 1994 Department of Health and Human Services report.(1) A few of the 90 different documented symptoms listed in the report as being caused by aspartame include: Headaches/migraines, dizziness, seizures, nausea, numbness, muscle spasms, weight gain, rashes, depression, fatigue, irritability, tachycardia, insomnia, vision problems, hearing loss, heart palpitations, breathing difficulties, anxiety attacks, slurred speech, loss of taste, tinnitus, vertigo, memory loss, and joint pain.

According to researchers and physicians studying the adverse effects of aspartame, the following chronic illnesses can be triggered or worsened by ingesting of aspartame:(2) Brain tumors, multiple sclerosis, epilepsy, chronic fatigue syndrome, Parkinson's disease, Alzheimer's, mental retardation, lymphoma, birth defects, fibro myalgia, and diabetes.

Aspartame is made up of three chemicals: Aspartic acid, phenylalanine, and methanol. The book, Prescription for Nutritional Healing, by James and Phyllis Balch, lists aspartame under the category of "chemical poison." As you shall see, that is exactly what it is.

ASPARTIC ACID (40% OF ASPARTAME)

Dr Russell L. Blaylock, a professor of Neurosurgery at the Medical University of Mississippi, recently published a book thoroughly detailing the damage that is caused by the ingestion of excessive aspartic acid from aspartame. [Ninety nine percent of monosodium glutamate 9MSG) is glutamic acid. The damage it causes is also documented in Blaylock's book.] Blaylock makes use of almost 500 scientific references to show how excess free excitatory amino acids such as aspartic acid and glutamic acid in our food supply are causing serious chronic neurological disorders and a myriad of other acute symptoms.(3)

SUMMARY OF HOW ASPARTATE (AND GLUTAMATE) CAUSE DAMAGE

Aspartate and glutamate act as neurotransmitters in the brain by facilitating the transmission of information from neuron to neuron. Too much aspartate or glutamate in the brain kills certain neurons by allowing the influx of too much calcium into the cells. This influx triggers excessive amounts of free radicals which kill the cells. The neural cell damage that can be caused by excessive aspartate and glutamate is why they are referred to as "excitotoxins." They "excite" or stimulate the neural cells to death.

Aspartic acid is an amino acid. Taken in its free form (unbound to proteins) it significantly raises the blood plasma level of aspartate and glutamate. The excess aspartate and glutamate in the blood plasma shortly after ingesting aspartame or products with free glutamic acid (glutamate precursor) leads to a high level of those neurotransmitters in certain areas of the brain.

The blood brain barrier (BBB) which normally protects the brain from excess glutamate and aspartate as well as toxins 1) is not fully developed during childhood, 2) does not fully protect all areas of the brain, 3) is damaged by numerous chronic and acute conditions, and 4) allows seepage of excess glutamate and aspartate into the brain even when intact.

The excess glutamate and aspartate slowly begin to destroy neurons. The large majority (75%+) of neural cells in a particular area of the brain are killed before any clinical symptoms of a chronic illness are noticed. A few of the many chronic illnesses that have been shown to be contributed to by long-term exposure excitatory amino acid damage include:

Multiple sclerosis (MS), ALS, memory loss, hormonal problems, hearing loss, epilepsy, Alzheimer's disease, Parkinson's disease, hypoglycaemia, AIDS dementia, brain lesions, and neuroendocrine disorders.

The risk to infants, children, pregnant women, the elderly, and persons with certain chronic health problems from excitotoxins are great. Even the Federation of American Societies For Experimental Biology (FASEB), which usually understates problems and mimics the FDA party line, recently stated in a review that "it is prudent to avoid the use of dietary supplements of L-glutamic acid by pregnant women, infants, and children. The Existence of evidence of potential endocrine responses, i.e., elevated cortisol and prolactin, and differential responses between males and females, would also suggest a neuroendocrine link and that supplemental L-glutamic acid should be avoided by women of childbearing age and individuals with affective disorders."(4) Aspartic acid from aspartame has the same deleterious effects on the body as glutamic acid.

The exact mechanism of acute reactions to excess free glutamate and aspartate is currently being debated. As reported to the FDA, those reactions include:(5)
Headaches/migraines, nausea, abdominal pains, fatigue (blocks sufficient glucose entry into brain), sleep problems, vision problems, anxiety attacks, depression, and asthma/chest tightness.

One common complaint of persons suffering from the effect of aspartame is **memory loss**. Ironically, in 1987, G.D. Searle, the manufacturer of aspartame, undertook a search for a drug to combat memory loss caused by excitatory amino acid damage. Blaylock is one of many scientists and physicians who are concerned about excitatory amino acid damage caused by ingestion of aspartame and MSG. A few of the many experts who have spoken out against the damage being caused by aspartate and glutamate include Adrienne Samuels, Ph.D., an experimental psychologist specializing in research design. Another is Olney, a professor in the department of psychiatry, School of Medicine, Washington University, a neuroscientist and researcher, and one of the world's foremost authorities on excitotoxins. (He informed Searle in 1971 that aspartic acid caused holes in the brain of mice.) Also included is Francis J. Waickman, M.D., a recipient of the Rinkel and Forman Awards, and Board certified in Paediatrics, Allergy, and Immunology.

Other concerned scientists include: John R. Hain, M.D., Board Certified Forensic Pathologist, and H.J. Roberts, M.D., FACP, FCCP, Diabetic Specialist, and selected by a national medical publication as "The Best Doctor in the US"

John Samuels is concerned, also. He compiled a list of scientific research sufficient to show the dangers of ingesting excess free glutamic and aspartic acid.

And there are many more who can be added to this long list.

PHENYLALANINE (50% OF ASPARTAME)

Phenylalanine is an amino acid normally found in the brain. Persons with the genetic disorder, phenylketonuria (PKU) cannot metabolise phenylalanine. This leads to dangerously high levels of phenylalanine in the brain (sometimes lethal). It has been shown that ingesting aspartame, especially along with carbohydrates can lead to excess levels of phenylalanine in the brain even in persons who do not have PKU. **This is not just a theory**, as many people who have eaten large amounts of aspartame over a long period of time and do not have PKU have been shown to have excessive levels of phenylalanine in the blood. Excessive levels of phenylalanine in the brain can cause the levels of serotonin in the brain to decrease, leading to emotional disorders such as **depression**. It was shown in human testing that phenylalanine levels of the blood were increased significantly in human subjects who chronically used aspartame.(6) Even a single use of aspartame raised the blood phenylalanine levels. In his testimony before the US Congress, Dr Louis J. Elsas showed that high blood phenylalanine can be concentrated in parts of the brain, and is especially dangerous for infants and foetuses. He also showed that phenylalanine is metabolised much more efficiently by rodents than by humans.(7)

One account of a case of extremely high phenylalanine levels caused by aspartame was recently published the the "Wednesday Journal" in an article entitled "An Aspartame Nightmare." John Cook began drinking 6 to 8 diet drinks every day. His symptoms started out as memory loss and frequent headaches. He began to crave more aspartame-sweetened drinks. His condition deteriorated so much that he experienced wide mood swings and violent rages. Even though he did not suffer from PKU, a blood test revealed a phenylalanine level of 80 mg/dl. He also showed abnormal brain function and brain damage. After he kicked his aspartame habit, his symptoms improved dramatically. (8)

As Blaylock points out in his book, early studies measuring phenylalanine build-up in the brain were flawed. Investigators who measured specific brain regions and not the average throughout the brain notice significant rises in phenylalanine levels. Specifically the hypothalamus, medulla oblongata, and corpus striatum areas of the brain had the largest increases in phenylalanine. Blaylock goes on to point out that excessive build-up of phenylalanine in the brain can cause schizophrenia or make one more susceptible to seizures.

Therefore, **long-term, excessive use of aspartame may provide a boost to sales of serotonin reuptake inhibitors such as Prozac** and drugs to control schizophrenia and seizures.

METHANOL (AKA WOOD ALCOHOL/POISON) (10% OF ASPARTAME)

Methanol/wood alcohol is a deadly poison. Some people may remember methanol as the poison that has caused some "skid row" alcoholics to end up blind or dead. Methanol is gradually released in the small intestine when the methyl group of aspartame encounter the enzyme chymotrypsin.

The absorption of methanol into the body is sped up considerably when free methanol is ingested. **Free methanol is created from aspartame when it is heated to above 86 Fahrenheit (30 Centigrade). This would occur when aspartame-containing product is improperly stored or when it is heated** (e.g., as part of a "food" product such as Jello).

Methanol breaks down into formic acid and formaldehyde in the body. Formaldehyde is a deadly neurotoxin. An EPA assessment of methanol states that methanol "is considered a cumulative poison due to the low rate of excretion once it is absorbed. In the body, methanol is oxidized to formaldehyde and formic acid; both of these metabolites are toxic." **The recommend a limit of consumption of 7.8 mg/day. A one-litre (approx. 1 quart) aspartame-sweetened beverage contains about 56 mg of methanol. Heavy users of aspartame-containing products consume as much as 250 mg of methanol daily or 32 times the EPA limit. (9)**

Symptoms from methanol poisoning include headaches, ear buzzing, dizziness, nausea, gastrointestinal disturbances, weakness, vertigo, chills, memory lapses, numbness and shooting pains in the extremities, behavioural disturbances, and neuritis. The most well known problems from methanol poisoning are vision problems including misty vision, progressive contraction of visual fields, blurring of vision, obscuration of vision, retinal damage, and blindness. Formaldehyde is a known carcinogen, causes retinal damage, interferes with DNA replication, causes birth defects. (10) Due to the lack of a couple of key enzymes, humans are many times more sensitive to the toxic effects of methanol than animals. Therefore, tests of aspartame or methanol on animals do not accurately reflect the danger for humans. As pointed out by Dr Woodrow C. Monte, Director of the Food Science and Nutrition Laboratory at Arizona State University, "There are no human or mammalian studies to evaluate the possible mutagenic, teratogenic, or carcinogenic effects of chronic administration of methyl alcohol."(11)

He was so concerned about the unresolved safety issues that he filed suit with the FDA requesting a hearing to address these issues. He asked the FDA to "slow down on this soft drink issue long enough to answer some of the important questions. It's not fair that you are leaving the full burden of proof on the few of us who are concerned and have such limited resources. You must remember that you are the American public's last defence. Once you allow usage (of aspartame) there is literally nothing I or my colleagues can do to reverse the course. Aspartame will then join saccharin, the sulfiting agents, and God knows how many other questionable compounds enjoined to insult the human constitution with governmental approval."(10) Shortly thereafter, the Commissioner of the FDA, Arthur Hull Hayes, Jr., approved the use of aspartame in carbonated beverages, he then left for a position with G.D. Searle's Public Relations firm. (11)

It has been pointed out that some fruit juices and alcoholic beverages contain small amounts of methanol. It is important to remember, however, that methanol never appears alone. In every case, ethanol is present, usually in much higher amounts. Ethanol is an antidote for methanol toxicity in humans. (9) The troops of Desert Storm were "treated" to large amounts of aspartame-sweetened beverages which had been heated to over 86 degrees F. in the Saudi Arabian sun. Many of them returned home with numerous disorders similar to what has been seen in persons who have been chemically poisoned by formaldehyde. The free methanol in the beverages may have been a contributing factor in these illnesses. Other breakdown products of aspartame such as DKP (discussed below) may also have been a factor.

In a 1993 act that can only be described as "unconscionable," the FDA approved aspartame as an ingredient in numerous food items that would always be heated to above 86 degrees F (30 degrees C).

DIKETOPIPERAZINE (DKP)

DKP is a by-product of aspartame metabolism. DKP has been implicated in the occurrence of brain tumors. Olney noticed that DKP, when nitrosated in the gut, produced a compound which was similar to N-nitrosourea, a powerful brain tumor causing chemical. Some authors have said that DKP is produced after aspartame ingestion. I am not sure if that is correct. It is definitely true that DKP is formed in liquid aspartame-containing products during prolonged storage.

G.D. Searle conducted animal experiments on the safety of DKP. The FDA found numerous experimental errors occurred, including "clerical errors, mixed-up animals, animals not getting drugs they were supposed to get, pathological specimens lost because of improper handling," and many other errors. (12) These sloppy laboratory procedures may explain why both the test and control animals had sixteen times more brain tumors than would be expected in experiments of this length. In an ironic twist, shortly after these experimental errors were discovered, the FDA used guidelines recommended by G.D. Searle to develop the Industry-wide FDA standards for Good Laboratory Practices. (11) DKP has also been implicated as a cause of uterine polyps and changes in blood cholesterol by FDA Toxicologist Dr Jacqueline Verrett in her testimony before the US Senate. (13)

AILMENTS RESULTING FROM ASPARTAME

The components of aspartame can lead to a wide variety of ailments. Some of these problems occur gradually, others are immediate, acute reactions. There is an enormous population of people who are suffering from symptoms contributed to by aspartame, yet they have no idea why herbs or drugs are not helping relieve their problems. There are other users of aspartame who appear not to be suffering immediate reactions to aspartame. But even these individuals are susceptible to the long-term damage caused by excitatory amino acids, phenylalanine, methanol, and DKP. A few of the many disorders that are of particular concern to me include the following.

Birth Defects.

Dr Diana Dow Edwards, a researcher was funded by Monsanto to study possible birth defects caused by the ingestion of aspartame. After preliminary data showed damaging information about aspartame, funding for the study was cut off. A Genetic Paediatrician at Emory University has testified that aspartame is causing birth defects.7360-367.

In the book, While Waiting: A Prenatal Guidebook by George R. Verrilli, M.D. and Anne Marie Mueser, it is stated that aspartame is suspected of causing brain damage in sensitive individuals. A foetus may be at risk for these effects. Some researchers have suggested that high doses of aspartame may be associated with problems ranging from dizziness and subtle brain changes to mental retardation.

Cancer (Brain Cancer).

In 1981, Satya Dubey, an FDA statistician, stated that the brain tumor data on aspartame was so "worrisome" that he could not recommend approval of NutraSweet. (14) In a two-year study conducted by the manufacturer of aspartame, twelve of the 320 rats fed a normal diet and aspartame developed brain tumors while none of the control rats had tumors. Five of the twelve tumors were in rats given a low dose of aspartame. (15) The approval of aspartame was a violation of the Delaney Amendment which was supposed to prevent cancer-causing substances such as methanol (formaldehyde) and DKP from entering our food supply. The late Dr Adrian Gross, an FDA toxicologist, testified before the US Congress that aspartame was capable of producing brain tumors. This made it illegal for the FDA to set an allowable daily intake at any level. He stated in his testimony that Searle's studies were "to a large extent unreliable" and that "at least one of those studies has established beyond any reasonable doubt that aspartame is capable of inducing brain tumors in experimental animals...." He concluded his testimony by asking, "What is the reason for the apparent refusal by the FDA to invoke for this food additive the so-called Delaney Amendment to the Food, Drug and Cosmetic Act? .... And if the FDA itself elects to violate the law, who is left to protect the health of the public?"(16)

In the mid-1970s it was discovered that the manufacturer of aspartame falsified studies in several ways. One of the techniques used was to cut tumors out of test animals and put them back in the study. Another technique used to falsify the studies was to list animals that had actually died as surviving the study. Thus, the data on brain tumors was likely worse than discussed above. In addition, a former employee of the manufacturer of aspartame, Raymond Schroeder told the FDA on July 13, 1977 that the particles of DKP were so large that the rats could discriminate between the DKP and their normal diet. (12)

It is interesting to note that the incidence of brain tumors in persons over 65 years of age has increase 67% between the years 1973 and 1990. Brain tumors in all age groups has jumped 10%. The greatest increase has come during the years 1985-1987. (17)

In his book, Aspartame (NutraSweet). Is it Safe? Roberts gives evidence that aspartame can cause a particularly dangerous form of cancer - primary lymphoma of the brain.

Diabetes.
The American Diabetes Association (ADA) is actually recommending this chemical poison to persons with diabetes. According to research conducted by H.J. Roberts, a diabetes specialist, a member of the ADA, and an authority on artificial sweeteners, aspartame:

1) Leads to the precipitation of clinical diabetes.
2) Causes poorer diabetic control in diabetics on insulin or oral drugs.
3) Leads to the aggravation of diabetic complications such as retinopathy, cataracts, neuropathy and gastroparesis.
4) Causes convulsions.

In a statement concerning the use of products containing aspartame by persons with diabetes and hypoglycaemia, Roberts says: "Unfortunately, many patients in my practice, and others seen in consultation, developed serious metabolic, neurologic and other complications that could be specifically attributed to using aspartame products. This was evidenced by:

"The loss of diabetic control, the intensification of hypoglycaemia, the occurrence of presumed 'insulin reactions' (including convulsions) that proved to be aspartame reactions, and the precipitation, aggravation or simulation of diabetic complications (especially impaired vision and neuropathy) while using these products.

"Dramatic improvement of such features after avoiding aspartame, and the prompt predictable recurrence of these problems when the patient resumed aspartame products, knowingly or inadvertently."

Roberts goes on to say:
"I regret the failure of other physicians and the American Diabetes Association (ADA) to sound appropriate warnings to patients and consumers based on these repeated findings which have been described in my corporate-neutral studies and publications."

Blaylock stated that excitotoxins such as that found in aspartame can precipitate diabetes in persons who are genetically susceptible to the disease. (5)

Emotional Disorders.
A double blind study of the effects of aspartame on persons with mood disorders was recently conducted by Dr Ralph G. Walton. **Since the study wasn't funded/controlled by the makers of aspartame, The NutraSweet Company refused to sell him the aspartame.** Walton was forced to obtain and certify it from an outside source.

The study showed a large increase in serious symptoms for persons taking aspartame. Since some of the symptoms were so serious, the Institutional Review Board had to stop the study. Three of the participants had said that they had been "poisoned" by aspartame. Walton concludes that "individuals with mood disorders are particularly sensitive to this artificial sweetener; its use in this population should be discouraged."(18) Aware that the experiment could not be repeated because of the danger to the test subjects, Walton was recently quoted as saying, "I know it causes seizures. I'm convinced also that it definitely causes behavioural changes. I'm very angry that this substance is on the market. I personally question the reliability and validity of any studies funded by the NutraSweet Company."(19)

There are numerous reported cases of low brain serotonin levels, depression and other emotional disorders that have been linked to aspartame and often are relieved by stopping the intake of aspartame.

Researchers have pointed out that increasing in phenylalanine levels in the brain, which can and does occur in persons without PKU, leads to a decreased level of the neurotransmitter, serotonin, which leads to a variety of emotional disorders. Dr William M. Pardridge of UCLA testified before the US Senate that a youth drinking four 16-ounce bottles of diet soda per day leads to an enormous increase in the phenylalanine level.

Epilepsy/Seizures.
With the large and growing number of seizures caused by aspartame, it is sad to see that the Epilepsy Foundation is promoting the "safety" of aspartame. At Massachusetts Institute of Technology, 80 people who had suffered seizures after ingesting aspartame were surveyed. Community Nutrition Institute concluded the following about the survey:

"These 80 cases meet the FDA's own definition of an imminent hazard to the public health, which requires the FDA to expeditiously remove a product from the market."

Both the Air Force's magazine Flying Safety and the Navy's magazine, Navy Physiology published articles warning about the many dangers of aspartame including the cumulative deleterious effects of methanol and the greater likelihood of birth defects. The articles note that the ingestion of aspartame can make pilots more susceptible to seizures and vertigo. Twenty articles sounding warnings about ingesting aspartame while flying have also appeared in the National Business Aircraft Association Digest (NBAA Digest 1993), Aviation Medical Bulletin (1988), The Aviation Consumer (1988), Canadian General Aviation News (1990), Pacific Flyer (1988), General Aviation News (1989), Aviation Safety Digest (1989), and Plane and Pilot (1990) and a paper warning about aspartame was presented at the 57th Annual Meeting of the Aerospace Medical Association (Gaffney 1986).

Recently, a hotline was set up for pilots suffering from acute reactions to aspartame ingestion. Over 600 pilots have reported symptoms including some who have reported suffering grand mal seizures in the cockpit due to aspartame. (21)

One of the original studies on aspartame was performed in 1969 by an independent scientist, Dr Harry Waisman. He studied the effects of aspartame on infant primates. Out of the seven infant monkeys, one died after 300 days and five others had grand mal seizures. Of course, these negative findings were not submitted to the FDA during the approval process. (22)

Why don't we hear about these things?

The reason many people do not hear about serious reactions to aspartame is twofold:
1) Lack of awareness by the general population. Aspartame-caused diseases are not reported in the newspapers like plane crashes. This is because these incidents occur one at a time in thousands of different locations across the US.
2) Most people do not associate their symptoms with the long-term use of aspartame. For the people who have killed a significant percentage of the brain cells and thereby caused a chronic illness, there is no way that they would normally associate such an illness with aspartame consumption. **How aspartame was approved is a lesson in how chemical and pharmaceutical companies can manipulate government agencies such as the FDA, "bribe" organizations such as the American Dietetic Association, and flood the scientific community with flawed and fraudulent industry-sponsored studies funded by the makers of aspartame.**

Erik Millstone, a researcher at the Science Policy Research Unit of Sussex University has compiled thousands of pages of evidence, some of which have been obtained using the freedom of information act 23, showing:

1. Laboratory tests were faked and dangers were concealed.
2. Tumors were removed from animals and animals that had died were "restored to life" in laboratory records.
3. False and misleading statements were made to the FDA.
4. The two US Attorneys given the task of bringing fraud charges against the aspartame manufacturer took positions with the manufacturer's law firm, letting the statute of limitations run out.
5. The Commissioner of the FDA overruled the objections of the FDA's own scientific board of inquiry. Shortly after that decision, he took a position with Burson-Marsteller, the firm in charge of public relations for G.D. Searle.

A Public Board of Inquiry (PBOI) was conducted in 1980. There were three scientists who reviewed the objections of Olney and Turner to the approval of aspartame. They voted unanimously against aspartame's approval. The FDA Commissioner, Dr Arthur Hull Hayes, Jr. then created a 5-person Scientific Commission to review the PBOI findings. After it became clear that the Commission would uphold the PBOI's decision by a vote of 3 to 2, another person was added to the Commission, creating a deadlocked vote. This allowed the FDA Commissioner to break the deadlock and approve aspartame for dry goods in 1981. Dr Jacqueline Verrett, the Senior Scientist in an FDA Bureau of Foods review team created in August 1977 to review the Bressler Report (a report that detailed G.D. Searle's abuses during the pre-approval testing) said:

"It was pretty obvious that somewhere along the line, the bureau officials were working up to a whitewash." In 1987, Verrett testified before the US Senate stating that the experiments conducted by Searle were a "disaster." She stated that her team was instructed not to comment on or be concerned with the overall validity of the studies. She stated that questions about birth defects have not been answered. She continued her testimony by discussing the fact that DKP has been shown to increase uterine polyps and change blood cholesterol and that increasing the temperature of the product leads to an increase in production of DKP. (13)

Revolving doors

The FDA and the manufacturers of aspartame have had a revolving door of employment for many years. In addition to the FDA Commissioner and two US Attorneys leaving to take positions with companies connected with G.D. Searle, four other FDA officials connected with the approval of aspartame took positions connected with the NutraSweet industry between 1979 and 1982 including the Deputy FDA Commissioner, the Special Assistant to the FDA Commissioner, the Associate Director of the Bureau of Foods and Toxicology and the Attorney involved with the Public Board of Inquiry. (24)

It is important to realize that this type of revolving-door activity has been going on for decades. The Townsend Letter for Doctors (11/92) reported on a study revealing that 37 of 49 top FDA officials who left the FDA took positions with companies they had regulated. They also reported that over 150 FDA officials owned stock in drug companies they were assigned to manage. Many organizations and universities receive large sums of money from companies connected to the NutraSweet Association, a group of companies promoting the use of aspartame. In January 1993, the American Dietetic Association received a US$75,000 grant from the NutraSweet Company. The American Dietetic Association has stated that the NutraSweet Company writes their "Facts" sheets. (25)

Many other "independent" organizations and researchers receive large sums of money from the manufacturers of aspartame. The American Diabetes Association has received a large amount of money from Nutrasweet, including money to run a cooking school in Chicago (presumably to teach diabetes how to use Nutrasweet in their cooking).

A researcher in New England who has pointed out the dangers of aspartame in the past is now a Monsanto consultant. Another researcher in the South-eastern US had testified about the dangers of aspartame on foetuses. An investigative reporter has discovered that he was told to keep his mouth shut to avoid causing the loss of a large grant from a diet cola manufacturer in the NutraSweet Association.

What is the FDA doing to protect the consumer from the dangers of aspartame? Less than nothing.

In 1992, the FDA approved aspartame for use in malt beverages, breakfast cereals, and refrigerated puddings and fillings. In 1993 the FDA approved aspartame for use in hard and soft candies, non-alcoholic favored beverages, tea beverages, fruit juices and concentrates, baked goods and baking mixes, and frostings, toppings and fillings for baked goods.

In 1991, the FDA banned the importation of stevia. The powder of the leaf has been used for hundreds of years as an alternative sweetener. It is used widely in Japan with no adverse effects. Scientists involved in reviewing stevia have declared it to be safe for human consumption - something which has been well known in many parts of the world where it is not banned. Everyone that I have spoken with in regards to this issue believes that stevia was banned to keep the product from taking hold in the US and cutting into sales of aspartame.(26)

What is the US Congress doing to protect the consumer from the dangers of aspartame? Nothing.

What is the US Administration (President) doing to protect the consumer from the dangers of aspartame? Nothing.

Aspartame consumption is not only a problem in the US. It is being sold in over 70 countries throughout the world.

ASPARTAME CAN BE FOUND IN:
- Instant breakfasts

- Breath mints
- Cereals
- Sugar-free chewing gum
- Cocoa mixes
- Coffee beverages
- Frozen desserts
- Gelatine desserts
- Juice beverages
- Laxatives
- Multivitamins
- Milk drinks
- Pharmaceuticals and supplements
- Shake mixes
- Soft drinks
- Tabletop sweeteners
- Tea beverages
- Instant teas and coffees
- Topping mixes
- Wine coolers
- Yoghurt

I have been told that aspartame has been found in products where it is not listed on the label. One must be particular careful of pharmaceuticals and supplements. I have been informed that even some supplements made by well-known supplement manufacturers such as Twinlabs contain aspartame.

The information I have related above is just the tip of the iceberg as far as damaging information about aspartame. In order for the reader to find out more, I have included some resources below.

BOOKS:

* Blaylock, Russell L., Excitotoxins: The Taste That Kills (Health Press, Santa Fe, New Mexico, c1994). One of the best books available on excitotoxins. Well worth reading!

* H. J. Roberts, M.D., Aspartame (NutraSweet), Is it Safe? Available from the Aspartame Consumer Safety Network.

* Sweet'ner Dearest, Available from the Aspartame Consumer Safety Network

* Mary Nash Stoddard, The Deadly Deception, Available from the Aspartame Consumer Safety Network.

* Barbara Mullarkey, Editor, Bittersweet Aspartame - A Diet Delusion,

* Available from the Aspartame Consumer Safety Network.

* The Aspartame Consumer Safety Network, The Aspartame Consumer Safety Network Synopsis.

* Dennis Remington, M.D. and Barbara Higa, R.D., The Bitter Truth About Artificial Sweetners, Available from the Aspartame Consumer Safety Network

ASPARTAME CONSUMER SAFETY NETWORK
PO Box 780634
Dallas, Texas 75378, USA.
Phone: (214) 352-4268

REFERENCES

(1) Department of Health and Human Services, Report on All Adverse Reactions in the Adverse Reaction Monitoring System, (February 25 and 28, 1994).

(2) Compiled by researchers, physicians, and artificial sweetener experts for Mission Possible, a group dedicated to warning consumers about aspartame.

(3) Excitotoxins: The Taste That Kills, by Russell L. Blaylock, M.D.

(4) Safety of Amino Acids, Life Sciences Research Office, FASEB, FDA Contract No. 223-88-2124, Task Order No. 8.

(5) FDA Adverse Reaction Monitoring System.

(6) Wurtman and Walker, "Dietary Phenylalanine and Brain Function," Proceedings of the First International Meeting on Dietary Phenylalanine and Brain Function., Washington, D.C., May 8, 1987.

(7) Hearing Before the Committee On Labor and Human Resources United States Senate, First Session on Examining the Health and Safety Concerns of Nutrasweet (Aspartame).

(8) Account of John Cook as published in Informed Consent Magazine. "How Safe Is Your Artificial Sweetner" by Barbara Mullarkey, September/October 1994.

(9) Woodrow C. Monte, Ph.D., R.D., "Aspartame: Methanol and the Public Health," Journal of Applied Nutrition, 36 (1): 42-53.

(10) US Court of Appeals for the District of Columbia Circuit, No. 84-1153 Community Nutrition Institute and Dr Woodrow Monte v. Dr Mark Novitch, Acting Commissioner, US FDA (9/24/85).

(11) Aspartame Time Line by Barbara Mullarkey as published in Informed Consent Magazine, May/June 1994.

(12) FDA Searle Investigation Task Force. "Final Report of Investigation of G.D. Searle Company." (March 24, 1976)

(13) Testimony of Dr Jacqueline Verrett, FDA Toxicologist before the US Senate Committee on Labor and Human Resources, (November 3, 1987).
(14) Internal FDA memorandum.
(15) Analysis prepared by Dr John Olney as a statement before the Aspartame Board of Inquire of the FDA. Also Excitotoxins by Russell Blaylock, M.D.
(16) Congressional Record SID835: 131 (August 1, 1985)
(17) National Cancer Institute SEER Program Data.
(18) Walton, Ralph G., Robert Hudak, Ruth Green-Waite "Adverse Reactions to Aspartame: Double-Blind Challenge in Patients from a Vulnerable Population," Biological Psychiatry, 1993:34:13-17.
(19) Barbara Mullarkey, "How Safe Is Your Artificial Sweetner," September/October 1994 issue of Informed Consent Magazine.
(20) US Air Force. "Aspartame Alert." Flying Safety, 48 (5): 20-21 (May 1992).
(21) Reported by the Aspartame Consumer Safety Network.
(22) Barbara Mullarkey, Bittersweet Aspartame, A Diet Delusion.
(23) Millstone, Eric "Sweet and Sour." The Ecologist, 25 (March/April 1994).
(24) Mary Nash Stoddard, Editor, "The Deadly Deception," Aspartame Consumer Safety Network.
(25) ADA Courier, January 1993, Volume 32, Number 1. (26) "FDA Rejects AHPA Stevia Petition" by Mark Blumenthal, Whole Foods, April 1994.

Hi All,

I'm sure you're already aware of some of the side affects of these poisons, thought I'd forward this article as it makes some disturbing observations.

Regards,
Cathy

Subject: ...Aspartame....

THE MULTIPLE SCLEROSIS FOUNDATION & FDA ARE SUING FOR COLLUSION WITH MONSANTO - Article written by Betty Martini

I have spent several days lecturing at the World Environmental Conference on "Aspartame: Marketed as 'NutraSweet', 'Equal', and 'Spoonful'". In the keynote address by the EPA, they announced that there was an epidemic of multiple sclerosis and systemic lupus, and they did not understand what toxin was causing this to be rampant across the United States. I explained I was there to lecture on exactly that subject.

When the temperature of aspartame exceeds 86 degrees F, the wood alcohol in aspartame coverts to formaldehyde and then to formic acid, which in turn causes metabolic acidosis. (Formic acid is the poison found in the sting of fire ants.) The methanol toxicity mimics multiple sclerosis; thus, people were being diagnosed with having multiple sclerosis in error. The multiple sclerosis is not a death sentence, where methanol toxicity is.

In the case of systemic lupus, we are finding it has become almost as rampant as multiple sclerosis, especially in Diet Coke and Diet Pepsi drinkers. Also, with methanol toxicity, the victims usually drink three to four 12-oz. cans of them per day, some even more. In the cases of systemic lupus, which is triggered by aspartame, the victim usually does not know that the aspartame is the culprit. The victim continues its use aggravating the lupus to such a degree, that sometimes it becomes life threatening.

When we get people off the aspartame, those with systemic lupus usually become asymptomatic. Unfortunately, we cannot reverse this disease. On the other hand, in the case of those diagnosed with Multiple Sclerosis (when in reality, the disease is methanol toxicity), most of the symptoms disappear. We have seen cases where their vision has returned and even their hearing has returned. This also applies to cases of tinnitus.

During a lecture I said, "If you are using aspartame [NutraSweet, Equal, Spoonful, etc.] and you suffer from fibromyalgia symptoms, spasms, shooting pains, numbness in your legs, cramps, vertigo, dizziness, headaches, tinnitus, joint pain, depression, anxiety attacks, slurred speech, blurred vision, or memory loss, you probably have ASPARTAME DISEASE!" People were jumping up during the lecture saying, "I've got this! Is it reversible?" It is rampant. Some of the speakers at my lecture even were suffering from these symptoms.

In one lecture attended by the Ambassador of Uganda, he told us that their sugar industry is adding aspartame! He continued by saying that one of the industry leader's son could no longer walk due in part to product usage! We have a very serious problem.

Even a stranger came up to Dr. Espisto, (one of my speakers) and myself and said, "Could you tell me why so many people seem to be coming down with MS?" During a visit to a hospice, a nurse said that six of her friends, who were heavy Diet Coke drinkers, had all been diagnosed with MS. This is beyond coincidence.

Here is the problem. There were Congressional Hearings when aspartame was originally included as a sweetener in 100 different products. Since this initial hearing, there have been two subsequent hearings, but to no avail. Nothing has been done. The drug and chemical lobbies have very deep pockets. Now there are over 5,000 products containing this chemical, and the PATENT HAS EXPIRED!!

At the time of this first hearing, people were going blind. The methanol in the Aspartame converts to formaldehyde in the retina of the eye. Formaldehyde is grouped in the same class of drugs as cyanide and arsenic - DEADLY POISONS!!! Unfortunately, it just takes longer to quietly kill, but it is killing people and causing all kinds of neurological problems.

Aspartame changes the brain's chemistry. It is the reason for severe seizures. This drug changes the dopamine level in the brain. Imagine what this drug does to patients suffering from Parkinson's Disease. This drug also causes birth defects. There is absolutely no reason to take this product.

It is NOT A DIET PRODUCT!! The Congressional record said, "It makes you crave Carbohydrates and will make you FAT." Dr. Roberts stated that when he got patients off aspartame, their average weight loss was 19 pounds per person. The formaldehyde stores in the fat cells, particularly in the hips and thighs.

Aspartame is especially deadly for diabetics. All physicians know what wood alcohol will do to a diabetic. We find that physicians believe that they have patients with retinopathy, when in fact, it is caused by the aspartame. The aspartame keeps the blood sugar level out of control, causing many patients to go into a coma. Unfortunately, many have died.

People were telling us at the conference of the American College of Physicians that they had relatives that had switched from saccharin to an aspartame product and how that relative had eventually gone into a coma. Their physicians could not get the blood sugar levels under control; thus the patients suffered acute memory loss and eventually coma and death.

Memory loss is due the fact that aspartic acid and phenylalanine are neurotoxic without the other amino acids found in protein, thus it goes past the blood brain barrier and deteriorates the neurons of the brain. Dr. Russell Blaylock, a prominent neurosurgeon of Jackson, Mississippi, said, "The ingredients stimulate the neurons of the brain to death, causing brain damage of varying degrees." Dr. Blaylock has written a book entitled EXCITOTOXINS: The Taste That Kills. (Health Press 1-800-643-2665).

Dr. H. J. Roberts, diabetic specialist and world expert on aspartame poisoning, has also written a book entitled Defense Against Alzheimer's Disease (1-800-814-9800). Dr. Roberts tells how aspartame poisoning is escalating Alzheimer's Disease, and indeed it is. As the hospice nurse told me, women are being admitted at 30 years of age with Alzheimer's Disease.

Dr. Blaylock and Dr. Roberts will be writing a position paper with some case histories and will post it on the Internet. According to the Conference of the American College of Physicians, "We are talking about a plague of neurological diseases caused by this deadly poison."

Dr. Roberts realized what was happening when aspartame was first marketed. He said, "His diabetic patients presented memory loss, confusion, and severe vision loss." At the Conference of the American College of Physicians, doctors admitted that they did not know. They had wondered why seizures were rampant (the phenylalanine in aspartame breaks down the seizure threshold and depletes serotonin, which causes manic depression, panic attacks, rage and violence).

Just before the Conference, I received a fax from Norway asking for a possible antidote for this poison because they are experiencing so many problems in their country. This "poison" is now available in 90 PLUS countries worldwide. Fortunately, we had speakers and ambassadors at the conference from different nations who have pledged their help. We ask that you help too.

Print this article out and warn everyone you know. Take anything that contains aspartame back to the store. Take the "NO ASPARTAME TEST" and send us your case history.

I assure you that MONSANTO, the creator of aspartame, knows how deadly it is. They fund the American Medical Association, American Dietetic Association, Congress, and the Conference of the American College of Physicians. The New York Times, November 15, 1996, ran an article on how the American Dietetic Association takes money from the food industry to endorse their products. Therefore, they cannot criticize any additives or tell about their link to MONSANTO.

How bad is this? We told a mother who had a child on NutraSweet to get off the product. The child was having grand mal seizures every day. The mother called her physician, who called the ADA, who told the doctor not to take the child off the NutraSweet. We are still trying to convince the mother that the aspartame is causing the seizures. Every time we get someone off of aspartame, the seizures stop. If the baby dies, you know whose fault it is, and what we are up against.

There are 92 documented symptoms of aspartame, from coma to death. The majority of them are all neurological, because the aspartame destroys the nervous system. Aspartame Disease is partially the cause to what is behind some of the mystery of the Desert Storm health problems. The burning tongue and other problems discussed in over 60 cases can be directly related to the consumption of aspartame product. Several thousand pallets of diet drinks were shipped to the Desert Storm troops. (Remember heat can liberate the methanol from the aspartame at 86 degrees F.) Diet drinks sat in the 120 degree F Arabian sun for weeks at a time on pallets. The servicemen and women drank them all day long. All of their symptoms are identical to aspartame poisoning.

Dr. Roberts says "Consuming aspartame at the time of conception can cause birth defects." According to Dr. Louis Elsas, Pediatrician and Professor of Genetics, at Emory University, in his testimony before Congress, " The phenylalanine concentrates in the placenta causing mental retardation. In the original lab tests, animals developed brain tumors, phenylalanine breaks down into DXP (a brain tumor agent.) When Dr. Espisto was lecturing on aspartame, one physician in the audience, a neurosurgeon, said, "When they remove brain tumors, they have found high levels of aspartame in them."

Although Stevia, a sweet food, NOT AN ADDITIVE, which helps in the metabolism of sugar, which would be ideal for diabetics, has now been approved as a dietary supplement by the FDA for years, the FDA has outlawed this sweet food because of their loyalty to MONSANTO. If it says "SUGAR FREE" on the label - DO NOT EVEN THINK ABOUT IT.

Senator Howard Hetzenbaum wrote a bill that would have warned all infants, pregnant mothers and children of the dangers of aspartame. The bill would have also instituted independent studies on the problems existing in the population (seizures, changes in brain chemistry, changes in neurological and behavioral symptoms). It was killed by the powerful drug and chemical lobbies, letting loose the hounds of disease and death on an unsuspecting public. Since the Conference of the American College of Physicians, we hope to have the help of several world leaders.

Again, please help us, too. There are a lot of people out there who must be warned, please let them know this information.

Women's Cancer Resource Center Laurie Moser, Assistant Director
1815 East 41st Street, Suite C Minneapolis, MN 55407-3425
1-800-908-8544 or 612-729-049

"You must do the things you think you cannot do" Eleanor Roosevelt

412-VACCINES

"Your failure to be informed does not make me a wacko." -- John Loeffler

I have no medical training, but I can read.

I find this topic severely distressing. Even those words fail to convey my feelings. Aggravating that distress is that I am not sure if it is simply a case of human madness and insanity initiated and motivated by callous greed, which would be the "best" case scenario, or if in fact there is something far more sinister than mere corporate greed implementing and sustaining this assault on humanity.

If vaccinations were an isolated or "stand alone" oddity, I may find it as just that, but as there are simultaneously literally hundreds of other similarly debilitating assaults on humanity, I am forced to doubt that it is more than an unfortunate set of circumstances. Pardon me if I see a pattern where others see nothing unusual, but hundreds of similar "hits" tend to make one look for a common cause. (This may be an ability identified by an intelligence test that identified me as a visual mathematician with ability to see patterns and make correlations,)

Unfortunately when it comes down to the debilitation or death of humanity, a common source is actually identifiable. Our commonly accepted "god" is obviously not only not helping us, but complicit. Writings in other sections of this book will give validation and also all the justification of this opinion. Jehovah or Enlil and the heirs are not sympathetic to humanity. God I hate it when a god is not on "our" side. Do the research. (The history of the "Genesis God" is elsewhere in this book.) It should become obvious that some non-human 3rd party is the author of unending schemes that should be obviously designed to create misery or destroy.

There can be absolutely no doubt, when the facts and case histories are laid open and exposed, that untold deaths, debilitation, and injury are a direct result of the vaccination "industry".

Regardless of what I think, it is imperative that you, the reader, put yourself in a position where you really can make an informed decision about the entire subject rather than just blindly accept and submit yourself and family to this practice.

Death and permanent injury are not denied by the "industry", rather they play down the unfortunates numbers and statistics as **just bad luck**, or a result of **"complications"** that are **unfortunate**. Put it this way: would you join a queue knowing at every so often one of you would be randomly taken out for immediate execution? Would you put your children in that line? Would you allow yourself or family to play with a loaded revolver?

As this is indeed such a painful issue (and I have had family issues relating to "bad reaction" to vaccines that resulted in our sane family doctor advising "no more vaccinations for the child") I must personally be brief in my comments. However the matter requires your most urgent attention and investigations. Hereunder are some pointers and information that are merely a start. The resources available for your consideration are larger than you can imagine.

In short we the people, are being wilfully and knowingly killed off. No one is being held accountable for what is genocide. Let's put aside emotional writing and get down to history.

****

1963. The mass **vaccination** campaigns of the 1950s and '60s may be causing hundreds of deaths a year because of a cancer-causing virus that contaminated the first polio vaccine, according to scientists. Known as SV40, the virus came from dead monkeys whose kidney cells were used to culture the first Salk vaccines. (Nice hey? – Bob) Doctors estimate that the virus was injected into tens of millions during the vaccination campaigns, including several million in Canada, **before being detected and screened out in 1963.** Those born between 1941 and 1961 are thought to be most at risk of having been infected.

Mid-1970's. The incidence of AIDS infections in Africa coincides exactly with the locations of the W.H.O. smallpox vaccination program in the mid-1970's (London Times, May 11, 1987). Some 14,000 Haitians then on UN secondment to Central Africa were also vaccinated in this campaign. Personnel actually conducting the vaccinations may have been completely unaware that the vaccine was anything other than what they were told.

1978. The Hepatitis B vaccine study appears to have been the initial means of planting the infection in New York City. The test protocol specified **non-monogamous males** only, and **homosexuals received a different vaccine from heterosexuals**. At least 25-50% of the first reported New York AIDS cases in 1981 had received the Hepatitis B test vaccine in 1978. **By 1984, 64% of the vaccine recipients had AIDS**, and the figures on the current infection rate for the participants of that study are held by the U.S. Department of Justice, and "unavailable."

****

"As a legislator, I believe mandated smallpox vaccines are very bad policy. The point is not that smallpox vaccines are necessarily a bad idea, but rather that intimately personal medical decisions should not be made by government. The real issue is individual medical choice. No single person, including the President of the United States, should ever be given the power to make a medical decision for potentially millions of Americans. Freedom over one's physical person is the most basic freedom of all, and people in a free society should be sovereign over their own bodies. When we give government the power to make medical decisions for us, we in essence accept that the state owns our bodies." (Ron Paul MD, LewRockwell.com)

Vaccinations - good or bad? updated April 2008

The idea of vaccination is that if you give the immune system a small "taste" of a bug (such as polio, whooping cough etc) it will make antibodies which will protect one against future exposures to the real thing. Good idea, but in practice it is not as simple.

My medical training tells me that all these issues should be resolved by logical argument. But in the modern world, all these arguments are tainted by **vested interest (primarily from drug companies)** and it is difficult to trust the data with which one is presented. Therefore one ends up working from either limited data, or untrustworthy data, or common sense and experience and ends up with a belief. So what you are getting below are my individual conclusions, but it is up to every parent to find what information they can and make up their own minds.

The evidence that vaccinations reduce incidence of disease is pretty thin. **Most infectious diseases have declined as a result of improved hygiene and nutrition.** Doctors believe that vaccinations work and are reluctant to diagnose a disease in a vaccinated child. **So for example since polio vaccine, polio is rarely diagnosed, but there has been an increase in aseptic meningitis.**

The medical profession, backed up by the pharmaceutical and chemical industry, are experts in cover-ups. When doctors find themselves in trouble they close ranks. Most people have seen cover-ups for themselves with **drug side effects (which kill huge numbers of people every year but are hushed up)**. I see cover ups in patients with pesticide poisoning, with problems from silicone breast implants and in Gulf War syndrome. Doubtless there are others and I know **vaccine damage is covered up and/or denied. I have seen too many children with serious health problems dating from vaccination for which there is no other explanation for their illness.** I have to believe the evidence of my own eyes.

Vaccines can cause harm

There is now strong evidence that part of Gulf War Syndrome was caused by multiple vaccinations. MMR has been linked with autism and there is still a case to be answered here. There are many cases of brain damaged children following triple vaccine (diptheria, pertussis, tetanus).

Vaccines may be causing harm in unseen ways

Polio vaccination may be the cause of the huge increase in post viral fatigue syndrome. Before polio vaccination, post viral syndrome was rare. This is because people caught polio (which occasionally results in paralysis) which is an enterovirus. They mounted a vigorous immune reaction against polio virus which gave them cross-immunity against all other enteroviruses including Epstein Barr (glandular fever), coxsackie B, ECHO etc. This protected them against post viral fatigue since this most commonly follows an enteroviral infection.

We now know that many cancers are caused by viral infection. Obvious examples include hepatitis B (primary liver cancer), cervical wart virus (cervical cancer) and AIDS (Kaposi's sarcoma). Chronic myeloid leukaemia is probably virally induced. How many other cancers could there be from which we are protected by proper exposure to a virus, but not protected by vaccination? Nobody knows the answer to this question. And certainly no studies are being done.

What is in a vaccine?

Not just bits of bacteria and viruses. No immune system is going to react vigorously against a few dead or half alive (attenuated) cells. To turn the immune system on a vaccine needs an immune adjuvant added. These include aluminium and mercury which are toxic in their own right. It may well be that autism following MMR is actually a mercury problem.

Vaccines are made from bugs which are grown in animal tissues including beef. There is evidence to suggest that the cases of new variant CJD in young people may be due to direct injection of prion from these tissue cultures.

So what is the alternative?

We should be tackling infectious disease by good hygiene and boosting the immune system.

Good hygiene

By good hygiene definitely I do not mean obsessively wiping down working surfaces with antiseptic wipes. Indeed this is counter productive because we need daily exposure to bacteria to train and programme the immune system. What I recommend is proper public health measures such as:

* Not pumping raw sewage into the seas for people to swim in.
* Not making animals travel hundreds of miles to slaughter houses so they crap themselves on the way and get covered with shit, contaminating meat subsequently. Please try to buy local produce, or organic produce which has animal care standards.
* Not keeping chickens so intensively that they need constant antibiotics to survive chronic salmonella.
* Moving towards more organic farming practices i.e. away from intensive farming, use of properly composted animal waste, using local suppliers etc.
* Sexually transmitted diseases are presently all too common. Take proper precautions.
* We should not be concentrating sick and ill people in large general hospitals. This means that antibiotic resistant organisms can develop and spread quickly from one patient to the next.
* There are many other ideas for good hygiene. It is important to think carefully for yourselves. Please do not assume that "hygienic" chemical solutions are the answer.

Boosting the immune system

Human beings live on a knife edge with their immune systems. The immune system has a delicate balancing act because it needs to be able to recognise bugs and attack them, it must recognise cancer cells and attack them, but it must recognise "self" i.e. human bacteria/bugs and human cells and and ignore them. It is already confused by chemicals.

We should not be thinking about getting rid of the bug. This will always be impossible simply because "nature abhors a vacuum" and if you get rid of one bug, another will take its place.

Therefore we should be thinking about individual resistance to disease i.e. making people so healthy through good diet, good micronutrient status (vitamins and minerals) and freedom from toxins (i.e. red herrings and obstacles) that the immune system can easily resist any bugs that do gain entry. For example, measles can cause eye damage, but not if there is good vitamin A status.

The trouble is that against all these arguments is the combined weight of the medical profession and pharmaceutical companies who financially drive government and control the Press telling us that vaccination is safe and desirable. Nowadays logical argument no longer prevails and policy is dictated by big business and cash.

So what vaccinations would I give my child today?

No DPT in the first few months of life (I would look for protection from breast feeding).
I would give polio because I do not want to risk paralysis and I believe good nutrition will protect my child from other severe enteroviral infections.
No MMR (I want my child to get these infections young when the immune system, with good nutrition, can deal efficiently with these infections). See MMR vaccination - should my child have it?
Good HealthKeeping is a website with information on obtaining single vaccines.
Once my child started running around outside I would give tetanus vaccination.
No BCG.
With a daughter, I would check rubella status as a teenager and use vaccination if she was negative: I do not want her to get rubella during pregnancy.

In conclusion
These, as I say, are my beliefs. They may well change in the future as I learn new things.

For a list of vaccine ingredients refer to RENSE.COM

Rense.com

Vaccine Ingredients -
Formaldehyde, Aspartame,
Mercury, Etc
11-11-4

This following list of common vaccines and their ingredients should shock anyone.

The numbers of microbes, antibiotics, chemicals, heavy metals and animal byproducts is staggering. Would you knowingly inject these materials into your children?

Acel-Immune DTaP - **Diphtheria-Tetanus-Pertussis** Wyeth-Ayerst 800.934.5556
* diphtheria and tetanus toxoids and acellular pertussis adsorbed, formaldehyde, aluminum hydroxide, aluminum phosphate, thimerosal, and polysorbate 80 (Tween-80) gelatin Act HIB

Haemophilus - **Influenza B** Connaught Laboratories 800.822.2463
* Haemophilus influenza Type B, polyribosylribitol phosphate ammonium sulfate, formalin, and sucrose

Attenuvax - **Measles** Merck & Co., Inc. 800-672-6372
* measles live virus neomycin sorbitol hydrolized gelatin, chick embryo

Biavax - **Rubella** Merck & Co., Inc. 800-672-6372
* rubella live virus neomycin sorbitol hydrolized gelatin, human diploid cells from aborted fetal tissue

BioThrax - Anthrax Adsorbed BioPort Corporation 517.327.1500
* nonencapsulated strain of Bacillus anthracis aluminum hydroxide, benzethonium chloride, and formaldehyde

DPT - **Diphtheria-Tetanus-Pertussis** GlaxoSmithKline 800.366.8900 x5231
* diphtheria and tetanus toxoids and acellular pertussis adsorbed, formaldehyde, aluminum phosphate, ammonium sulfate, and thimerosal, washed sheep RBCs

Dryvax - **Smallpox** (not licensed d/t expiration) Wyeth-Ayerst 800.934.5556
* live vaccinia virus, with "some microbial contaminants," according to the Working Group on Civilian Biodefense polymyxcin B sulfate, streptomycin sulfate, chlortetracycline hydrochloride, and neomycin sulfate glycerin, and phenol -a compound obtained by distillation of coal tar vesicle fluid from calf skins Engerix-B

Recombinant **Hepatitis B** GlaxoSmithKline 800.366.8900 x5231
* genetic sequence of the hepatitis B virus that codes for the surface antigen (HbSAg), cloned into GMO yeast, aluminum hydroxide, and thimerosal

Fluvirin Medeva Pharmaceuticals 888.MEDEVA 716.274.5300
* influenza virus, neomycin, polymyxin, beta-propiolactone, chick embryonic fluid

FluShield Wyeth-Ayerst 800.934.5556
* trivalent influenza virus, types A&B gentamicin sulphate formadehyde, thimerosal, and polysorbate 80 (Tween-80) chick embryonic fluid

Havrix - Hepatitis A GlaxoSmithKline 800.366.8900 x5231
* hepatitis A virus, formalin, aluminum hydroxide, 2-phenoxyethanol, and polysorbate 20 residual MRC5 proteins -human diploid cells from aborted fetal tissue

HiB Titer - Haemophilus Influenza B Wyeth-Ayerst 800.934.5556
* haemophilus influenza B, polyribosylribitol phosphate, yeast, ammonium sulfate, thimerosal, and chemically defined yeast-based medium

Imovax Connaught Laboratories 800.822.2463

* rabies virus adsorbed, neomycin sulfate, phenol, red indicator human albumin, human diploid cells from aborted fetal tissue

IPOL Connaught Laboratories 800.822.2463
* 3 types of polio viruses neomycin, streptomycin, and polymyxin B formaldehyde, and 2-phenoxyethenol continuous line of monkey kidney cells

JE-VAX - Japanese Ancephalitis Aventis Pasteur USA 800.VACCINE
* Nakayama-NIH strain of Japanese encephalitis virus, inactivated formaldehyde, polysorbate 80 (Tween-80), and thimerosal mouse serum proteins, and gelatin

LYMErix - Lyme GlaxoSmithKline 888-825-5249

* recombinant protein (OspA) from the outer surface of the spirochete Borrelia burgdorferi kanamycin aluminum hydroxide, 2-phenoxyethenol, phosphate buffered saline

MMR - **Measles-Mumps-Rubella** Merck & Co., Inc. 800.672.6372
* measles, mumps, rubella live virus, neomycin sorbitol, hydrolized gelatin, chick embryonic fluid, and human diploid cells from aborted fetal tissue

M-R-Vax - **Measles-Rubella** Merck & Co., Inc. 800.672.6372
* measles, rubella live virus neomycin sorbitol hydrolized gelatin, chick embryonic fluid, and human diploid cells from aborted fetal tissue

Menomune - **Meningococcal** Connaught Laboratories 800.822.2463
* freeze-dried polysaccharide antigens from Neisseria meningitidis bacteria, thimerosal, and lactose

Meruvax I - **Mumps** Merck & Co., Inc. 800.672.6372
* mumps live virus neomycin sorbitol hydrolized gelatin

NYVAC - (new **smallpox** batch, not licensed) Aventis Pasteur USA 800.VACCINE
* highly-attenuated vaccinia virus, polymyxcin B, sulfate, streptomycin sulfate, chlortetracycline hydrochloride, and neomycin sulfate glycerin, and phenol -a compound obtained by distillation of coal tar vesicle fluid from calf skins

Orimune - **Oral Polio** Wyeth-Ayerst 800.934.5556
* 3 types of polio viruses, attenuated neomycin, streptomycin sorbitol monkey kidney cells and calf serum

Pneumovax - **Streptococcus** Pneumoniae Merck & Co., Inc. 800.672.6372
* capsular polysaccharides from polyvalent (23 types), pneumococcal bacteria, phenol,

Prevnar **Pneumococcal** - 7-Valent Conjugate Vaccine Wyeth Lederle 800.934.5556
* saccharides from capsular Streptococcus pneumoniae antigens (7 serotypes) individually conjugated to

diphtheria CRM 197 protein aluminum phosphate, ammonium sulfate, soy protein, yeast

RabAvert - **Rabies** Chiron Behring GmbH & Company 510.655.8729
* fixed-virus strain, Flury LEP neomycin, chlortetracycline, and amphotericin B, potassium glutamate, and sucrose human albumin, bovine gelatin and serum "from source countries known to be free of bovine spongioform encephalopathy," and chicken protein

Rabies Vaccine Adsorbed GlaxoSmithKline 800.366.8900 x5231
*rabies virus adsorbed, beta-propiolactone, aluminum phosphate, thimerosal, and phenol, red rhesus monkey fetal lung cells

Recombivax - Recombinant **Hepatitis B** Merck & Co., Inc. 800.672.6372
* genetic sequence of the hepatitis B virus that codes for the surface antigen (HbSAg), cloned into GMO yeast, aluminum hydroxide, and thimerosal

RotaShield - Oral Tetravalent Rotavirus (recalled) Wyeth-Ayerst 800.934.5556
* 1 rhesus monkey rotavirus, 3 rhesus-human reassortant live viruses neomycin sulfate, amphotericin B potassium monophosphate, potassium diphosphate, sucrose, and monosodium glutamate (MSG) rhesus monkey fetal diploid cells, and bovine fetal serum smallpox (not licensed due to expiration)

40-yr old stuff "found" in Swiftwater, PA freezer Aventis Pasteur USA 800.VACCINE
* live vaccinia virus, with "some microbial contaminants," according to the Working Group on Civilian Biodefense polymyxcin B sulfate, streptomycin sulfate, chlortetracycline hydrochloride, and neomycin sulfate glycerin, and phenol -a compound obtained by distillation of coal tar vesicle fluid from calf skins

Smallpox (new, not licensed) Acambis, Inc. 617.494.1339 in partnership with Baxter BioScience
* highly-attenuated vaccinia virus, polymyxcin B sulfate, streptomycin sulfate, chlortetracycline hydrochloride, and neomycin sulfate glycerin, and phenol -a compound obtained by distillation of coal tar vesicle fluid from calf skins

TheraCys **BCG** (intravesicle -not licensed in US for tuberculosis) Aventis Pasteur USA 800.VACCINE

* live attenuated strain of Mycobacterium bovis monosodium glutamate (MSG), and polysorbate 80 (Tween-80)

Tripedia - **Diphtheria-Tetanus-Pertussis** Aventis Pasteur USA 800.VACCINE
*Corynebacterium diphtheriae and Clostridium tetani toxoids and acellular Bordetella pertussis adsorbed aluminum potassium sulfate, formaldehyde, thimerosal, and polysorbate 80 (Tween-80) gelatin, bovine extract

US-sourced Typhim Vi - **Typhoid** Aventis Pasteur USA SA 800.VACCINE
* cell surface Vi polysaccharide from Salmonella typhi Ty2 strain, aspartame, phenol, and polydimethylsiloxane (silicone)

Varivax - **Chickenpox** Merck & Co., Inc. 800.672.6372
* varicella live virus neomycin phosphate, sucrose, and monosodium glutamate (MSG) processed gelatin, fetal bovine serum, guinea pig embryo cells, albumin from human blood, and human diploid cells from aborted fetal tissue

YF-VAX - **Yellow Fever** Aventis Pasteur USA 800.VACCINE
* 17D strain of yellow fever virus sorbitol chick embryo, and gelatin

http://www.informedchoice.info/cocktail.html
Vaccine Liberation Information
NOTE: THIMEROSAL = MERCURY
http://www.vaclib.org/pdf/exemption.htm

Vaccinations: Good, Bad or Just Plain Ugly

The FDA and other "watchdog" government agencies seldom are called to account for erroneous or irresponsible decisions. In the Dow Chemical silicone breast implant suit, the government was recently awarded $9.8 million for medical expenses paid out through Medicare and Medicaid. It didn't seem to matter that another agency, the FDA, of the same government had previously approved the use and sale of these implants and is currently considering whether to allow them to be sold again.

Further, these same agencies show definite bias when it comes to evaluating the risks associated with drugs. A good example is the fact that the agencies are constantly pushing for vaccinations and flu shots. For some reason, however, they neglect to tell the public that **the preservative in these flu shots and vaccines is mercury.**

IS THERE SUCH A THING AS HEALTHY MERCURY?

When it comes to other sources of mercury, though, they are extremely vigilant. They have issued repeated warnings on the consumption of various fish, including tuna, shark, swordfish, and mahi-mahi, because of possible mercury contamination. And since mercury is particularly harmful to nerve cells, government health authorities have stressed that infants and small children shouldn't be fed these foods, and pregnant and nursing mothers should avoid eating tuna also.

However the facts state that most canned tuna contains less mercury contamination than tuna steaks, which come from larger tuna. It's hard to tell how much, if any, mercury these products contain. Smaller fish are safer, and so are fish like sole, sardines, herring, bass, catfish, salmon and shellfish.

Although the EPA (Environmental Protection Agency) has determined that the maximum allowable daily exposure to mercury is 0.1 microgram per kilogram of body weight, the new flu vaccine for babies, called Fluzone, contains 25 micrograms of mercury per 0.5 ml dose.

Practically all vaccines contain mercury and aluminum. And vaccines are not "safer" sources of these toxic minerals. It doesn't matter if the mercury comes from fish or from a vaccine. The potential for neurological damage remains the same. But for some reason, even though we're warned about fish consumption, vaccines and flu shots are strongly encouraged and, in many instances, even required by law. It shouldn't come as any surprise that more babies seem to be developing autism problems, and the risk of developing Alzheimer's disease is steadily increasing.

ALZHEIMER'S LINKED TO FLU SHOTS

In the year 2000, there were approximately 5 million people in the U.S. with Alzheimer's, and it has become the fourth leading cause of death in individuals over the age of 75. By the year 2010, it is estimated that over 7 million individuals will have the disease, and by 2025, 22 million will develop Alzheimer's.

As the general population continues to consume more contaminated food, water, and medicines, these predictions may very well prove accurate. One expert at the 1997 National Vaccine Information Center (NVIC) International Vaccine Conference stated that anyone who had five consecutive flu vaccine shots increased their risk of developing Alzheimer's disease by a factor of 10 over someone who received only two or fewer shots.

A powerful herb to prevent alzheimer's

It's worth mentioning, while we're on the Alzheimer's topic, that the elderly in India have the lowest incidence of Alzheimer's disease in the world. Only 1 percent of the elderly in India suffer from Alzheimer's. In contrast, the Alzheimer's Association in this country says that 10% of our population over 65 years old has the disease, and half of those over 85 have Alzheimer's. Researchers have theorized that the low incidence of Alzheimer's among the Indian population could be due to their increased consumption of the spice turmeric, a component of curry. Animal studies have supported this theory.

Studies have shown that when either turmeric or curcumin, (a major component of turmeric) was added to the diets of animals bred to develop Alzheimer's, the brain damage was significantly lessened. [Neurobiol Aging 01;22(6):993-1005] [J Neurosci 01:21(21):8370-8377]

Turmeric has been shown to have very strong antioxidant properties that can be very effective at normal dietary doses. This spice may be one of the easiest and least expensive methods of combating the growing epidemic of Alzheimer's disease.

Better than a flu shot

When it comes to beating the flu, **selenium** can increase your odds. Selenium is a necessary mineral for the production of antioxidants within the body. New animal research from the University of North Carolina has found that a dietary deficiency of selenium may cause a harmless strain of the flu virus to mutate into a virulent pathogen.

When selenium-deficient mice were given a known flu virus and compared to mice with normal selenium levels, researchers found that the selenium-deficient animals experienced far more serious symptoms, such as lung damage. Based on this new research, other researchers are wondering if the more potent viruses, such as HIV, also mutated in environments where there were selenium deficiencies. It makes sense when you consider the well-known fact that most of the worldwide flu outbreaks originate in China, where large segments of the population are selenium-deficient.

Whether you decide to get flu shots or other vaccinations is a personal choice but as you weigh the pros and cons of such a decision, **don't be naive enough to think any of our government agencies have your best interests as their top priority. It could be a fatal mistake.**

THE VACCINATION
By Patricia Crutchfield

His trusting eyes looked up at me
He smiled his sweetest smile
What a precious gift from God he was
My son my first born child,

The nurse came in and weighed him
Put a thermometer briefly in his ear
Then she told me to take off his diaper
And expose his plump little rear.

I did as I was instructed
For I knew the procedure by now
It's time for his next vaccination
This time I won't flinch, I vow.

The syringes and vial of the serums
Lay benignly on her sterile steel tray
And though I try to watch her,
I find myself turning away.

His scream at the prick of the needle
Sends a bolt of pure terror through me
It's animal like pitch was not normal
And I turned around quickly to see.

His beautiful body went rigid
Then spasmed again and again
What's happening to my poor baby?

And what can I do to help him?

I could sense the nurse's pure panic
As she called out to the doctor to come
The seconds that passed seems like hours
And where is that screaming coming from?

I open my eyes in a room filled with light
The silence a deafening roar
My husband is standing beside me
He says everything fine, but his tears tell me more

I try to sit up, but I'm weary
Another needle pierces my arm
I drift off once again into darkness
But my mind beats a steady alarm.

Two days and two nights I am sedated
Until now no one tells me why
Then the doctor appears with my husband
And immediately I start to cry

My most precious gift has been taken
He'll never again be mine to hold
His body once so warm and loving
Now lays on a slab icy cold

I'm sorry says the good doctor
A reaction we couldn't foresee
Please accept my sincerest condolence
I guess it was just meant to be

Our son now plays with the angels
And my heart breaks anew everyday
Its the angels who tickle his tummy
And it's in their arms not mine, he will lay

A statistic, one in seventeen hundred
That's what they say of my son
But I say one child is too many
To die from a vaccination

So mothers do not be so trusting
Hear me before it's to late
Don't lose your child to the "program"
Investigate before you vaccinate

Families Raise Concern Over Mercury In Vaccines
Debate Continues Over Past Use Of Thimerosal
POSTED: 1:37 p.m. EST November 4, 2002
11:08 p.m. EST November 4, 2002 DURHAM COUNTY, N.C.

A growing contingent of parents believes a mercury-based preservative in those vaccines may have done more harm than good. In 1999, at the request of the Food and Drug Administration, drug companies agreed to begin removing a controversial preservative called thimerosal from vaccines. Some families believe the removal comes too late. Jackson Bono is a happy, curious 13-year-old challenged by a myriad of medical and developmental problems. Jackson has trouble speaking and focusing and works with a tutor.

"The toll it takes on a family is remarkable," said Scott Bono, Jackson's father. Like most parents, Scott and Laura Bono had their son vaccinated when he was a baby. They now blame his problems on thimerosal and its main ingredient, mercury.
"Little did we ever suspect that the very immunizations that were to protect him from childhood diseases were poisoning him with mercury," Scott Bono said. Thimerosal kills harmful bacteria and has been in vaccines for decades. In the early 1990s, the number of recommended childhood vaccines increased. Over the last decade the national autism rate has risen drastically. In North Carolina, the rate has more than quadrupled, according to the state Department of Public Instruction.

Some people see a connection. If you add up the amount of mercury in baby vaccines with thimerosal, the levels exceed those considered safe for adults by the FDA. The Bonos said Jackson was a normal, healthy baby until he received a bundle of vaccines when he was 16 months old. They said, soon after, he stopped talking and making eye contact. Jackson developed autistic tendencies, like spinning uncontrollably. He also suffered severe allergies, seizures and stomach trouble.

"It was a cruel tragedy that happened with our son," Laura Bono said. Dr. Samuel Katz, chairman emeritus of paediatrics at Duke, is considered

one of the foremost authorities on vaccines in the country. He raises doubts that thimerosal ever hurt children. "Whenever we have a problem, we like to know whose fault is it. Unfortunately, vaccines have become an easy target," he said. Katz said, "The evidence to support these claims is lacking." However, in 1999, he recommended drug companies take thimerosal out of vaccines. A 2001 report from the National Institute of Medicine also concluded the evidence does not support the claims. Researchers conceded, "the hypothesis is biologically plausible."

"Given that its mercury and we know that mercury has no beneficial effects, my statement to the FDA was that there's really no reason to use something like thimerosal," said Michael Aschner, a Wake Forest University neurobiologist. Aschner has studied mercury for 20 years. Research from the University of Calgary backs up his work and found mercury can destroy brain cells. Aschner points out that the ethylmercury in thimerosal is different from the damaging methylmercury found in some fish. He feels the issue clearly deserves much more study.

"If you do it in a dish, ethylmercury does cause significant effects, toxic effects. There's no question about it," Ascher said. "But, again, what you have to be careful of is how you translate what you see in a dish into a human being." The biggest obstacle parents of special needs children face in making the thimerosal argument is the fact that millions of children, a vast majority, got the same vaccine and never got sick.

"Why is it that all people who smoke don't get cancer? The body reacts differently to different antagonists," Salisbury attorney Bill Graham said. Graham represents 40 families who believe thimerosal hurt their children. He believes evidence is mounting that federal regulators knew that thimerosal could be harmful long before drug companies felt pressure to remove it from vaccines. A study sanctioned by the Centers for Disease Control and Prevention shows infants immunized with thimerosal vaccines were 2.5 times more likely to develop neurological disorders, but it was never released. Instead, the study continued and the results changed. Graham questions why vaccines were never recalled.

"Do you think that thimerosal vaccines that are potentially harmful could still be out there? They could be. They could be on the shelf right now," Graham said. "I really think the thimerosal issue has become a feeding frenzy. It's like the sharks with blood in the water," Katz said. The Bonos said they do not want blood. They want families like theirs to be heard for Jackson's sake, and others like him. "He's lost his childhood and he may not ever be what he should have been," Laura Bono said. Parents like the

Bonos can file claims with the National Vaccine Injury Compensation Program. Because of the debate over thimerosal, the federal government has put all the claims on hold until further studies are completed. There was no recall of thimerosal vaccines, so it is possible some could still be on shelves. Anyone with concerns should talk to their child's pediatrician and ask for thimerosal-free vaccines. Both sides of the debate stress the importance of immunizing children.
Reporter: Cullen Browder
Photographer: Gil Hollingsworth
OnLine Producer: Michelle Singer

FOR VAST AMOUNTS OF INFORMATION CHECK:
http://www.vaccinetruth.org/

413-Gems of wisdom

All truth goes through three stages. First it is ridiculed. Then it is violently opposed. Finally, it is accepted as self-evident."

(Schopenhauer)
****

"Condemnation without investigation is the height of ignorance." Albert Einstein

....

"In the field of vaccination, medical training is simple indoctrination."
****

Inoculations are the true weapons of mass destruction causing an epidemic of GENOCIDE
Rebecca Carley, MD
Court Qualified Expert in VIDS (Vaccine Induced Diseases)
http://www.drcarley.com

....

..discussing vaccination with a doctor is like discussing vegetarianism with a butcher...........(George Bernard Shaw)

What good fortune for those in power that the people do not think." ~Adolf Hitler

"When you once see something as false which you have accepted as true, as natural, as human, then you can never go back to it" - J. Krishnamurti

It also gives us a very special, secret pleasure to see how unaware the people around us are of what is really happening to them." ~Adolf Hitler

" Fear of disease, fear of microorganisms, fear of the unknown, is the tool of the clever that keeps the weak in line"
~ Tim O'Shea, DC

What a strange religion medicine makes. It's the only religion that is federally backed, and even amid scientific controversy, cannot be questioned openly without persecution or ridicule."

Why Doctors do not understand the evils of vaccinations....

"It is difficult to get a man to understand something when his salary depends upon his not understanding it!" Upton Sinclair

No one has ever successfully proven that any child has ever benefited from an injection of rotting matter combined with nerve and brain destroying poisons, the actual ingredients of vaccines. – Dewey

What is the name of the test that can be given to determine if a child can safely receive a vaccine?
It's called a breath test. You hold a mirror in front of the child and if condensation appears, they are still alive and cannot "safely" receive a vaccine. - steve

"Uneducated people believe what they are told...Educated people question what they are told"

"People do not like to think. If one thinks, one must reach conclusions. Conclusions are not always pleasant."
-Helen Keller

You can't wake a person who is pretending to be asleep. ~ Navajo Proverb
The art of medicine consists of amusing the patient while nature cures the disease—Voltaire

"A truth's initial commotion is directly proportional to how deeply the lie was believed... When a well-packaged web of lies has been sold gradually to the masses over generations, the truth will seem utterly preposterous and its speaker, a raving lunatic." --Dresden James

"He's the best physician that knows the worthlessness of most medicines."
"God heals and the Doctor takes the fee." - Benjamin Franklin, (1706-1790)

For us to bombard a newborn baby with a whole battery of vaccines as, in effect, their very first immunologic experience I think is reckless beyond measure. I would say it borders on the criminal. Dr. Moscowitz

If you think that something is right just because everyone believes it, then you are not thinking" - Vievienne Westwood

Knowledge makes a man unfit to be a slave."
Frederick Douglass

Men occasionally stumble on the truth, but most of them pick themselves up and hurry off as if nothing had happened.
Winston Churchill

"First they ignore you, then they laugh at you, then they fight you, then you win." ~Ghandi

"Your failure to be informed does not make me a wacko." -- John Loeffler

I have no medical training, but I can read.

"The great tragedy of science - a beautiful hypothesis slain by an ugly fact." - Thomas Huxley

"If you think you're too small to be effective, you've never been in bed with a mosquito." - Betty Reese

"I know that most men, including those at ease with the problems of the greatest complexity, can seldom accept even the simplest and most obvious truth if it be such as would oblige them to admit the falsity of conclusions which they have delighted in explaining to colleagues, which they have proudly taught to others, and which they have woven, thread by thread, into the fabric of their lives."- Leo Tolstoy—

Right is right, even if everyone is against it; and wrong is wrong, even if everyone is for it --William Penn

414-WHERE IS THE GOLD?

BECAUSE IT DOES NOT REALLY SEEM TO BE HERE ANYMORE. IT'S GONE.

In October of 2009 the Chinese received a shipment of gold bars.Gold is regularly exchanges between countries to pay debts and to settle the so-called balance of trade. Most gold is exchanged and stored in vaults under the supervision of a special organization based in London, the London Bullion Market Association (or LBMA). When the shipment was received, the Chinese government asked that special tests be performed to guarantee the purity and weight of the gold bars. Inthis test, four small holed are drilled into the gold bars and the metal is then analyzed.

Officials were shocked to learn that the bars were fake. They contained cores of tungsten with only a outer coating of real gold.What's more, these gold bars, containing serial numbers for tracking,originated in the US and had been stored in Fort Knox for years. There were reportedly between 5,600 to 5,700 bars, weighing 400 oz. each, in the shipment!

http://www.daily.pk/fake-gold-bars-in-bank-of-england-and-fort-knox-14477/

•

Pakistan Daily

Fake gold bars in Bank of England and Fort Knox
Written by (Author) World Jan 11, 2010

It's one thing to counterfeit a twenty or hundred dollar bill. The amount of financial damage is usually limited to a specific region and only affects dozens of people and thousands of dollars. Secret Service agents quickly notify the banks on how to recognize these phony bills and retail outlets usually have procedures in place (such as special pens to test the paper) to stop their proliferation.

But what about gold? This is the most sacred of all commodities because it is thought to be the most trusted, reliable and valuable means of saving wealth.

A recent discovery — in October of 2009 — has been suppressed by the main stream media but has been circulating among the "big money" brokers and financial kingpins and is just now being revealed to the public. It involves the gold in Fort Knox — the US Treasury gold — that is the equity of our national wealth. In short, millions (with an "m") of gold bars are fake!

Who did this? Apparently our own government.

Background

In October of 2009 the Chinese received a shipment of gold bars. Gold is regularly exchanges between countries to pay debts and to settle the so-called balance of trade. Most gold is exchanged and stored in vaults under the supervision of a special organization based in London, the London Bullion Market Association (or LBMA). When the shipment was received, the Chinese government asked that special tests be performed to guarantee the purity and weight of the gold bars. In this test, four small holed are drilled into the gold bars and the metal is then analyzed.

Officials were shocked to learn that the bars were fake. They contained cores of tungsten with only a outer coating of real gold. What's more, these gold bars, containing serial numbers for tracking, originated in the

US and had been stored in Fort Knox for years. There were reportedly between 5,600 to 5,700 bars, weighing 400 oz. each, in the shipment!

At first many gold experts assumed the fake gold originated in China, the world's best knock-off producers. The Chinese were quick to investigate and issued a statement that implicated the US in the scheme.

What the Chinese uncovered:
Roughly 15 years ago — during the Clinton Administration [think Robert Rubin, Sir Alan Greenspan and Lawrence Summers] — between 1.3 and 1.5 million 400 oz tungsten blanks were allegedly manufactured by a very high-end, sophisticated refiner in the USA [more than 16 Thousand metric tonnes]. Subsequently, 640,000 of these tungsten blanks received their gold plating and WERE shipped to Ft. Knox and remain there to this day.

According to the Chinese investigation, the balance of this 1.3 million to 1.5 million 400 oz tungsten cache was also gold plated and then allegedly "sold" into the international market. Apparently, the global market is literally "stuffed full of 400 oz salted bars". Perhaps as much as 600-billion dollars worth.

An obscure news item originally published in the N.Y. Post [written by Jennifer Anderson] in late Jan. 04 perhaps makes sense now.

DA investigating NYMEX executive ,Manhattan, New York, –Feb. 2, 2004.
A top executive at the New York Mercantile Exchange is being investigated by the Manhattan district attorney. Sources close to the exchange said that Stuart Smith, senior vice president of operations at the exchange, was served with a search warrant by the district attorney's office last week. Details of the investigation have not been disclosed, but a NYMEX spokeswoman said it was unrelated to any of the exchange's markets. She declined to comment further other than to say that charges had not been brought. A spokeswoman for the Manhattan district attorney's office also declined comment."

The offices of the Senior Vice President of Operations — NYMEX — is exactly where you would go to find the records [serial number and smelter of origin] for EVERY GOLD BAR ever PHYSICALLY settled on the exchange. They are required to keep these records. These precise records would show the lineage of all the physical gold settled on the exchange and hence "prove" that the amount of gold in question could not have possibly come from the U.S. mining operations — because the amounts in question coming from U.S. smelters would undoubtedly be vastly bigger than domestic mine production.

No one knows whatever happened to Stuart Smith. After his offices were raided he took "administrative leave" from the NYMEX and he has never been heard from since. Amazingly, there never was any follow up on in the media on the original story as well as ZERO developments ever stemming from D.A. Morgenthau's office who executed the search warrant.

Are we to believe that NYMEX offices were raided, the Sr. V.P. of operations then takes leave — all for nothing?

The revelations of fake gold bars also explains another highly unusual story that also happened in 2004:
LONDON, April 14, 2004 (Reuters) — NM Rothschild & Sons Ltd., the London-based unit of investment bank Rothschild [ROT.UL], will withdraw from trading commodities, including gold, in London as it reviews its operations, it said on Wednesday.

Interestingly, GATA's Bill Murphy speculated about this back in 2004; "Why is Rothschild leaving the gold business at this time my colleagues and I conjectured today? Just a guess on my part, but [I] suspect something is amiss. They know a big scandal is coming and they don't want to be a part of it… [The] Rothschild wants out before the proverbial "S" hits the fan." — BILL MURPHY, LEMETROPOLE, 4-18-2004

What is the GATA?
The Gold Antitrust Action Committee (GATA) is an organisation which has been nipping at the heels of the US Treasury Federal Reserve for several years now. The basis of GATA's accusations is that these institutions, in coordination with other complicit central banks and the large gold-trading investment banks in the US, have been manipulating the price of gold for decades.

What is the GLD?GLD is a short form for Good London Delivery. The London Bullion Market Association (LBMA) has defined "good delivery" as a delivery from an entity which is listed on their delivery list or meets the standards for said list and whose bars have passed testing requirements established by the associatin and updated from time to time. The bars have to be pure for AU in an area of 995.0 to 999.9 per 1000. Weight, Shape, Appearance, Marks and Weight Stamps are regulated as follows:

Weight: minimum 350 fine ounces AU; maximum 430 fine ounces AU, gross weight of a bar is expressed in troy ounces, in multiples of 0.025, rounded down to the nearest 0.025 of an troy ounce.

Dimensions: the recommended dimensions for a Good Delivery gold bar are: Top Surface: 255 x 81 mm; Bottom Surface: 236 x 57 mm; Thickness: 37 mm.

Fineness: the minimum 995.0 parts per thousand fine gold. Marks: Serial number; Assay stamp of refiner; Fineness (to four significant figures); Year of manufacture (expressed in four digits).

After reviewing their prospectus yet again, it becomes pretty clear that GLD was established to purposefully deflect investment dollars away from legitimate gold pursuits and to create a stealth, cesspool / catch-all, slush-fund and a likely destination for many of these fake tungsten bars where they would never see the light of day — hidden behind the following legalese "shield" from the law:

[Excerpt from the GLD prospectus on page 11]
"Gold bars allocated to the Trust in connection with the creation of a Basket may not meet the London Good Delivery Standards and, if a Basket is issued against such gold, the Trust may suffer a loss. Neither the Trustee nor the Custodian independently confirms the fineness of the gold bars allocated to the Trust in connection with the creation of a Basket. The gold bars allocated to the Trust by the Custodian may be different from the reported fineness or weight required by the LBMA's standards for gold bars delivered in settlement of a gold trade, or the London Good Delivery Standards, the standards required by the Trust. If the Trustee nevertheless issues a Basket against such gold, and if the Custodian fails to satisfy its obligation to credit the Trust the amount of any deficiency, the Trust may suffer a loss."

The Federal Reserve knows but is apparently part of the scheme Earlier this year GATA filed a second Freedom of Information Act (FOIA) request with the Federal Reserve System for documents from 1990 to date having to do with gold swaps, gold swapped, or proposed gold swaps.

On Aug. 5, The Federal Reserve responded to this FOIA request by adding two more documents to those disclosed to GATA in April 2008 from the earlier FOIA request. These documents totaled 173 pages, many parts of which were redacted (blacked out). The Fed's response also noted that there were 137 pages of documents not disclosed that were alleged to be exempt from disclosure.

GATA appealed this determination on Aug. 20. The appeal asked for more information to substantiate the legitimacy of the claimed exemptions from disclosure and an explanation on why some documents, such as one posted on the Federal Reserve Web site that discusses gold swaps, were not included in the Aug. 5 document release.

In a Sept. 17, 2009, letter on Federal Reserve System letterhead, Federal Reserve governor Kevin M. Warsh completely denied GATA's appeal. The entire text of this letter can be examined at http://www.gata. org/files/ GATAFedRespon" The first paragraph on the third page is the most revealing. "In connection with your appeal, I have confirmed that the information withheld under exemption 4 consists of confidential commercial or financial information relating to the operations of the Federal Reserve Banks that was obtained within the meaning of exemption 4. This includes information relating to swap arrangements with foreign banks on behalf of the Federal Reserve System and is not the type of information that is customarily disclosed to the public. This information was properly withheld from you."

above statement is an admission that the Federal Reserve has been involved with the fake gold bar swaps and that it refuses to disclose any information about its activities!

The above statement is an admission that the Federal Reserve has been involved with the fake gold bar swaps and that it refuses to disclose any information about its activities!

Why use tungsten?

If you are going to print fake money you need to have the special paper, otherwise the bills don't feel right and can be easily detected by special pens that most merchants and banks use. Likewise, if you are going to fake gold bars you had better be sure they have the same weight and properties of real gold.

In early 2008 millions of dollars in gold at the central bank of Ethiopia turned out to be fake. What were supposed to be bars of solid gold turned out to be nothing more than gold-plated steel. They tried to sell the stuff to South Africa and it was sent back when the South Africans noticed this little problem. The problem with making good-quality fake gold is that gold is remarkably dense. It's almost twice the density of lead, and two-and-a-half times more dense than steel. You don't usually notice this because small gold rings and the like don't weigh enough to make it obvious, but if you've ever held a larger bar of gold, it's absolutely unmistakable: The stuff is very, very heavy.

The standard gold bar for bank-to-bank trade, known as a "London good delivery bar" weighs 400 troy ounces (over thirty-three pounds), yet is no bigger than a paperback novel. A bar of steel the same size would weigh only thirteen and a half pounds.

According to gold expert, Theo Gray, the problem is that there are very few metals that are as dense as gold, and with only two exceptions they all cost as much or more than gold.

The first exception is depleted uranium, which is cheap if you're a government, but hard for individuals to get. It's also radioactive, which could be a bit of an issue.

The second exception is a real winner:
tungsten. Tungsten is vastly cheaper than gold (maybe $30 dollars a pound compared to $12,000 a pound for gold right now). And remarkably, it has exactly the same density as gold, to three decimal places. The main differences are that it's the wrong color, and that it's much, much harder than gold. (Very pure gold is quite soft, you can dent it with a fingernail.)

A top-of-the-line fake gold bar should match the color, surface hardness, density, chemical, and nuclear properties of gold perfectly. To do this, you could could start with a tungsten slug about 1/8-inch smaller in each dimension than the gold bar you want, then cast a 1/16-inch layer of real pure gold all around it. This bar would feel right in the hand, it would have a dead ring when knocked as gold should, it would test right chemically, it would weigh *exactly* the right amount, and though I don't know this for sure, I think it would also pass an x-ray fluorescence scan, the 1/16″ layer of pure gold being enough to stop the x-rays from reaching any tungsten. You'd pretty much have to drill it to find out it's fake.

Such a top-quality fake London good delivery bar would cost about $50,000 to produce because it's got a lot of real gold in it, but you'd still make a nice profit considering that a real one is worth closer to $400,000.

What's going to happen now?
Politicians like Ron Paul have been demanding that the Federal Reserve be more transparent and open up their records for public scrutiny. But the

Fed has consistently refused, stating that these disclosures would undermine its operation. Yes, it certainly would!

The Jane Burgermeister Archive Article Archive Rothschilds Implicated in Fake Gold Bar Scandal

Rothschilds Implicated in Fake Gold Bar Scandal

Sunday, 17 January 2010 15:00

News - Latest News

0diggsdigg
Fake gold bars in Bank of England and Fort Knox

Written by (Author) World Jan 11, 2010

(Well this sure as heck causes some alarm bells to ring in my mind. By itself it is of little or no consequence, but when one ties it in with literally thousands of other "situations" it makes one think, "yes there is a connection, but I wouldn't have seen it just looking at one isolated matter. Maybe I wouldn't have correlated 20 things, but there is a limit to my acceptance. Now I want to know. WHAT IS GOING ON HERE?"

Then there is the unanswered case of all the gold bullion that went missing when "9/11" happened. Here is a clip off the Internet:)

Precious Metals Stored Beneath the World Trade Center

One of the less noted of the possible motives for the attack was the creation of diversion in order to steal hundreds of millions of dollars worth of precious metals.

*By September of 2003, **9-11 Research** had published the following story about the discrepancy between the value of precious metals reportedly stored in the Comex vaults beneath WTC 4 and the value reportedly recovered in late 2001 following the attack. (The September, 2003 version of the page is archived on archive.org.)*

e x c e r p t
title: Missing Gold

authors: Jim Hoffman

Missing Gold

A King's Ransom in Precious Metals Seems to have Disappeared

The basement of 4 World Trade Center housed vaults used to store gold and silver bullion. Published articles about precious metals recovered from the World Trade Center ruins in the aftermath of the attack mention less than $300 million worth of gold. All such reports appear to refer to a removal operation conducted in late October of 2001. On Nov. 1, Mayor Rudolph Giuliani announced that "more than $230 million" worth of gold and silver bars that had been stored in a bomb-proof vault had been recovered. A New York Times article contained: 2

Two Brinks trucks were at ground zero on Wednesday to start hauling away the $200 million in gold and silver that the Bank of Nova Scotia had stored in a vault under the trade center ... A team of 30 firefighters and police officers are helping to move the metals, a task that can be measured practically down to the flake but that has been rounded off at 379,036 ounces of gold and 29,942,619 ounces of silver ..

Another article gave a figure of $650 million to the value of gold in the 4 WTC vault.

Unknown to most people at the time, $650 million in gold and silver was being kept in a special vault four floors beneath Four World Trade Center. The gold and silver were recently recovered.

An article in the TimesOnline gives the following rundown of precious metals that were being stored in WTC vaults belonging to Comex. 4

- *Comex metals trading - 3,800 gold bars weighing 12 tonnes and worth more than $100 million*
- *Comex clients - 800,000 ounces of gold with a value of about $220 million*
- *Comex clients - 102 million ounces of silver, worth $430 million*
- *Bank of Nova Scotia - $200 million of gold*

The TimesOnline *article is not clear as to whether the $200 million in gold reported by the Bank of Nova Scotia was part of the $220 million in gold held by Comex for clients. If so, the total is $750 million; otherwise $950 million.*

There appear to be no reports of precious metals discovered between November of 2001 and the completion of excavation several months later.

It would seem that at least the better part of a billion dollars worth of precious metals went missing. It is not plausible that whatever destroyed the towers vaporized gold and silver, which are heavy malleable metals that are extremely unlikely to participate in chemical reactions with other materials.

References
1Thanksgiving at Ground Zero,National Real Estate Investor,
2Below Ground Zero, Silver and Gold,New York Times,11/1/2001
3Cache of Gold Found at WTC Two truckloads retrieved through a tunnel in rubble2,,
4Crushed towers give up cache of gold ingots,TimesOnline,11/1/02
5,Reuters and New York Daily News,

site: 911research.wtc7.net page:
911research.wtc7.net/wtc/evidence/gold.html

Tons of gold vanished out of the World Trade Center after 9/11, who took it?

Even the most brain-washed person can not answer what happened to tons of gold and silver that was stored in the lower vaults of the World Trade Center after the 9/11 disaster. Everyone has to know that bin Laden didn't take it. Who ever took the gold is totally responsible of the events of 9/11 and they are not Arabs, they are Americans.
I first heard about the missing gold on a news report on 9/14, when rescue workers looking for survivors found all the vault doors open and empty. I knew then that this was an inside job, because common sense says that terrorists couldn't move that much gold prior to the buildings collapse.
With more people talking about the missing gold, Giuliani reported on 11/1, that some of the gold was recovered. Still over a billion dollars worth was still unaccounted for. Those who still believe that bin Laden did it, how did he get away with the gold unseen? Those who laugh at conspiracies find the gold.

•

Additional Details
I realize that 9/11 was not about gold. It was staged to frighten American citizens out of a lot of their freedoms and constitutional rights, which is all in place with the Patriot Act and Home Land Security. I'm not surprised that so few people know what when on in the World Trade Towers and why there would be so much gold and silver there.
I used the analysis of the gold because behind every scam put on US

citizens, money is involved. Those who are really interested in the truth will seek it out themselves as I did, the rest will find comfort in ignorance.

It hurts my soul that we have been killing people in another country all these years that had nothing to do with 9/11 and we cheer those who are committing this murder.

Nothing related to 9/11 makes even common sense, unfortunately too many Americans are afraid to accept the truth about the country that we live in.

415-The Population Control Agenda

"To die will be an awfully big adventure." – Sir James Barrie

(Article off Internet) by Bob Lee • Sunday January 20, 2002 at 07:26 PM

Stay armed, stay free!
Rense.com

The Population Control
Agenda - A Timeline
From Bob Lee <rboblee@home.com>
http://www.fortunecity.com/roswell/daniken/443/populationcontrolagenda2. htm
Robert E. Lee, M.S., M.S.W., L.C.S.W.
Author "AIDS: An Explosion of the Biological Time-bomb?" c2000
Author "AIDS in America: Our Chances, Our Choices" c1987
website: http://www.bhc.edu/eas tcampus/leeb/aids/index.html
11-15-00

1963 The mass vaccination campaigns of the 1950s and '60s may be causing hundreds of deaths a year because of a cancer-causing virus that contaminated the first polio vaccine, according to scientists. Known as SV40, the virus came from dead monkeys whose kidney cells were used to culture the first Salk vaccines. Doctors estimate that the virus was injected into tens of millions during the vaccination campaigns, including several million in Canada, before being detected and screened out in 1963. Those born between 1941 and 1961 are thought to be most at risk of having been infected.

...Humanity must drastically scale down its industrial activities on Earth, change its consumption lifestyles, stabilize and then reduce the size of the human population by humane means, and protect and restore wild ecosystems and the remaining wildlife on the planet." The Wildlands Project

Chemtrails Lab Analysis -Virulent Bio-Toxin Soup

"Inasmuch as ye have done it unto the least of these my brethren, ye have done it unto Me." (Matthew 25:40)

A History Timeline Of Population Control

By Robert Howard

The Tartars had the idea of infecting the enemy by catapulting bodies infected with bubonic plague over the walls of the city of Kaffa. Some historians believe that this event was the cause of the epidemic of plague that swept across medieval Europe killing 25 million.

1763 The British during the French-Indian War. The Native Americans greatly outnumbered the British and were suspected of being on the side of the French. As an "act of good will" the British give blankets to the Indians, but the blankets came from a hospital that was treating smallpox victims and consequently smallpox raged through the Native American community and devastated their numbers.

1814 Andrew Jackson, whose portrait appears on the U.S. $20 bill today, supervised the mutilation of 800 or more Creek Indian corpses, the bodies of men, women and children that his troops had massacred, cutting off their noses to count and preserve a record of the dead, slicing long strips of flesh from their bodies to tan and turn into bridle reins.
(Historian Ward Churchill, A LITTLE MATTER OF GENOCIDE; HOLOCAUST AND DENIAL IN THE AMERICAS, 1492 TO THE PRESENT
(San Francisco: City Lights Books, 1997).
ISBN 0-87286-323-9. pg.186)
U.S. Presidents And The Masonic Power Structure

1911, Turkey established gun control. From 1915-1917, 1.5 million Armenians, unable to defend themselves, were rounded up and exterminated.

1918 The modern history of Biological Warfare (BW) starts in 1918 with the Japanese formation of a special section of the Army (Unit 731) dedicated to BW. The thought at the time was "Science and Technology are the Key's to Winning War and BW is the most cost effective."

1918-1919 Flu pandemic that killed over 20 million worldwide and over 500,000 here in the US.

1920's Unlike the malignant twists of nature, ranging from bubonic plague through to potato blight, which have killed masses throughout the ages, both the beef and pituitary hormone CJD crises were manmade. Scrapie, the sheep equivalent of BSE and CJD, has been around for more than two centuries. Somewhat differently, human spongiform encephalopathy was unheard of before two German physicians, Creutzfeldt and Jakob, independently reported the initial cases in the 1920s. BSE, too, was unheard of until a decade after cattle began to be fed the protein-rich remains of scrapie-infected sheep to accelerate their growth.

1929 The Soviet Union established gun control. From 1929 to 1953, approximately 20 million dissidents, unable to defend themselves, were rounded up and exterminated.

1930's Less known to the public is that fluoride also accumulates in bones. "The teeth are windows to what's happening in the bones," explained Paul Connett, Professor of Chemistry at St Lawrence University, New York, to these reporters. In recent years, pediatric bone specialists have expressed alarm about an increase in stress fractures among young people in the US. Connett and other scientists are concerned that fluoride-linked to bone damage in studies since the 1930s-may be a contributing factor. Link

1931 Japan expanded its territory by taking over part of Manchuria and Unit 731 moved in to secure "an endless supply of human experiment materials." Essentially all prisoners of war were available for Biological Warfare (BW) experiments.

1931 Dr. Cornelius Rhoads, under the auspices of the Rockefeller Institute for Medical Investigations, infects human subjects with cancer cells. He later goes on to establish the U.S. Army Biological Warfare facilities in Maryland, Utah, and Panama, and is named to the U.S. Atomic Energy Commission. While there, he begins a series of radiation exposure experiments on American soldiers and civilian hospital patients. Link

1932 The Infamous Tuskegee Study In recent history, we have seen the influence of occult population control advocates here in America. Nowhere is that influence better demonstrated than in the Tuskegee Study, a scientific research program in which 400 syphilis-infected black men were recruited by the U.S. Public Health Service back in 1932. The participants were all told that they would be treated for their infections, but instead of treating their illness, all medicines were withheld. The black men were then actively prevented from obtaining treatment elsewhere as their bodies, and the bodies of their wives and children, were systematically ravaged by disease. The evil men who conceived that Nazi-style study justified their atrocity by alleging that scientists needed

to learn how untreated syphilis progressed in the human body. For a period of forty years, between 1932 and 1972, the genocidal Tuskegee Study continued. It was not until 1972, when one newspaper finally had the courage to break the story to the public, that the Tuskegee Study was finally terminated.
The Population Control Agenda

1934 The original lesson about the infectious nature of these brain diseases mad cow disease" or bovine spongiform encephalopathy (BSE) in cattle, and Creutzfeldt-Jakob disease (CJD) in humans) came from a 1934 vaccine catastrophe in the UK which brought scrapie, or "mad sheep disease", to almost 5,000 out of 18,000 lambs within two years of their immunization against louping-ill virus infection. Tracing back, scientists discovered that the vaccine serum was prepared from a number of lambs whose dams had subsequently developed scrapie, but the significance of scrapie passing vertically from ewes to their lambs, and horizontally from lamb to lamb by virtue of the vaccine injections, was kept from international eyes by a series of egotistical carry-ons which prevented the data from reaching the pages of the scientific literature for a further 15 years.

1935 The Pellagra Incident. After millions of individuals die from Pellagra over a span of two decades, the U.S. Public Health Service finally acts to stem the disease. The director of the agency admits it had known for at least 20 years that Pellagra is caused by a niacin deficiency but failed to act since most of the deaths occured within poverty-striken black populations.
Link

1935 China established gun control. From 1948 to 1952, 20 million political dissidents, unable to defend themselves, were rounded up and exterminated.

1938 Germany established gun control in 1938 and from 1939 to 1945, 6 to 7 million Jews, gypsies, homosexuals, the mentally ill, and 12 million Christians who were unable to defend themselves, were rounded up and exterminated.

1939 Margaret Sanger organized her "Negro project," a program designed to eliminate members of what she believed to be an "inferior race." Margaret Sanger justified her proposal because she believed that: "The masses of Negroes ...particularly in the South, still breed carelessly and disastrously, with the result that the increase among Negroes, even more

than among whites, is from that portion of the population least intelligent and fit..."
The Population Control Agenda

1940 Four hundred prisoners in Chicago are infected with Malaria in order to study the effects of new and experimental drugs to combat the disease. Nazi doctors later on trial at Nuremberg cite this American study to defend their own actions during the Holocaust. Link

1941 Japanese planes sprayed bubonic plague over parts of China. At least 5 separate instances of this occurring have been documented. In 1942 "bacterial bombs" were deployed on mainland China but these attacks were determined to be ineffective.

1942 The United States (US) becomes aware of the Japanese efforts in Biological Warfare (BW) and decided to start its own program. These acts were not the only atrocities committed, however. The Japanese released thousands of plague infested rats prior to their surrender, with unknown consequences. They also tested on American POW's during the war and the U.S. Government apparently knew about it, but did nothing (perhaps a worse atrocity). These people killed over 3000 POWs, including many Americans, in a variety of grisly experiments. What they did instead was to offer immunity to would-be war criminals in exchange for the information the Japanese learned from these experiments!

1942 Chemical Warfare Services begins mustard gas experiments on approximately 4,000 servicemen. The experiments continue until 1945 and made use of Seventh Day Adventists who chose to become human guinea pigs rather than serve on active duty. Link

1942 Great Britain was also developing a program in Biological Warfare (BW). The program focused on anthrax spores and their viability and "range of spread" when delivered with a conventional bomb. The fateful **Gruinard Island** off the coast of Scotland was chosen as the site for this testing. It was thought that it was far enough off the coast as too prevent any contamination of the mainland, which later turned out to be false. The data gathered from these experiments was used by both Great Britain and the U.S. to develop bombs that were better able to effectively disperse spores.

1943 After an outbreak of anthrax in sheep and cattle in 1943 on the coast of Scotland that directly faced Gruinard, the British decided to stop testing. A tragic consequence of this testing is that even today Gruinard

Island is contaminated with Bacillus anthracis spores. The original idea for decontamination was to start a brushfire that burned off the top of the soil and killed all traces of the organisms. Unfortunately, the spores unexpectedly embedded themselves in the soil so total decontamination of the island was/is impossible. As long as no ground is disturbed, we are supposedly safe, but birds that travel back and forth from mainland to island probably don't know this!

1943 Planning began in1943 with the appointment of a special New York State Health Department committee to study the advisability of adding **fluoride** to Newburgh's drinking water. The chairman of the committee was, again, Dr Harold C. Hodge, then chief of fluoride toxicity studies for the Manhattan Project. Subsequent members of the committee included Henry L. Barnett, a captain in the Project's Medical Section, and John W. Fertig, in 1944 with the Office of Scientific Research and Development-the super-secret Pentagon group which sired the Manhattan Project. Their military affiliations were kept secret. Hodge was described as a pharmacologist, Barnett as a pediatrician. Placed in charge of the Newburgh project was David B. Ast, chief dental officer of the New York State Health Department. Ast had participated in a key secret wartime conference on fluoride, held by the Manhattan Project in January 1944, and later worked with Dr Hodge on the Project's investigation of human injury in the New Jersey incident, according to once-secret memos.

1944 A Manhattan Project memorandum of 29 April 1944 states: "Clinical evidence suggests that uranium hexafluoride may have a rather marked central nervous system effect... It seems most likely that the F [code for fluoride] component rather than the T [code for uranium] is the causative factor." The memo, from a captain in the medical corps, is stamped SECRET and is addressed to Colonel Stafford Warren, head of the Manhattan Project's Medical Section. Colonel Warren is asked to approve a program of animal research on CNS effects. "Since work with these compounds is essential, it will be necessary to know in advance what mental effects may occur after exposure... This is important not only to protect a given individual, but also to prevent a confused workman from injuring others by improperly performing his duties. The author of the 1944 CNS research proposal attached to the 29 April memo was Dr Harold C. Hodge-at the time, chief of fluoride toxicology studies for the University of Rochester division of the Manhattan Project.

1944 When a severe pollution incident occurred downwind of the E.I. DuPont de Nemours Company chemical factory in Deepwater, New Jersey. The factory was then producing millions of pounds of fluoride for

the Manhattan Project whose scientists were racing to produce the world's first atomic bomb. The farms downwind in Gloucester and Salem counties were famous for their high-quality produce. Their peaches went directly to the Waldorf Astoria Hotel in New York City; their tomatoes were bought up by Campbell's Soup. But in the summer of 1944 the farmers began reporting that their crops were blighted: "Something is burning up the peach crops around here." They said that poultry died after an all-night thunderstorm, and that farm workers who ate produce they'd picked would sometimes vomit all night and into the next day. "I remember our horses looked sick and were too stiff to work," Mildred Giordano, a teenager at the time, told these reporters. Some cows were so crippled that they could not stand up; they could only graze by crawling on their bellies. The account was confirmed in taped interviews with Philip Sadtler (shortly before he died), of Sadtler Laboratories of Philadelphia, one of the nation's oldest chemical consulting firms. Sadtler had personally conducted the initial investigation of the damage. The farmers were stonewalled in their search for information about fluoride's effects on their health, and **their complaints have long since been forgotten**. But they unknowingly left their imprint on history: their complaints of injury to their health reverberated through the corridors of power in Washington and triggered intensive, secret, bomb program research on the health effects of fluoride.
Link

1944 U.S. Navy uses human subjects to test gas masks and clothing. Individuals were locked in a gas chamber and exposed to mustard gas and lewisite.
Link

1945 May. Newburgh's water was fluoridated, and over the next 10 years its residents were studied by the New York State Health Department.

1945-1955 Much of the original proof that fluoride is safe for humans in low doses was generated by A-bomb program scientists who had been secretly ordered to provide "evidence useful in litigation" against defense contractors for fluoride injury to citizens. The first lawsuits against the American A-bomb program were not over radiation, but over fluoride damage, the documents show. Human studies were required. Bomb program researchers played a leading role in the design and implementation of the most extensive US study of the health effects of fluoridating public drinking water, conducted in Newburgh, New York, from 1945 to 1955. Then, in a classified operation code-named "Program F", they secretly gathered and analyzed blood and tissue samples from

Newburgh citizens with the cooperation of New York State Health Department personnel. The original, secret version (obtained by these reporters) of a study published by Program F scientists in the August 1948 Journal of the American Dental Association1 shows that evidence of adverse health effects from fluoride was censored by the US Atomic Energy Commission (AEC)-considered the most powerful of Cold War agencies-for reasons of "national security". The bomb program's fluoride safety studies were conducted at the University of Rochester-site of one of the most notorious human radiation experiments of the Cold War, in which unsuspecting hospital patients were injected with toxic doses of radioactive plutonium. The fluoride studies were conducted with the same ethical mindset, in which "national security" was paramount.
Link
Hitler claimed to have gotten his inspiration for the "final solution" from the extermination of Native Americans in the U.S.

1947 By then, as the 1950s dawned, mad sheep disease was shown in the United States to jump the species barrier when a scrapie-infected food supplement brought a similar brain illness to farm-raised mink in 1947. By this stage, the medico-scientific fraternity was intensely preoccupied with another incurable brain illness, kuru, which had reached epidemic proportions amongst the Fore people living in the highlands of New Guinea. Anthropologists from the University of Adelaide unraveled a chain of events to trace the origin of kuru back to the reverent consumption of deceased tribal members' bodies. Kuru was essentially eradicated by New Guinean authorities acting in 1959 on the anthropological clue to outlaw the eating of human flesh.

1947 The CIA begins its study of LSD as a potential weapon for use by American intelligence. Human subjects (both civilian and military) are used with and without their knowledge.

1949 U.S. Army begins 20 years of simulated germ warfare attacks against American cities, conducting at least 239 open air tests.

1950 Sept. 20-26. One of the biggest experiments involved the use of Serratia marcescens and bacillus globigi being sprayed over 117 square miles of the San Francisco area, causing pneumonia-like infections in many of the residents. The family of one elderly man who died in the test sued the government, but lost. To this day, syraceus is a leading cause of death among the elderly in the San Francisco area.

1950 Department of Defense begins plans to detonate nuclear weapons in desert areas and monitor downwind residents for medical problems and mortality rates

1953 U.S. military releases clouds of zinc cadmium sulfide gas over Winnipeg, St. Louis, Minneapolis, Fort Wayne, the Monocacy River Valley in Maryland, and Leesburg, Virginia. Their intent is to determine how efficiently they could disperse chemical agents.

1953 CIA initiates Project **MKULTRA**. This is an eleven year research program designed to produce and test drugs and biological agents that would be used for mind control and behavior modification. Six of the sub-projects involved testing the agents on unwitting human beings.

1955 Another case was the joint Army-CIA BW test in 1955, still classified, in which an undisclosed bacteria was released in the Tampa Bay region of Florida, causing a dramatic increase in whooping cough infections, including twelve deaths.

1956 The Soviet Union accused the U.S. of using biological weapons in Korea, which lead them to threaten future use of Chemical and Biological weapons. This changed the focus of the U.S. program to a more defensive one. Before this, the bulk of the research was based at Ft. Detrick and used "surrogate biological agents" to model more deadly organisms. Most of the offensive tests were based on "secret spraying" of organisms over populated areas. This program was (supposedly) shut down in 1969.

1956 Cambodia established gun control. From 1975 to 1977, one million "educated" people, unable to defend themselves, were rounded up and exterminated.
Other experiments included tests on Minneapolis that were disguised as "smoke screen tests" because residents were told a harmless smoke was being tested so that cities might be 'hidden' from radar guided missiles.

1956 U.S. military releases mosquitoes infected with Yellow Fever over Savannah, Ga and Avon Park, Fl. Following each test, Army agents posing as public health officials test victims for effects.

1958 LSD is tested on 95 volunteers at the Army's Chemical Warfare Laboratories for its effect on intelligence.

1960 The Army Assistant Chief-of-Staff for Intelligence (ACSI) authorizes field testing of LSD in Europe and the Far East. Testing of the

European population is code named Project THIRD CHANCE; testing of the Asian population is code named Project DERBY HAT.

1963 The mass **vaccination** campaigns of the 1950s and '60s may be causing hundreds of deaths a year because of a cancer-causing virus that contaminated the first polio vaccine, according to scientists. Known as SV40, the virus came from dead monkeys whose kidney cells were used to culture the first Salk vaccines. Doctors estimate that the virus was injected into tens of millions during the vaccination campaigns, including several million in Canada, before being detected and screened out in 1963. Those born between 1941 and 1961 are thought to be most at risk of having been infected.

1964 Guatemala established gun control. From 1964 to 1981, 100,000 Mayan Indians, unable to defend themselves, were rounded up and exterminated.

1965 Aspartame is the technical name for the brand names, NutraSweet, Equal, Spoonful, and Equal-Measure. Aspartame was discovered by accident in 1965, when James Schlatter, a chemist of G.D. Searle Company was testing an anti-ulcer drug. Aspartame was approved for dry goods in 1981 and for carbonated beverages in 1983.

1965 Project CIA and Department of Defense begin Project MKSEARCH, a program to develop a capability to manipulate human behavior through the use of mind-altering drugs.

1965 Prisoners at the Holmesburg State Prison in Philadelphia are subjected to **dioxin**, the highly toxic chemical component of **Agent Orange** used in Viet Nam. The men are later studied for development of cancer, which indicates that Agent Orange had been a suspected carcinogen all along.

1966 CIA initiates Project MKOFTEN, a program to test the toxicological effects of certain drugs on humans and animals.

1966 July 7-10, The virus Bacillus subtilis was released throughout the New York subway system, conducted by the U.S. Army's Special Operations Division. Due to the vast number of people exposed it would virtually impossible to identify, let alone prove, and specific health problems resulting directly from this test.

1967 CIA and Department of Defense implement Project MKNAOMI, successor to MKULTRA and designed to maintain, stockpile and test biological and chemical weapons.

1968 CIA experiments with the possibility of poisoning drinking water by injecting chemicals into the water supply of the FDA in Washington, D.C.

1968 - 69 The Hong Kong flu, which was influenza A type H3N2, killed over 30,000 people in the U.S. alone. That was a fortuitous learning event for some because it taught them that the flu could still conceivably be used to wipe out a population. But at the same time, it pointed out the need to precondition the populace so that those who might normally be resistant could be rendered susceptible. Hence the development of the vaccine program and the aerial spraying procedures to condition the population. The purpose of the chemicals in the chemtrails is to help the viral envelope fuse with lung cells, permitting easier penetration and infection.

1969 At a House Appropriations hearing, the Defense Department's Biological Warfare (BW) division requested funds to develop through gene-splicing a new disease that would both resist and break down a victim's immune system. "Within the next 5 to 10 years it would probably be possible to make a new infective micro-organism which could differ in certain important respects from any known disease-causing organisms. Most important of these is that it might be refractory to the immunological and therapeutic processes upon which we depend to maintain our relative freedom from infectious diseases." The funds were approved. AIDS appeared within the requested time frame, and has the exact characteristics specified.

1970 Uganda established gun control. From 1971 to 1979, 300,000 Christians, unable to defend themselves, were rounded up and exterminated.

1972 The World Health Organization (W.H.O.) published a similar proposal: "An attempt should be made to ascertain whether viruses can in fact exert selective effects on immune function, e.g., by ...affecting T cell function as opposed to B cell function. The possibility should also be looked into that the immune response to the virus itself may be impaired if the infecting virus damages more or less selectively the cells responding to the viral antigens." (Bulletin of the W.H.O., vol. 47, p 257-274.) This is a clinical description of the function of the AIDS virus.

1972 It was discovered that black children as young as age five were having psychosurgery performed on them at the University of Mississippi in Jackson in order to control "hyperactive" and "aggressive" behavior. Their brains were being implanted with electrodes that were heated up to melt areas of the brain that regulate emotion and intellect. When we first opposed these experiments, and eventually stopped them, we did so despite resistance from organized psychiatry and the research community.

Mid-1970's The incidence of AIDS infections in Africa coincides exactly with the locations of the W.H.O. smallpox vaccination program in the mid-1970's (London Times, May 11, 1987). Some 14,000 Haitians then on UN secondment to Central Africa were also vaccinated in this campaign. Personnel actually conducting the vaccinations may have been completely unaware that the vaccine was anything other than what they were told.

1975 Two neuroscientists, Laura and (the late) Eli Manuelides, from Yale University in the US, went on to illustrate by 1975 that injections of human blood, like injections of brain taken from kuru and CJD victims, transmitted the disease across the species barrier to laboratory animals. Their prophetic, but unheeded, message implied that blood was the vehicle that carried the agent of CJD around the body until it chanced upon an hospitable residence like the brain. This meant that the blood route was the key to the transmission of CJD from a primary host to a secondary host.

1975 The virus section of Fort Detrick's Center for Biological Warfare Research is renamed the Fredrick Cancer Research Facilities and placed under the supervision of the National Cancer Institute (NCI) . It is here that a special virus cancer program is initiated by the U.S. Navy, purportedly to develop cancer-causing viruses. It is also here that retrovirologists isolate a virus to which no immunity exists. It is lat er named HTLV (Human T-cell Leukemia Virus).

1976 Nobel Prize went to American scientist Carleton Gajdusek for his experiments demonstrating that injections of kuru brain (1967) and CJD brain (1969) reproduced similar illnesses in chimpanzees.
A striking feature of AIDS is that it's ethno-selective. The rate of infection is twice as high among Blacks, Latinos and Native Americans as among whites, with death coming two to three times as swiftly. And over 80% of the children with AIDS and 90% of infants born with it are among these minorities. "Ethnic weapons" that would strike certain racial

groups more heavily than others have been a long-standing U.S. Army BW objective. (Harris and Paxman, p 265)
The "discovery" of the AIDS virus (HTLV3) was announced by Dr. Robert Gallo at the National Cancer Institute, which is on the grounds of Fort Detrick, Maryland, a primary U.S. Army biological warfare research facility.

1977 Senate hearings on Health and Scientific Research confirm that 239 populated areas had been contaminated with biological agents between 1949 and 1969. Some of the areas included San Francisco, Washington, D.C., Key West, Panama City, Minneapolis, and St. Louis.

1978 The Hepatitis B vaccine study appears to have been the initial means of planting the infection in New York City. The test protocol specified non-monogamous males only, and homosexuals received a different vaccine from heterosexuals. At least 25-50% of the first reported New York AIDS cases in 1981 had received the Hepatitis B test vaccine in 1978. By 1984, 64% of the vaccine recipients had AIDS, and the figures on the current infection rate for the participants of that study are held by the U.S. Department of Justice, and "unavailable."

1978-1987 Even as the understanding of spongiform encephalopathy increased, various human pituitary hormone programs in countries such as Australia, France, New Zealand, the United Kingdom and United States were attracting hefty government sponsorships. Few of the programs' stalwarts caught on to the implications of the Manuelides' experiments, and unsuccessful attempts between the years of 1978 and 1987 to filter the CJD agent out of the pituitary hormones being injected into unsuspecting short-statured children and infertile women were left to one of this era's rare visionaries, British scrapie expert Alan Dickinson. At about the same time, a British Royal Commission on Environmental Pollution in 1979 raised the possibility that the unregulated cycling of protein-rich sheep remains back into animal feed might spread scrapie to cattle, as it had done to farm mink in the US three decades beforehand, via the oral route.

1979 There was an explosion at a Soviet plant in Sverdlosk and an outbreak of anthrax followed. At the time, all accusations of BW research were vigorously denied by Soviet officials, with the explanation that anthrax outbreaks can occur naturally and that the explosion was merely a coincidence.

1979 June , a well-dressed, articulate stranger visited the office of the Elberton Granite Finishing Company and announced that he wanted to build an edifice to transmit a message to mankind. He identified himself as R. C. Christian, but it soon became apparent that was not his real name. He said that he represented a group of men who wanted to offer direction to humanity, but to date, almost two decades later, no one knows who R. C. Christian really was, or the names of those he represented. Several things are apparent. The messages engraved on the Georgia Guidestones deal with four major fields: (1) Governance and the establishment of a world government, (2) Population and reproduction control, (3) The environment and man's relationship to nature, and (4) Spirituality. In the public library in Elberton, I found a book written by the man who called himself R.C. Christian. I discovered that the monument he commissioned had been erected in recognition of Thomas Paine and the occult philosophy he espoused. Indeed, the Georgia Guidestones are used for occult ceremonies and mystic celebrations to this very day. Tragically, only one religious leader in the area had the courage to speak out against the American Stonehenge, and he has recently relocated his ministry.
The Georgia Guidestones

1981 Aspartame was invented by the G D Searle Co. acquired by Monsanto in 1985. For 16 years FDA refused to approve it until 1981 when Commissioner Arthur Hayes overruled the objections of a Public Board of Inquiry and the protests of the American Soft Drink Association and blessed it. The tests submitted by Searle were so bad the Department of Justice, initiated prosecution of Searle for fraud. Then the defense lawyers hired the prosecutors, Sam Skinner and Wm. Conlon, and the case expired when the statute of limitations ran out. Aspartame/Nutrasweet, a toxin that blinds, drops intelligence, eradicates memory, grows brain tumors and other cancers, brings fatigue. Depression, ADD, panic, rage, paranoia, diabetes, seizures, suicide and death. This toxin is supported by unlimited advertising and the manufacturers pay off the American Diatetics Association, the American Diabetics Association, the AMA, and whomever else, to convince us its safe as rain. These lies are backed by a Federal Bureaucracy knowing it may kill your child, but the bureaucrat who approved the poison got a fat job as have many of his successors. Suppose this government watchdog, ignoring thousands of consumer complaints, has become an Attack Dog protecting corporate corruption. This is the bitter reality of Aspartame/Nutrasweet, Monsanto, the FDA, Coca Cola, Pepsi, and the hundreds of food, drink and drug makers who add to their products a known poison Conceived in Fraud and Dedicated to the Proposition that

Profit is all that Matters! (They're Poisoning Our Kids - Aspartame Warning The Facts From Betty Martin <mailto:Mission-Possible-USA@altavista.netMission-Possible-USA@altavista.net< /FONT>

Dr Miguel A. Baret of the Dominican Republic removed milk from 360 children's diets, because cow's milk has a specific protein that can cause diabetes, especially in children. They drank juice laced with aspartame instead and many developed "abnormal restlessness, lack of concentration, irritability and depression." When Dr Baret removed it: "The results were astonishing. Their symptoms disappeared in 4-6 days in ALL of them!" Thank you, Dr Baret, for this study showing what aspartame does to the brains of our kids!

1984, July 4. The first detailed charges regarding AIDS as a BW weapon were published in the Patriot newspaper in New Delhi, India. It is hard to say where the investigations of this story in the Indian press might have led, if they had not been sidetracked by two major domestic disasters shortly thereafter: the assassination of Indira Gandhi on Oct. 31 and the Bhopal Union Carbide plant "accident" that killed several thousand and injured over 200,000 on Dec. 3.

1985 The first of the fatal legacies of this form of medical madness emerged with four cases of CJD in human pituitary growth hormone-treated children.

1986 According to the Proceedings of the National Academy of Sciences (83:4007-4011), HIV and VISNA are highly similar and share all structural elements, except for a small segment which is nearly identical to HTLV. This leads to speculation that HTLV and VISNA may have been linked to produce a new retrovirus to which no natural immunity exists.

1987 Dr Louis Elsas, Professor of Pediatrics & Genetics at Emory University, testified before Congress; "Aspartame is in fact a well known neurotoxin and teratogen [triggers birth defects] which in some undefined dose will irreversibly in the developing child or fetal brain, produce adverse effects I am particularly angry at this type of advertising that is promoting the sale of a neurotoxin in the childhood age group." [Nov 2, 1987]

Neurosurgeon Russell Blaylock, MD, **declares Aspartame is a toxin like arsenic and cyanide that causes confusion, disorientation, seizures, cancer, pancreatic, uterine, ovarian and brain tumors and**

leads to Alzheimer's. Read Excitotoxins, the Taste That Kills [505-474-0303]. Hear Dr. Blaylock's radio interview on http://www.dorway.com/' Courageous whistleblowers like these have spoken in three congressional hearings, but industry's lobbying and political action keep the poison in the foods of the world. Our recourse as consumers is personal communication since the media is paid by advertising to push Nutrasweet/Equal/Diet Coke, etc..

1987 Department of Defense admits that, despite a treaty banning research and development of biological agents, it continues to operate research facilities at 127 facilities and universities around the nation.

1989 During which time the number of French children at risk of growth-hormone-related CJD had practically doubled, the first French children fulfilled that tragic legacy. In 1993, those responsible for this travesty were threatened with manslaughter charges. By 1997, France had half of the world's 100-plus cases of pituitary hormone-related CJD.

1990 More than 1500 six-month old black and Hispanic babies in Los Angeles are given an "experimental" measles vaccine that had never been licensed for use in the United States. CDC later admits that parents were never informed that the vaccine being injected to their children was experimental.

1991 Although the general elitism of human-pituitary programs restricted this brand of medical madness to North America, Europe and Australasian, Third World children and women did not altogether escape the insanity of applying Frankenstein medicine to social conditions. A medical report in 1991 linked the CJD death of a young Brazilian man, like those of five youthful New Zealand men and women, with a childhood treatment involving pituitary growth hormone obtained from the US. Unfortunately, the fate of women in Mexico City whose breasts were injected with US pituitary hormones in an appalling experiment to increase the volume of milk in lactating mothers (some already pregnant again) will probably never be known.

1992 It was discovered the federal violence initiative--the federal government's agency-wide plan to go into America's inner cities to experiment on children in the hope of finding genetic and biological causes for violence. We opposed this program as racist and abusive of children. Our efforts led to the cancellation of this program. It also led the chief sponsor of the program, psychiatrist Frederick Goodwin, to resign from his post as director of NIMH and to leave a career in the government. The fenfluramine studies at Columbia and Queens College are part of the violence initiative. They were created under its umbrella before it was cancelled. They confirm our fears that while the public aspects of the violence initiative were withdrawn, the actual individual projects continue unabated.

1992 Boris Yeltsin confirmed that anthrax was being researched at Sverdlosk and vowed to stop all "Soviet" BW research. Unfortunately, defectors have contradicted Yeltsin and there are rumors that although the 'official government' statement and ideal may be an elimination of biological weapons, the military is still actively pursuing a BW program on its own.

1993 The FDA approved aspartame as an ingredient in numerous food items that would always be heated to above 86°degrees F (30°Degrees C). An act that can only be described as "unconscionable"

1994 One has only to learn what really happened to the Christians in Rwanda between April and July of 1994 to imagine what may lie in store for Christians here in America at some time in the not-too-distant future. After the Christian Tutsis had been disarmed by governmental decree in the early 1990s, Hutu-led military forces began to systematically massacre the defenseless Christians. The massacre began in April 1994 and continued until July 1994. Using machetes rather than bullets, the Hutu forces were able to create a state of abject fear and terror within the helpless Christian population as they systematically butchered hundreds of thousands of them. The United Nations immediately convened hearings on the genocide taking place in Rwanda, but Madeline Albright, the American Ambassador to the United Nations, argued strenuously that neighboring African nations should not be allowed to intervene until the "civil war had come to an end." In reality, of course, there was no civil war since those being slaughtered had no weapons with which to defend themselves; it was simply a matter of mass murder. In addition to blocking intervention by neighboring nations, Madeline Albright also insisted that the word "genocide" must not be used, and that the United Nations forces stationed in Rwanda were not to be allowed to intervene. In the three months that followed, between one-half and three-quarters of a million Christians were systematically dismembered, hacked to death, and slaughtered in the bloody carnage that ensued. Tens of thousands of Christians were murdered in their churches; tens of thousands more were murdered in their hospitals and in their schools. On several occasions, United Nations soldiers stationed in Rwanda actually handed over helpless Christians under their protection to members of the Hutu militia.

They then stood by as their screaming charges were unceremoniously hacked to pieces. At the end of the carnage, in late July 1994, the American government rewarded the Hutu murderers with millions of dollars in foreign aid. Strangely, the American press has remained silent (Subversio n Of The Free Press By The CIA) about the fact that almost all of those who were slaughtered were Christians, and it was the policies of our government that were primarily responsible for blocking efforts by neighboring African countries to intervene. The Population Control Agenda

1994 With a technique called "gene tracking," Dr. Garth Nicolson at the MD Anderson Cancer Center in Houston, TX discovers that many returning Desert Storm veterans are infected with an altered strain of Mycoplasma incognitus, a microbe commonly used in the production of biological weapons. Incorporated into its molecular structure is 40 percent of the HIV protein coat, indicating that it had been man-made.

1995 Dr Phyllis Mullenix, former head of toxicology at Forsyth Dental Center in Boston and now a critic of fluoridation. Animal studies which Mullenix and co-workers conducted at Forsyth in the early 1990s indicated that fluoride was a powerful central nervous system (CNS) toxin and might adversely affect human brain functioning even at low doses. (New epidemiological evidence from China adds support, showing a correlation between low-dose fluoride exposure and diminished IQ in children.) Mullenix's results were published in 1995 in a reputable peer-reviewed scientific journal.

1995 The University of Rochester's classified fluoride studies, code-named "Program F", were started during the war and continued up until the early 1950s. They were conducted at its Atomic Energy Project (AEP), a top-secret facility funded by the AEC and housed at Strong Memorial Hospital. It was there that one of the most notorious human radiation experiments of the Cold War took place, in which unsuspecting hospital patients were injected with toxic doses of radioactive plutonium. Revelation of this experiment-in a Pulitzer Prize&endash;winning account by Eileen Welsome-led to a 1995 US presidential investigation and a multimillion-dollar cash settlement for victims.

1995 U.S. Government admits that it had offered Japanese war criminals and scientists who had performed human medical experiments salaries and immunity from prosecution in exchange for data on biological warfare research.

1995 Dr. Garth Nicolson, uncovers evidence that the biological agents used during the Gulf War had been manufactured in Houston, TX and Boca Raton, Fl and tested on prisoners in the Texas Department of Corrections.

1996, 27 June without public notice, the FDA removed all restrictions from aspartame allowing it to be used in everything, including all heated and baked goods. The truth about aspartame's toxicity is far different than what the NutraSweet Company would have you readers believe.

1996 A new scientific paper dealing with a meta-analysis of 23 different scientific studies on the relationship between first-trimester abortions and breast cancer was published in a British medical journal. That study clearly demonstrated a higher incidence of breast cancer in women who had hadfirst-trimester abortions. In response to that publication, the American Medical Association (AMA), the American Cancer Society (ACS), and pro-abortion/population-control advocates joined together in an unholy alliance to attack the conclusions of the authors, and to block all efforts to disseminate that information to American physicians.
The Population Control Agenda

1997 October. Speaking from Washington, DC, Nobel Prize winner for discovering the role of molecules known as "prions" in the invariably fatal brain illnesses such as "mad cow disease" or bovine spongiform encephalopathy (BSE) in cattle, and Creutzfeldt-Jakob disease (CJD) in humans, Dr Stanley Prusiner from the University of California predicted that the first drug therapy, which would not necessarily be a cure for BSE or CJD, was at least five years away. At the same time, on the opposite side of the Atlantic, the post-mortem of Chris Warne, a 36-year-old fitness fanatic from Derbyshire, England, revealed that he was the 21st victim of the new variant of CJD which had spread from BSE-infected cattle to humans via the food chain. Only 18 months earlier, a British House of Commons admission that BSE-infected meat had probably caused the CJD deaths of 10 youthful Britons left the British meat industry in tatters. Since then, the history of BSE has gradually unfolded to reveal a brain-dead imperialism, one which, while blinded by its own arrogant greed to inflate market profits, has treated public and, indeed, world health with gay abandon. Formerly a rare disease which affected less than one per million in most countries, one worst-case scenario predicts that BSE-infected meat will push the incidence of CJD in humans to claim 10,000 British lives by the year 2000, and a further 10 million by the year 2010. Another predicts that half the British people, some 30 million, will be left brain-dead by CJD. As Chris Warne's

mother commented, her son was a health-conscious sportsman, but "after winning medals in March, by July he couldn't stand on his feet, and by October he was gone".
Volume 5, #1

Researchers at the US Army Medical Research Institute of Infectious Diseases (or USAMRIID) at Fort Detrick in Frederick MD have reconstructed and modified the H1N1 Spanish Flu virus, making it far more deadly than it ever was back when it was responsible for the 1918-1919 flu pandemic that killed over 20 million worldwide and over 500,000 here in the US.

FRANCE is facing a new health scandal following allegations that the prestigious Pasteur Institute willfully ignored warnings that up to 600 children were being injected with cancerous hormones.

The disclosures that children may have been put at risk emerged during a new investigation into the link between the growth hormones and Creutzfeldt-Jakob disease (CJD), which caused a major scandal when acknowledged in the Nineties . So far, 74 children have died of CJD after being treated with growth hormones extracted from the bodies of victims of neurological illnesses in the Eighties.

That scandal and the contaminated blood affair in which 4,000 people were infected with HIV through unscreened blood transfusions - despite the availability of an American test forthe virus - have contributed to France's acute sensitivity on health issues. When the contaminated blood affair came to trial in 1998 , it was alleged that authorization for the American test was withheld to give the institute time to develop a rival French test.

The latest claims suggest that in 1985 the institute sold a batch of growth hormone to French hospitals without waiting for safety checks which showed the batch to have cancer marker cells at five times the permitted limit for use. The institute is further alleged to have made no efforts to withdraw the batch once it was aware of the risk.

References

Cole, Leonard A. Clouds of secrecy:
the army's germ warfare tests over populated areas,
Rowman & Littlefield, Totowa, N.J. , 1988

Hersh, Seymour M. Chemical and biological warfare:
America's hidden arsenal,
Bobbs-Merrill,
Indianapolis, 1968

Murphy, Sean.
No fire, no thunder: the threat of chemical and biological weapons,
Monthly Review Press,
New York , 1984

Piller, Charles.
Gene wars: military control over the new genetic technologies,
Beech Tree Books,
New York, 1988

Spiers, Edward M. Chemical and Biological Weapons:
A Study in Proliferation,
St. Martin's Press,
New York, 1994

A Higher Form of Killing:
The Secret Story of Chemical and Biological Warfare
by R. Harris and J. Paxman,
p 266, Hill and Wang, pubs.

Covert Action Information Bulletin #28 ($5),
Box 50272,
Washington, D.C. 20004;

Bio-Attack Alert ($20), Dr. Robert Strecker,
1501 Colorado Blvd.,
Los Angeles, CA 90041;

Radio Free America #16 by Dave Emery and Nip Tuck
(3 tapes, $10),
Davkore Co.,
1300-D Space Park Way,
Mountain View, CA 94043.

Critique - Exposing Consensus Reality,
P.O. Box 11368, Santa Rosa, CA 95406.
$15.00 for three issues (one year).

Project Paperclip by Clarence Lasby,
Atheneum 214, NY, and Gehlen:

Spy of the Century by E.H. Cookridge,
Random House.

DESIGNER DISEASES: AIDS As Biological & Psychological Warfare
by Waves Forrest

Contaminated Early Polio Vaccinations Linked To Cancer Epidemic
By Robert Matthews & Adrian Humphreys
The Sunday Telegraph and National Post Ontario,
Canada 5-18-99

On Line References

History of Biological Warfare
Eugenics
http://www.holisticmed.com/aspartame/
http://www.nexusmagazine.com/
The Population Control Agenda
Get links to over 200 sites on aspartame

US Government Experiments On Children During The Cold War

Chemically Induced Compliance: The Drugging of Kids and Garbage Science in America

http://user.mc.net/dougp/ahm/cor1.htmEugenics

The Illuminati Wants to Reduce the Population with 90%

http://www.healthnewsnet.com/humanexperiments.html

Silent Weapons For A Quiet War

Another Dark Chemtrail Hypothesis

Subversion of the free Press by the CIA

Comment
From Alan Cantwell
11-15-00

1976 Nobel Prize went to American scientist Carleton Gajdusek for his experiments demonstrating that injections of kuru brain (1967) and CJD brain (1969) reproduced similar illnesses in chimpanzees.

A striking feature of AIDS is that it's ethno-selective. The rate of infection is twice as high among Blacks, Latinos and Native Americans as among whites, with death coming two to three times as swiftly. And over 80% of the children with AIDS and 90% of infants born with it are among these minorities. "Ethnic weapons" that would strike certain racial groups more heavily than others have been a long-standing U.S. Army BW objective. (Harris and Paxman, p 265)

The "discovery" of the AIDS virus (HTLV3) was announced by Dr. Robert Gallo at the National Cancer Institute, which is on the grounds of Fort Detrick, Maryland, a primary U.S. Army biological warfare research facility.

I would quibble about substantiating evidence for the last paragraph (Gallo, HIV 1976 timeline).

NEWS

dompost.co.nz

ACC in legal battle over $250,000 bill from flu jab

Kiran Chug

A WOMAN who fell seriously ill after having a flu jab has been left with a medical bill of more than $250,000, which the Accident Compensation Corporation is refusing to pay.

Allison Cottle is embroiled in a legal battle with ACC as she prepares a High Court challenge to its decision not to pay, which a judge says is legally sound.

Mrs Cottle, who has permanent New Zealand residency but lives in Cleveland, in the United States, contracted a life-threatening illness a month after having the flu jab in March 2007.

A week after the vaccination, she was on holiday in the US, where she was diagnosed with Guillain-Barre syndrome, which causes spinal paralysis, after suffering a suspected stroke.

Yesterday, she told *The Dominion Post* of the unfairness of the decision made by ACC and the impact it was having.

"It's very stressful. My husband is unwell and we're just working through the matter ourselves."

Mrs Cottle did not want to comment further while the matter was before the courts.

Her lawyer, John Miller, said appeal papers had been filed in the High Court against the earlier decision made by Judge Beattie.

"It's your nightmare scenario of getting ill in the US and having to have an emergency treatment."

Court documents show Mrs Cottle went on holiday to the US a week after receiving the flu jab known as Vaxigrip in March.

After suffering a suspected stroke, she was taken to hospital, where the seriousness of her condition saw her admitted to intensive care and not given medical clearance to return home until July.

She was faced with a medical bill of US$180,482 (NZ$250,000), which ACC said it would not pay.

When Mrs Cottle asked for the decision to be reviewed, ACC was ordered to pay for the treatment.

However, it has now successfully challenged that review through the courts, with Judge Beattie finding the law was on ACC's side as the legislation stated that it should not pay for acute treatment received overseas. In his judgment, Judge Beattie said Mrs Cottle could not have known the flu jab would cause her to develop Guillain-Barre syndrome when she left for her holiday in the US, or that it would have such a "deleterious effect" on her well-being.

Although, legally, ACC must not pay for treatment received overseas, Judge Beattie said the case exposed a possible gap in the legislation.

In his decision, he described Mrs Cottle's situation as a "classic example" of one where ACC should have discretionary power to pay for her treatment.

A spokesman for ACC said the judge had found that ACC had interpreted the law correctly when making its decision.

Mr Miller said he had previously represented a client who developed a similar syndrome to Mrs Cottle after having a flu jab.

That client won her case for ACC to cover her treatment, as she had developed the illness in New Zealand.

He was unsure whether Mrs Cottle had travel insurance when she went on holiday.

So much for the assurances of the safety of this (and any other) vaccinations. This article from the Wellington NZ Dominion Post of 10 September 2010. Frankly I am surprised it was published, even though on page 10.

416-ME – HERE AND NOW

"Then I woke up or was fully aware – with full awareness of all the levels referred to. It was SO QUIET. SO peaceful. Then I knew "this is dead" – but I was fully awake/aware/alive; but I knew I was dead. Hastily I 'submerged' back into the previous level so as to re-emerge to mortality – DEAD AGAIN."

"I know something – now I know nothing No memories except swirling embracing all devouring 'evil'. I was devoured. I was hopelessly and forever trapped in a recycling evil dream of vast unpleasantness."
(Bob's Legacy – Book 1)

The trouble is that I know all of this and more. And I know that not only do people not know, but also don't care.

I become increasingly confused by the claims or allegations that the individual, as, or in the form that the individual experiences in this (temporal) mortality, will endure forever or eternally as presumably the same unchanged individual. Unchanged that is apart from the obviously necessary physical adjustments that would need to be made for the physical star stuff body to so endure forever. Whatever "forever" means. Frankly I think the word meaningless in this concept.

It would imply that the personality, the memories, all that makes us an individual "us" or in fact "me", is destined to survive "forever". I think it paradoxical that "commencing" from here and now we are alleged to have no end of existence, yet to have had a beginning, a childhood, an incremental development right up to this point in time, this current "now". What if we died at mortal age 5, or whatever age? Is that to be the "start"?

I analyze myself, my very being, and I conclude that in reality all I ever experience is always simply and just an ever-changing state of **"now"**. All else is in the mind. I have no memory or claimed memory of anything before my alleged birth into this star stuff world. I find the claim for everlasting existence for Bob Maddison as Bob Maddison one fraught with untold numbers of problems and lack of rationality or logic. On the other hand I find no difficulty with the concept of individual "me" intelligence existing "forever". That component is indeed part of the very quantum soup that is the fabric of the universe, whereas "Bob Maddison" is a mere transient identity that seemingly had a beginning, was given a name and a form (body) and will in the manner of a transient being, simply terminate its physical mortal existence. The star stuff will recycle, of that there is no doubt, and the energy that experiences "me, here and now" must likewise return to its source.

It has often been questioned as to how exactly an eternal "Bob Maddison" by any name would exist, and what such a one would do. The movie "Zardoz" postulates that after a few centuries boredom sets in and one deteriorates into an "apathetic" who thenceforth does nothing at all. I heard the question somewhere, "what does one do when one has bungy jumped all around the universe?" I tend to think the same. Indeed what would one do imprisoned in a "Bob Maddison" persona eternally?

Existing in a "here and now" state however ultimately is all that we can experience or know, and when one puts the mental operations into gear that becomes self-evident. All else is memory, hearsay from external sources, or conjecture. "Here and now" is perhaps the one ultimate reality.

So with this in mind, I imagine in the ever-changing flow of the "here and now" I experience and know as Bob Maddison, there will be a change at some stage. The "me", the eternal fabric of the universe stuff, that god stuff within this body, will experience a putting aside of the "Bob Maddison" identity and experience. Bob Maddison will die, and "me" will return for duty or reassignment to the life force that gave it.

After all, would one really want to live eternally, plagued with all the mental baggage one accumulates in even a brief experience such as mortality? Imagine after untold millions of years meeting someone, and recalling "old times". Does that even make sense? Do the Einstein thing and just use your imagination and mentally put yourself one million years into the future in that claimed eternal individual life of yourself as that Joe or Jane Doe. Would the concepts of "whatever happened to…." Or "how are your family/children" etc have any meaning? If one spends enough time thinking about this situation, it becomes obviously untenable. Or how about those who claim we will spend eternity at the feet of "god" singing his praises? Think hard about that one. Mentioned earlier in these writings were the questions about the future for those with wasted, lost, or shortened mortal lives. The more one thinks about it and broadens the thinking, the more problems with reason and logic one finds.

So here I find the problem. Yet the facts seem certain and beyond dispute that this "being" that is currently "us" will go in two separate directions when the "here and now" determines the time is right.

FORM AND IDENTITY.

Another reason for my confusion or bewilderment comes from contemplations as to our "true" form and nature. More so if it is indeed to endure for "all time". Then of course there is the matter of our true or real identity.

The person known in this world as "Bob Maddison" is a creation *of* this world. I had a beginning here and I will have and end of existence here. No doubt about that. My name was changed when my mother remarried in the 1950's, and no one left alive on this planet knows who my mortal father was. My beginnings on earth are now shrouded in mist, and lost because of well kept secrets and the death of all generations prior to mine. However working just with what I do know, it is established that "Bob Maddison" is a mortal and temporal earther life form. Memory indicates that once he had virtually **no personality** and **no memories** that precedes specific establishable memorised events in his ever-changing "now". He was once in effect a *"blank"*. He may well still be such.

Likewise his mortal body is purely a creation of this world, this planet. I have extensively dealt with mortal form elsewhere, and reiterate that this body seems *designed along specific lines and for specific identifiable limited purposes.* It is truly a limited utility vehicle, designed to work, learn, ingest, excrete, procreate and *do little else.* It serves adequately for this life, planet and conditions only. Its physical brain will cease practical functionality after a specific time, and at a given age will start to shrink.

It begs the question. What then is our true form? Further, what is the individual's true identity. In reality, do we as individuals even have a permanent or long term (eternal?) identity or "name"? I doubt it would be, as in my mortality, "Bob Maddison", as that name begs to be purely incidental, variable, and a mere quirk of fortune.

Surely the name as is manifest on this earth as "Bob Maddison", or by any other earth given name for an *earth given body* could not be an "eternal" name. No, they must surely be temporary in nature only. How rash to presume that "me" who only became "Bob Maddison" by reason of mortal birth as a blank programmable mortal, will now for all eternity, starting from planet earth "dot", will remain as that same identity. It defies reason to claim such, and would indicate the required amount of "brain work" has not been used to fully understand the claim and the deficiencies that will reveal themselves. The *god* stuff that animates this mortal earth flesh body must and does have "elsewhere its beginnings", as the verse so succinctly states. My point that must be ***clearly*** understood is that it was NOT known as "Bob" in any (or "if any") pre-mortal state. If you fail to understand that, then finish reading this now. Or re-read it from the beginning.

Likewise I totally doubt that its "form" was, and indeed is, probably much like that of mortal human life either. I think it somewhat arrogant to think or thoughtlessly assume that this mortal shape would thereby be the acme of physical development given the literally millions of diverse forms that currently exist on planet earth alone. Perhaps then, as a truly multidimensional being or entity, it may be more correct to assume or presume that there is no *absolute* form, but that such an entity may well be what is now known as a *"shape shifter or changer"*. (As in and from the "movies")(Note also that often in "dream state" we do not seem to actually be "embodied" at all, but function perfectly, apparently disembodied, as a "formless" intelligence, seeing everything.)

Such an *attribute* would accommodate and make acceptable (or understandable) claims by visionaries (etc.) that sightings, contact, and other events with various forms of beings/entities, (non-earther mortals) occur, exist or have been experienced. (This also has been covered in my writings.)

That at least would not be illogical, and at the same time would easily accommodate claims and reports of various forms of beings/entities by visionaries and such as claim sightings or contact with non-human mortal beings. We should give this **serious** thought, but most people will not.

It is clear to me then, that the whole matter of the nature, form, and identity of our true selves **needs some very serious questioning,** rather than blindly accepting the claims made by those obviously not qualified to make such claims or give such assurances.

EVER ONWARDS

So here we all are, on little old backwater planet earth not really knowing who or what we are as individuals, and no comprehension of the nature of man the species. (Meanwhile we watch the T.V., drink the tap water, use the toothpaste, eat the supermarket food, take vaccines, go to church and not think, and think we are "free". Etc.) We are mostly merely processed people, passing mortality with **no** "merit" badges.

Generally anything beyond that is just a worse and total confusion. The acceptable common paradigms of "god, man and the universe" and "destinies" areas are the almost exclusive domains of the ignorant and blind who vainly attempt to lead the unlearned blind.

Yet clues exist for us. For instance individually we fastidiously record absolutely every minutiae surrounding our personal existence. We can replay the most obscure of memories even from dreams, as well as mentally re-live, re-taste, and re-experience the most trivial and seemingly inconsequential of things. **We are unbelievable recording devices**. Yet we have no pre-mortal memories. (No, we will not re-visit reincarnation etc here again as its covered elsewhere.) One may ask if this is purely for our own individual requirements or benefits, **or is "something" feeding off or downloading these memory records**.

I have considerable respect for the whole "Anunnaki" story and clearly see therein the origins of the bible's Genesis. **It is entirely possible that the record of modern homo-sapiens-sapiens origin at the hands of an alien geneticist bent on creating a species of dumbded down, were procreating slave labour could be correct. Upon extensive analysis I find it answers more questions than it poses**. Even the historical timing is right. Where I had mental indigestion was figuring out the full story and consequences if that record is in fact correct. What are the implications for all the "Bob Maddisons" that make our species?

How did this "Anunnaki, Enki, Enlil" et al fit into the framework of any "real" issue of who we really and ultimately are? In other words, was there the ancient claim that "Enki" manufactured the current human race, engineering an intelligent species of mortals, irreconcilable with the acceptance (knowledge) of that species certainly having a body of star stuff earth, while being animated and driven by the "god stuff" energy?

A kind of mentally shattering concept to get one's head around. The god of Christendom certainly has no part in this model. Unless one comes to the logical conclusion that "he" or "they" (the gods) are certainly there, only not one but several, and they now have revealed individual names. Jehovah by whatever spelling (such matters not, regardless of claims otherwise) becomes unmasked as the human hating Enlil. Even that revelation sheds unbelievable light on biblical records of its god's dealing (frequently violently destructive) with human species. It also sheds light on "genesis god's" insecurity, incessant demands for loyalty, severe punishment, harsh laws, intolerance to all not obedient to the letter, and obvious lack of love for mankind. Also tells us why Moses had to ask this god who he was, because Moses had evidently not dealt with him before.

The impositions of such as Jehovah, Enlil etc has been dealt with in these writings, and my opinions have not changed since that was written on 16 October 2006.

Refer: -WHAT OF GOD? WHICH, WHO, AND WHERE?

So for whatever reason, by whatsoever means, perhaps we merely accepted to do a tour of duty on this earth, we have entered a mortal shell, driving it via the interface of the brain, and find ourselves on this earth, wiped clean and devoid of all knowledge of anything else. Yet we record all here, thus logically this is not a new talent. Even if it is an "Enki, Enlil" creation or manufacturing thing and we are serving as working brutes, then this does not mean, infer, or imply that the "me" within each of us is not sourced or have it's origin somewhere else. That is what was the source of my mental indigestion. It was solved with the realization that there was effectively "two creations". Somewhat like the 2 creation accounts that are clumsily revealed and recorded in the early chapters of Genesis. (As one religion states, "one account in Genesis was a physical creation, the other was a 'spiritual' creation of mankind *before coming to earth"*.)

Perhaps also there are no untold zillions of years of eternal future, but always and only just a "me, here and now", and the "here and now" changes one's identity, form, and goodness knows what else. All the untold "here and now" states would indicate one's continuing to experience, observe, and record all things. As an old "Bob Maddison" identity drops off (subsequent to a mortal death) and is no longer current or important, it needs be put aside, forgotten. Assuming individual survival, I think it must be so, else one would remain forever limited, anchored eternally to whatever blind prejudices, ignorance or conceit possessed in this one brief mortal experience. That would also infer this life is the paramount cycle, and that seems wrong, conceited, and illogical.

The main issue to understand is simply that although the identity of Bob Maddison as I currently know and experience him may be lost, it does not matter. Understand that you and I have effortlessly and without trauma experienced the loss of our earlier mortal selves frequently. As we change we "put aside", outgrow the older perhaps less understanding and developed self. As we change that self becomes forgotten as it is no longer the then current "me, here and now" us, but more like a "former us". I became somewhat aware of our true changing nature some decades ago when I recognised that I was no longer the same Bob Maddison as I could recall from various past events or times. Entire attitudes, belief systems, even personality become so changed as to challenge the belief that it is in fact still the same person. People change, they really do. This is one reason why I am so set against state imposed death sentences etc.

We all exist only as "me, here and now" and can only at best recall what or who we once were earlier in our current incarnation. It is no loss at all, as it is still the real "me" that progresses along the "here and now" path. I shall always be "me, here and now" even if I forever forget the experience of a life as Bob Maddison for that brief mortal span. I shall forever be alert, intelligent, and experiencing. I doubt I would ever wish to eternally (whatever that means) remember and thus experience all the events of such a brief and ultimately shallow life experience as is afforded by mortality. How could one possibly start so-called “eternal life” with the base of such an insignificant and generally meaningless life as that of a Bob Maddison. No doubt as I pass in death and am forgotten to all, likewise I shall forget this life and experiences just as I wake up and soon forget the contents and experiences of the dream body accompanying me on this journey.

Of course none of this is claimed as correct, true, or real in part or in total, but serves as a statement of current thinking of the current Bob Maddison. Who knows, what is real or not, so complex is the mind even with its human limitations. Maybe the only real thing is the mental exercise one gets from all the contemplations and reasoning.

(First printed and published in “The Human Disaster” November 2010. Reviewed and edited April 2013.)

SUMMARY OF SOME LIES, OMISSIONS, MISINFORMATION:

- The god of Christianity is not the universal creator at all, but a human hating alien. (Come on, as he is not one of us he is then by definition, alien)
- Humans did not "evolve" from a lower species on earth over millions of years. We appeared less than 500000 yrs ago, as an already domesticated species.
- We have never been "separated" from god and need no "forgiveness" from any but ourselves.
- Royalty, governments, and priesthoods were created solely to **control** the vast numbers of humans.
- Schooling is not for education at all, but to teach and inculcate fear, obedience, and subservience, and to memorize materials to make students "useful".
- Vaccines are dangerous, debilitating, and deadly concoctions with no proven benefit but with a litany of medical disasters, death, and disability to show.
- Gun control calls are not about a "law abiding" community as people can always get or make weapons; it is solely about control of the masses, and easier extermination.
- Fluoride in water supplies has nothing to do with dental hygiene, but is a proven means of sterilization, inducing lethargy, obedience and compliance of entire populations. The same chemical base is in the psychotic control drugs.
- Religion is not beneficial whereas spirituality is such. Religion is solely about control of the masses and also making it easier to get people to war and killing.
- The "Bible" is not the word of god at all, but a human record of human perceptions, interpretations, and thoughts.
- The "Big Bang" theory is just a theory or best guess of the education system's luminaries. Other explanations make far more sense, just like evolution.
- There is vast evidence for the progressive debilitation of the human race, to be culminated with a vast culling. Even the U.N. supports a goal of a mere <2 billion population.

- Civilization did not begin a mere 6000 or so years ago, that is merely the dating of the earliest of extant records.
- There is **unimpeachable** evidence that the entire Apollo Lunar claims are false, supplied compliments of the claimant NASA. NASA refuse to comment.
- The Christian religions all follow an erroneous paradigm in teaching that humans "have" or will "become" a spirit. In fact we really are "spirits" that have a body for a short time on earth. How did they get that so wrong?
- Sovereign governments no longer control their own countries any more than elected governments are there to serve its people. Control is in the hand of a hidden few.
- The entire story and history of human habitation on earth has been falsified and misrepresented to us by "experts" or authorities who wave credentials that really only evidence that they have learned, memorized their dogma well.
- The amount of evidence from varied sources would indicate and evidence that the earth has been involved in numerous "extinction events" within the period of human habitation thereon.
- As schooling does not equate to education, the vast majority of humans no longer have the power to think "outside of the square" to which they have been **assigned**.
- The "Media" in all its varieties are the tools of the few who run the show, to misinform, spread propaganda, control the paradigm, and provide the distraction, the "bread and circuses" to amuse the masses.
- There is now and always has been an ongoing war against the entire human race. We will lose. It is not our earth.

Mankind, and all life forms are multidimensional entities and not mere animated gross elements of this earth.

Bob's note: (re; Jehovah, Yahweh etc.)

(Relative to article 303 Pages 424-457)

I think that "The Name" is more likely a "title", and used by whoever was the leading or acting chief or general "god" in this particular area (geographical) at any time.

IF THERE WAS A SUPERIOR OR SUPREME "GOD" (not as in the case of this "family" patriarch, Anu) but a "God" to Anu, Enlil, Enki, et al, then such would have been known and recorded. For example Thoth was the designated chronicler of the Anunnaki etc.

It seems that no such record of a "God" (to them) exists from any source whatsoever, except for the record of the Hebrew, who so designated a supreme "god" for and by them alone.

Thus the god as later recorded by the Hebrew, was and obviously is NOT a universal "God" or even known or recognized by anyone apart from them. As I mention, if such a one existed as a fact, it would surely be known and recorded by the Anunnaki. As there is no such recording, I am left to conclude that "The Fabric Of the Universe" is the sole Omnific source of all things. (As discussed in my essays etc herein.)

The initial essay dated 16 October 2006 is contained in its entirety from page 40 in Book One of this revised edition.

www.ingramcontent.com/pod-product-compliance
Lightning Source LLC
Chambersburg PA
CBHW031816170526
45157CB00001B/72

* 9 7 8 1 3 0 4 6 1 6 9 8 2 *